FRONTIER SCIENCE

Frontier Science

*Northern Canada, Military Research,
and the Cold War, 1945–1970*

MATTHEW S. WISEMAN

UNIVERSITY OF TORONTO PRESS
Toronto Buffalo London

© University of Toronto Press 2024
Toronto Buffalo London
utorontopress.com
Printed in Canada

ISBN 978-1-4875-0419-9 (cloth)   ISBN 978-1-4875-1963-6 (EPUB)
                                 ISBN 978-1-4875-1962-9 (PDF)

Library and Archives Canada Cataloguing in Publication

Title: Frontier science : Northern Canada, military research, and
   the Cold War, 1945–1970 / Matthew S. Wiseman.
Names: Wiseman, Matthew S., 1986– author.
Description: Includes bibliographical references and index.
Identifiers: Canadiana (print) 20230567029 | Canadiana (ebook) 20230567053 |
   ISBN 9781487504199 (cloth) | ISBN 9781487519636 (EPUB) |
   ISBN 9781487519629 (PDF)
Subjects: LCSH: Military research – Canada, Northern – History – 20th
   century. | LCSH: Cold War.
Classification: LCC U395.C2 W57 2024 | DDC 355/.07071 – dc23

Cover design: Val Cooke
Cover image: Princess Patricia's Canadian Light Infantry soldier on Exercise
Sweetbriar, 1950. Library and Archives Canada / Department of National
Defence fonds / e010750870.

We wish to acknowledge the land on which the University of Toronto
Press operates. This land is the traditional territory of the Wendat, the
Anishnaabeg, the Haudenosaunee, the Métis, and the Mississaugas of the
Credit First Nation.

University of Toronto Press acknowledges the financial support of the
Government of Canada, the Canada Council for the Arts, and the Ontario
Arts Council, an agency of the Government of Ontario, for its publishing
activities.

 Canada Council   Conseil des Arts
for the Arts   du Canada

ONTARIO ARTS COUNCIL
CONSEIL DES ARTS DE L'ONTARIO
an Ontario government agency
un organisme du gouvernement de l'Ontario

Funded by the   Financé par le
Government   gouvernement
of Canada   du Canada

MIX
Paper | Supporting
responsible forestry
FSC® C016245

For my parents, Fay and Shane

Contents

List of Illustrations ix
Acknowledgments xi
Abbreviations and Acronyms xv

Introduction: Scientists, War, and Canada's Northern Frontier 3
1 Fort Churchill and Defence Research Northern Laboratory 36
2 Acclimatization, Cold Tolerance, and Biochemical Experimentation 67
3 Entomology, Insect Control, and Biological Warfare 96
4 The Changing Science of Arctic Warfare 123
5 Operation Hazen and the International Geophysical Year 146
6 Nuclear Fallout and the Northern Radiation Study 178
Conclusion: Reflections on Northern Canada, Military Research, and the Cold War 208

Notes 223
Bibliography 267
Index 287

Illustrations

0.1 Defence Research Board coat of arms, ca. 1950s–60s 14
0.2 Defence Research Board recruitment poster circulated to Canadian universities and academic scientists, ca. 1958–60 31
1.1 Map of the location of the former Fort Churchill military garrison 45
1.2 Defence Research Northern Laboratory, ca. 1955–64 60
2.1 A group of medical scientists and Inuuk guides in front of a temporary clinic hut near Coral Harbour, 1948 79
2.2 A White soldier and an Inuk man undergo scientific tests inside Defence Research Northern Laboratory, 1954 87
3.1 Map of Northern Insect Survey bases, 1949 103
3.2 Entomologists perform a radioactive experiment with mosquito larvae near Fort Churchill, 1952 109
4.1 Princess Patricia's Canadian Light Infantry soldier on Exercise Sweetbriar, 1950 130
4.2 Soldiers undergoing tests for physiological responses to cold temperatures and windchill at Fort Churchill, ca. 1953–4 136
4.3 MGM-18 Lacrosse missile in cold-weather trials near Fort Churchill, ca. 1958–9 141
5.1 Map of Lake Hazen, Ellesmere Island, 1960 149
5.2 Operation Hazen field party stand in front of a temporary shelter, Ellesmere Island, ca. 1957–8 163
5.3 Governor General Georges Vanier waving to a crowd outside RCAF Fort Churchill, 1961 173
6.1 Radiation Protection Division instructions for collecting human bone samples, 1964 199
6.2 Map of radioactive fallout sampling locations in subarctic and Arctic Canada, 1966 202

Acknowledgments

The research for this book began in the basement archive room of the Laurier Centre for Military, Strategic and Disarmament Studies (LCMSDS) – now Laurier Centre for the Study of Canada – at Wilfrid Laurier University (WLU). I completed my doctorate in history at WLU in 2017, where I studied under the supervision of Kevin Spooner and worked part-time as a research associate with Terry Copp and Mark Humphries at the "Centre." While compiling archival papers for *The Selected Works of George R. Lindsey: Operational Research, Strategic Studies, and Canadian Defence in the Cold War* (2019), I came across several references to the Defence Research Board (DRB) and military-sponsored science carried out in northern Canada and the Arctic between the 1940s and the 1970s. Intrigued by what I had read in Lindsey's papers, I travelled to Library and Archives Canada in Ottawa and retrieved several boxes filled with newspaper clippings that had been collected and scrapbooked by the DRB's public relations staff during the Cold War. I read about cold-weather acclimatization studies, frostbite, hypothermia, freeze-dried foods, entomology, meteorology, and several other topics of apparent military interest to the Canadian armed services. Much of the information pointed to Fort Churchill and Defence Research Northern Laboratory, and I returned to Waterloo convinced that I had stumbled upon the topic for my doctoral project.

Studying the North from a southern location challenged me to comprehend and reconcile the temporal and spatial distance underpinning my research. I spent two weeks in Churchill parsing archival documents and speaking with anyone willing to reflect on the past; in doing so, I barely scratched the surface of life for residents living on the shore of Hudson Bay in Manitoba's northeastern corner. My research trip was extremely fruitful, nonetheless. I received a Northern Research Fund from the Churchill Northern Studies Centre (CNSC), a scientific research establishment located thirty minutes from town on the site of the decommissioned Churchill Research Range (CRR). The research grant paid a portion of my travel costs and gave me access to accommodation, meals, and a rental pickup truck for the duration of my stay. Living at the tail end of Launch Road also gave me a glimpse into the residual effects of the Cold War. The road east from town toward the range site passes the Churchill airport, once the site of the Fort Churchill military base, long since decommissioned and demolished. Farther down the winding gravel road rests the shipwrecked *Ithaca*, on the seabed 19 kilometres east of Churchill in Bird Cove. As the coastal road veers inland, a former radar site of the Mid-Canada Line pops its double-domed white roof through a cluster of spruce trees atop a roadside hill. The abandoned radar site sits empty and locked, electronics and metal parts rusting on the inside and out. The road ends at the range launch site, itself standing in the long shadow of the Cold War, the remnants of past science and a Churchill of old.

In sharp contrast, CNSC is an eco-friendly research facility bustling with scientists and visitors year-round. The staff and students welcomed me with curiosity and jest, somewhat surprised (and perhaps wary) that a "social scientist" was among them. I spent much of my time immersed in a small but plentiful library, reading a rich selection of books, periodicals, government records, and local sources about the history of science in Churchill and northern Canada. The collection took me by surprise. Here was an unexpected supplement to the rich archival documents maintained by the volunteer staff at Churchill's public library, where I read local newspaper clippings and viewed community records and photographs that are unavailable elsewhere.

Churchill's vibrant history is alive and well because residents record and maintain both oral and written records out of care and respect for the community they call home. No one treated me as an outsider. Lorraine Brandson, the curator of Itsanitaq Museum, gave me access to books and resources on Churchill's natural and cultural heritage; William Erickson discussed his research on aurora borealis for the National Research Council at the former CRR; and LeeAnn Fishback, CNSC Scientific Coordinator, discussed the role and importance of science in northern Manitoba, gave me access to the CRR site, and connected me with people in the community. I owe an immense debt of gratitude to everyone who shared tidbits, stories, and reflections while encouraging my research. It is my sincere hope that this book, even in a small way, begins to return the kindness and generosity that you all showed me.

Equally important to this book was the tremendous support and guidance that I received from countless colleagues and peers. Kevin Spooner supported my decision to study the DRB's Arctic research program, read and revised chapters during the research and writing process, and mentored me throughout my entire degree. His dedication to my professional growth is unwavering. He is always the first to submit reference letters on my behalf and continues to connect me with people and opportunities well after my graduation. He also helped gather my superb committee, which included Roger Sarty, Ryan Touhey, Alistair Edgar, and Andrew Burtch. I thank you all for nurturing the research on which this book is based and for your steadfast encouragement along the way.

I could not have asked for a better and more enriching educational experience. Geoff Keelan, Kirk Goodlet, Kellen Kurschinksi, and Caitlin McWilliams welcomed me as a friend and colleague at the LCMSDS. We shared research stories and discussed our mutual interests in history over drinks and food. I was also fortunate to meet Trevor Ford, Lyndsay Rosenthal, Matt Baker, Kyle Falcon, Kandace Bogaert, Alex Souchen, Brittany Dunn, Eliza Richardson, Eric Story, Katrina Pasierbek, and Alec Maavara during my time at the LCMSDS. My wider cohort and peer group in the Tri-University graduate program included Whitney Wood, Marjorie Hopkins, Joe Buscemi, Erin Schuurs, Ian Muller, Joshua Tavenor, Christopher Bowles, Russ Freure, Alice Glaze, and Anne Vermeyden. You all listened happily to my strange and bizarre anecdotes about the Cold War as we punched through the gauntlet, and I will remain forever grateful for the time we spent together as graduate students, colleagues, and friends.

The research for this book continued after graduation. I was fortunate to receive three successive postdoctoral fellowships after leaving WLU, which enabled me to continue and expand my research and also ensured I had the time to revise the original project into the final product that you now read. I completed a Social Sciences and Humanities Research Council of Canada (SSHRC) Postdoctoral Fellowship with Timothy Sayle and John English in the Bill Graham Centre for Contemporary International History at the University of Toronto, an Associated Medical Services History of Medicine and Healthcare Postdoctoral Fellowship with Shelley McKellar in the Department of History at Western University, and a Banting Postdoctoral Fellowship with Jane Nicholas in the Department of History at St. Jerome's University. Each opportunity provided additional resources and funding for new research, and I am most thankful for the professional support and guidance of my terrific postdoctoral mentors. You are all giants in your respective fields, and I cannot thank you enough for your wisdom and commitment to my success.

I have been fortunate to remain at the University of Waterloo, "across the creek" at main campus, where my colleagues in the Department of History have welcomed me and encouraged my studies. Among others, I extend my thanks to Daniel Gorman, Andrew Hunt, Katherine Bruce-Lockhart,

Geoffrey Hayes, Alexander Statiev, and Robyn Wilkinson for giving me opportunities, answering my questions, and helping me learn and grow as a historian and teacher. Your collegiality is dear to me, and I am grateful for the opportunity to work together as colleagues moving forward.

Historical research takes time, resources, and a generous amount of assistance. Several national, provincial, and university archives hold DRB records, attesting to the challenges and exciting possibilities of studying the history of science in Canada. I did my best to thank all archivists and curators whom I had the pleasure to meet, but I want to acknowledge their collective efforts here by expressing my gratitude in writing. Archivists in Canada, Britain, and the United States made the research for this book possible: thank you all for answering queries, retrieving documents, and reviewing access-to-information requests. Financial assistance covered most of the travel costs associated with the research conducted for this study. I wish to acknowledge financial support from SSHRC, WLU, LCMSDS, CNSC, University of Toronto, Western University, and the University of Waterloo. I also benefited from a strong editorial network, including Len Husband and the entire editorial acquisitions and support team at University of Toronto Press. Len stuck with this project from day one, and I thank him here for his professionalism, guidance, and commitment to seeing this book through to publication. On a similar note, I want to thank the anonymous peer reviewers who read the manuscript and provided helpful comments and suggestions on earlier drafts, Matthew Kudelka for judiciously copy editing the book, and Megan Raffa for illustrating the maps.

The one constant in my life is family. My parents Fay and Shane, siblings Jennifer and Mark, brother-in-law Chris, and sister-in-law Hayley all experienced the highs and lows of this research project with me from the start. My partner, Alycia, entered my life later in the writing process. We now have a son, Benjamin, and I cannot wait to see what life holds for us next. Thank you all from the bottom of my heart. You may now read and reminisce – you're welcome.

Note on Permissions

Portions of the book have appeared in earlier work. I am grateful for permission to incorporate material from the following publications:

Matthew S. Wiseman. "The Origins and Early History of Canada's Cold War Scientific Intelligence, 1946–65." *International Journal* 77, no. 3 (2022): 7–25.
Matthew S. Wiseman. "Unlocking the 'Eskimo Secret': Defence Science in the Cold War Canadian Arctic, 1947–1954." *Journal of the Canadian Historical Association* 26, no. 1 (2015): 191–223.

Abbreviations and Acronyms

AEC	Atomic Energy Commission (US)
AGARD	Advisory Group on Aeronautical Research and Development, NATO
AINA	Arctic Institute of North America
BW	biological warfare
CAF	Canadian Armed Forces
CAORE	Canadian Army Operational Research Establishment
CIA	Central Intelligence Agency (US)
CIB	Canadian Infantry Brigade
CIIA	Canadian Institute of International Affairs
COPUOS	Committee on the Peaceful Uses of Outer Space (UN)
CRR	Churchill Research Range
Cs-137	cesium-137
CW	chemical warfare
DDT	dichloro-diphenyl-trichloroethane
DEW Line	Distant Early Warning Line
DND	Department of National Defence
DRB	Defence Research Board (Canada)
DRCL	Defence Research Chemical Laboratory (Canada)
DRKL	Defence Research Kingston Laboratory (Canada)
DRML	Defence Research Medical Laboratories (Canada)
DRNL	Defence Research Northern Laboratory (Canada)
DRTE	Defence Research Telecommunications Establishment (Canada)
HBC	Hudson's Bay Company
HMCS	Her Majesty's Canadian Ship
IAEA	International Atomic Energy Agency
ICBM	intercontinental ballistic missile
ICR	Institute of Cancer Research (United Kingdom)
ICRP	International Committee on Radiological Protection

IGY	International Geophysical Year
IPY	International Polar Year
JAWS	Joint Arctic Weather Stations (Canada–US)
JIB	Joint Intelligence Bureau (Canada–US)
JSES	Joint Services Experimental Testing Station (Canada–US)
KGB	Committee for State Security (Soviet Union)
MAD	mutual assured destruction
MCC	Military Cooperation Committee (Canada–US)
MCL	Mid-Canada Line
MSF	Mobile Striking Force
NASA	National Aeronautics and Space Administration (US)
NATO	North Atlantic Treaty Organization
NCO	non-commissioned officer
NDHQ	National Defence Headquarters
NFB	National Film Board (Canada)
NORAD	North American Air/Aerospace Defense Command
NRC	National Research Council (Canada)
NWMP	North-West Mounted Police
NWT	Northwest Territories
OR	operational research
ORS	Operations Research Section (Canada)
P-32	radioactive phosphorous
PJBD	Permanent Joint Board on Defence (Canada–US)
RAF	Royal Air Force
RCAF	Royal Canadian Air Force
RCMP	Royal Canadian Mounted Police
RCN	Royal Canadian Navy
RDB	Research and Development Board (US)
RDX	Research Department eXplosive
RMS	Royal Mail Ship
RN	Royal Navy
RPD	Radiation Protection Division (Canada)
SAC	Strategic Air Command
SES	Suffield Experimental Station
SPRI	Scott Polar Research Institute
Sr-90	strontium-90
UK	United Kingdom
UN	United Nations
UNSCEAR	UN Scientific Committee on the Effects of Atomic Radiation
US	United States
USAF	United States Air Force
USCGC	United States Coast Guard Cutter

USN	United States Navy
USS	United States Ship
USSR	Union of Soviet Socialist Republics
WP	Warsaw Pact

FRONTIER SCIENCE

Introduction: Scientists, War, and Canada's Northern Frontier

For ten days in late 1951 or early 1952, exact dates unknown, a group of scientists working in Toronto observed the slow and painful death of six shivering dogs subjected to a research experiment. The scientists shaved the fur of seven dogs and exposed the group, without anaesthetic, to a cold room at −30 degrees Celsius until death. One dog survived. Of the seven test subjects, three received approximately one pound of food twice daily. The other four starved, surviving solely on water and any limited exchange of body heat. "The animals lay huddled in a corner of their cage and shivered almost continuously throughout their confinement," wrote the scientists in their research report. "They looked cold and miserable with frosted whiskers, but each time an observer entered the cold room, the dogs would rise to their feet and wag their tails. They did not seem to object strenuously to this treatment and the longest survivor, at the end of ten days, would hop back into his cage at a command."[1] The scientists recorded the rectal temperature of each dog every two hours, took blood sugar estimates twice daily, and recorded terminal body changes upon death. The animals endured this treatment throughout day and night, surviving anywhere from five hours to an astonishing ten days. The lone survivor was one of the three fed dogs. He escaped the fate of the others only because the scientists removed him from the cold room after ending the experiment for "humanitarian reasons."

If not for the "unpleasant nature of the experiment," the research might have continued. In fact, the scientists apologized for the "short" sample size in their report. They provided no indication of the recruitment process, nor did they record the age or breed of the selected dogs. Instead, they described the stimulus for the experiment and listed their methods and observations. The research was a study of animal behaviour as well as body response during exposure to severe cold, and both observations were important for a much wider scientific research agenda.

In the year preceding the experiment, lead scientist Wilfred G. Bigelow had created the pacemaker, an important medical device and one of two pioneering achievements that later earned him an appointment to the Order of Canada in 1981 and an induction into the Canadian Medical Hall of Fame in 1997. His other significant achievement was developing the use of hypothermia as a medical procedure. Bigelow recognized that lowering the body's core temperature reduced the amount of oxygen required by the patient, which made open heart surgery safer. Neither medical breakthrough might have been possible if not for experimental research performed on animals – research funded by, and conducted for, Canada's Department of National Defence (DND).

During the 1950s, Bigelow was one of hundreds of Canadian scientists with ties to the Defence Research Board (DRB), the branch of Canada's defence department that provided scientific and technical support to the armed services. He chaired the DRB's Panel on Frostbite and Immersion Foot and conducted cold-related experiments at Defence Research Medical Laboratories (DRML), a newly established military research facility located on the premises of Royal Canadian Air Force Station Downsview in northwest Toronto.[2] Hundreds of animals went under the knife at DRML. Bigelow cooled no fewer than 176 dogs over a two-year period alone, carrying on a long tradition in Canadian medical science of using animals in advance of human trials.[3] As Matthew Klingle has shown, lauded research physicians Frederick Banting and Charles Best sacrificed several dogs in the 1920s to obtain organs for experimental injections during insulin research trials conducted at the University of Toronto.[4] Bigelow's research had different objectives, but he too experimented on dogs in place of human subjects. He attempted cardiac surgery in thirty-nine cases, causing the death to nineteen subjects. In a follow-up experiment, he cooled and operated on thirteen rhesus monkeys and six groundhogs.[5] Seven of the thirteen monkeys died, one during surgery and six of infection or pneumonia in the weeks that followed. Five of the six survivors became subjects for related experiments, and the lone survivor ended up as a mascot in the Royal Canadian Navy (RCN). All six groundhogs survived the procedure, but only one avoided infection and death, ultimately becoming the subject of other studies.

Bigelow experimented on animals to develop a safe technique for rewarming the human body from severe hypothermia. Acquainting the armed services with a proven method of resuscitation was important for protecting the soldiers, sailors, and aviators called upon to serve their country in cold and wet conditions. Senior government authorities in Ottawa made this clear in December 1952 when medical scientists funded by the DRB considered testing Bigelow's rewarming technique on human subjects.[6] Bigelow's work was especially valuable in the field of Arctic research, a broad area of particular importance to the men who championed science to protect and promote Canadian security interests during the earliest decades of the Cold War. Between the late 1940s and the early 1970s, scientists employed by or under contract to the DRB conducted a wide range of experiments designed to study and improve the physical and psychological capabilities of military personnel for potential operations in northern Canada and the Arctic. Canadian officials had little fear of a full-scale enemy invasion in the North at the beginning of the Cold War, but over time, technological advances in nuclear weaponry and long-range bombers sparked calls for a calculated and coordinated strategic defence plan.[7] Prior to the advent of ballistic missiles in the mid-1950s, Canadian soldiers had trained for land-based incursions in the North to prevent enemy forces from establishing forward operating bases on the fringes of the North American continent.[8] Scientists employed by or under contract to the DRB studied military problems in the North and generated knowledge deemed useful for helping the armed services deploy and operate in northern conditions, the better to defend Canada in the atomic age.[9] Military research functioned accordingly, ultimately providing information and ideas thought vital to ensuring Canadian security and national defence during an active and tense period in world affairs.

But what motivated Bigelow to freeze dogs to death for military research purposes? Did he and his co-researchers share the military and strategic defence concerns of their government sponsors, or did scientific curiosity and research funding drive their work? Did other Canadian scientists embrace military research to support the armed services, and, if so, what were the consequences of the militarization of Canadian science in the early decades of the Cold War? This book examines the complicated but important history of military-sponsored Arctic research in post-1945 Canada, using science as a lens, with the goal of reconstructing and interpreting the role of scientists and scientific expertise in shaping how the Canadian state understood and approached its military concerns about northern Canada and the Arctic.

Many scientists besides Bigelow conducted military-sponsored research for the federal government with the aim of improving the operational capabilities of Canadian personnel in the northern regions of the country. Bigelow's contribution was unique. Others studied the North and the people living there for non-military purposes, motivated by international research and the perceived paternalistic responsibilities of northern administration and governance. For all of these scientists, the common denominator was the Cold War. Whether performing or championing military or non-military research, the scientists and government officials discussed in this book carried out or supported Arctic research activities in response to the political and military atmosphere of the early post-war period. For some, science was a tool for defending Canada and protecting Canadian citizens from real and potential threats; for others, the Cold War provided a research opportunity, and they consciously supported the interests of the state while achieving individual and collective goals for professional development and the advancement of scientific knowledge.

Bigelow belonged to both camps. He was born into a medical family in Brandon, Manitoba. His father, Wilfred Abram, founded the first private medical clinic in Canada, and his mother, Grace Ann, was a nurse and midwife. He earned his medical degree from the University of Toronto in 1938 before serving as a captain in the Royal Canadian Army Medical Corps and performing battlefield surgeries on the front lines during the Second World War. He spent a year at Johns Hopkins Medical School after the war, later returning to Canada on appointments to the surgical staff of Toronto General Hospital in 1947 and to the Department of Surgery at the University of Toronto the following year.[10]

Bigelow's wartime experience treating soldiers motivated him to continue his military research and study medical problems for the Canadian armed services after the war – a common pattern among DRB-funded scientists. "Injuries from cold have played a major role in many military campaigns," he and D.R. Webster wrote in an article they co-authored for a special civil defence issue of *Canadian Medical Association Journal*. "History has many examples of the bitter price paid by those who entered these campaigns unprepared, or were overtaken by winter with insufficient clothing, food and warmth."[11] Bigelow and Webster wrote these words in 1952 while Canadian soldiers were fighting under the UN banner in Korea; both were aware of the rising incidence of frostbite, immersion foot, and other cold-induced injuries suffered by the multinational forces during the winter battles in the Korean peninsula's hilly northeast. Medical science, these two men believed, was a necessary complement for military preparations, and they gladly worked for Canada's defence department to serve their country in "peacetime."[12]

Under the looming spectre of nuclear war, military-funded research in post-1945 Canada consistently promised civilian benefits as well. "It is conceivable that an atomic bomb attack upon a large city during very cold weather would produce casualties suffering from hypothermia or dangerous reduction in body temperature," Bigelow and Webster wrote, tying their research to wider societal concerns over civil defence.[13] As the first fully controlled laboratory study on the subject of cold-body resuscitation published in the open medical literature, their work laid the foundation for subsequent international research and eventually the application of hypothermia for treating patients.[14] At the time, however, neither author published the full details of their work for the Canadian defence department. Scientists, physician researchers, engineers, and technicians (and occasionally research assistants and graduate students) who conducted full- or part-time research for the DRB, and by extension the Canadian armed services, swore an oath of secrecy and agreed to contractual stipulations restricting the oral and written communication of research.[15]

The Cold War fostered secrecy at several levels of society. Medical scientists conformed to the unusually harsh restrictions and rules of professional conduct. In doing so, they maintained national security, thereby protecting liberal-democratic values and ensuring their continued access to government coffers. Only years later did Bigelow acknowledge that funding from the defence department had supported the cold-room experiments on animals. "Happily, the Defence Research Board of Canada saw merit in this project," he wrote upon reflection in 1984. "They were the first to give financial support to our research, primarily to learn something about the physiology of hypothermia. They had a special interest, of course, in accidental exposure to hypothermia and in a technique for resuscitation."[16] The civilian and military applications of the research reflected a common theme among Canadian scientists and medical researchers who performed experimental work on behalf of the armed services during the first half of the Cold War: the threat against their country and way of life was real, and it motivated and justified the continuation and expansion of military-sponsored research.

Canada and the Cold War

From the end of the Second World War in September 1945 until the fall of the Berlin Wall in November 1989, global leaders feared the outbreak of a nuclear war between the world's two superpowers: the United States and the Soviet Union. In the end, the globe escaped that cataclysm, but in the interim, the stalemate between the superpowers – inflamed by espionage, technological competition, proxy battles, and psychological warfare – became known as the Cold War. Historians quibble over the exact dates that marked the beginning and end of this significant period in modern world history. This book concentrates on the years between 1945 and 1970. I locate the start of Canada's Cold War in mid-February 1946, when Igor Gouzenko's defection from the Soviet Union became public.

On the evening of 5 September 1945, three weeks after Japan surrendered following the atomic bombing of Hiroshima and Nagasaki, Gouzenko, a cipher clerk at the Soviet Embassy in Ottawa, fled from work with 109 documents detailing the activities of two Soviet spy rings in Canada.[17] After being turned away by the *Ottawa Journal* and the Justice Department, and narrowly escaping a team of Soviet agents, who ransacked their apartment in search of the incriminating documents, Gouzenko and his wife, Svetlana, together with their infant son, received protection from the RCMP.[18] Gouzenko's defection sparked a series of secret interrogations and proceedings that exposed Moscow's espionage efforts and shattered any remnants of the wartime alliance between the Western partners and the Soviet Union.

The Gouzenko affair helped usher in a new era of international tensions that saw Canada firmly committed to the post-war interests of its two closest allies, Britain and the US. From the mid-1950s through to the end of the 1980s, the armed forces of the two superpowers and their military blocs, NATO and the Warsaw Pact, brought the world to the brink of annihilation as they fought each other on the political, economic, diplomatic, and cultural fronts. The US Central Intelligence Agency (CIA) went to war against its Soviet counterpart, the Committee for State Security (KGB), and the two states clashed in occasional proxy wars on the ground. In October 1962, the world watched in fear as US President John F. Kennedy and Soviet Premier Nikita Khrushchev, each with the power to unleash unimaginable destruction upon the world, flirted with a thermonuclear war over the installation of nuclear-armed Soviet missiles in Cuba. The same year as the Cuban Missile Crisis, *Army Information Digest*, the official magazine of the US Army, defined the term "Cold War" as the "use of political, economic, technological, sociological, and military measures, short of overt armed conflict involving regular military forces, to achieve national objectives."[19] In this battle of one state against the other – *us versus them* – Ottawa was decisively tied to its friendly neighbour in Washington. Canadians had fought and died for the Allied cause during the Second World War, and they emerged from that global conflict unwilling to sacrifice or lose the freedoms that victory had secured.

The Cold War amounted to an ideological battle in which the capitalist United States and its Western partners were pitted against the communist Soviet Union and its Eastern Bloc. The differences and commonalities between the superpowers – political, economic, technological, cultural – spurred a bitter rivalry that would dominate world affairs for nearly five decades. But historians caution against characterizing the Cold War merely as a stand-off between two powerful states with opposing political orthodoxies. The years between 1940 and 1962 saw more than forty countries in Africa, the Middle East, and Asia gain independence, marking the dawn of a new era for more than 800 million people around the globe and the rise of new challenges for their national leaders.[20] Events outside North America and Europe, beyond the frame of the Cold War, would shape the post-1945 world, regardless of the thoughts and actions of government officials in Washington and Moscow. Nevertheless, the imperialist undertones that drove superpower diplomacy motivated foreign policy elites in the United States and the Soviet Union to define strength in terms of numbers. In their efforts to persuade or coerce people around the world to choose either capitalism or communism, both states fought beyond their borders, thus taking the Cold War global.

The stark realities of the superpower struggle for world dominance and ideological supremacy become even clearer when one considers the powerful mark the Cold War left *within* specific borders. For senior DND officials in Ottawa, the military and strategic threat in the immediate post-war period cast a spotlight on the Arctic. Geography placed Canada between the two superpowers, dangerously exposed to long-range bombers, and the country's northern defences were particularly weak. A strategic doctrine of mutual assured destruction (MAD) encouraged both the US and the Soviet Union to bolster their military presence in the Arctic and position nuclear weapons close to the North Pole, and this set the stage for a techno-military stand-off that had immediate and lasting consequences for peoples, communities, and resources located near and above the Arctic Circle. This Cold War battle for technological supremacy and military domination in the Far North affected Canada and Canadians. Stuck in between two colossus superpowers, Canada was at the mercy of two rival states and trained its eyes on both sides of the divide. Ottawa responded by pouring substantial federal funds into the Arctic sciences in the hope that novel research findings would define and help strengthen Canada's military commitment to North Atlantic security in a vast and largely unprotected region. To those ends, military-sponsored scientists and engineers turned their attention north, pursuing Arctic warfare projects in the name of security and national defence. The following pages trace the motivations and consequences of that pursuit.

Cold War Science and Northern Canada

Scientific expertise played an important role in the Cold War consciousness of Western democracies. Citizens in the United States, Britain, Canada, and elsewhere, under the influence of government propaganda, tended to view the Cold War as a battle over culture. Communist political ideology, totalitarianism, and social conformity represented a threat to the individual freedoms that Canadians had fought and died to protect during the Second World War. In the tense and uncertain political atmosphere of the early Cold War, fresh off helping to secure victory over Germany and Japan, scientists in the West embraced new fields of study and research opportunities, secured new sources of funding, and promoted the view that they were essential to the functioning of a healthy democracy.

In this vein, *Frontier Science* examines the circumstances and ideas behind the militarization of Arctic research in Canada in the early decades of the Cold War. I adopt Laura McEnaney's definition of militarization as a process that constitutes "the gradual encroachment of military ideas, values, and structures into the civilian domain."[21] Driven by the politics of national security and a strong belief in the advantages of scientific research, senior officials in the defence department championed science as a means to prepare the armed services in peacetime and to bolster Canada's contribution to North Atlantic security. But despite escalating tensions between the superpowers and real concerns over security and national defence in the northern regions of the country, senior officials in Ottawa sponsored Arctic research for reasons other than to protect Canadian strategic interests. Scientific knowledge about northern Canada and the Arctic also provided military planners with appreciations deemed useful for responding to the evolving threat of atomic, biological, and chemical warfare. Military considerations in Ottawa, London, and Washington were clear: gaining knowledge about scientific and technical advances in the Soviet Union was essential for understanding and defending against the communist enemy.

In response to real and evolving security and defence issues in the North, military officials and government scientists in Canada sponsored and performed Arctic research to generate knowledge useful for assessing the high-latitude functionality of the Soviet military. Theoretically, developing knowledge about Canada's northern geography and the effects of extreme climate on soldiers would provide practical and strategic information thought necessary to understand the complexities of northern warfare and predict the capabilities of Soviet soldiers, weapons, and military equipment in Arctic and subarctic conditions. Northern Canada was sovereign territory, partly accessible by rail and air, and considered sparsely populated. DND officials and defence scientists in Ottawa approached the North as a natural laboratory, a space free of military conflict and ideal for replicating the climate and terrain conditions faced by enemy forces on the Soviet side of the Arctic. Unable to peer through the Iron Curtain, Canadian scientists tested personnel and equipment in a region considered ripe for military research.

This perceived need for scientific knowledge about the Soviet Union and Soviet military capabilities in cold, high-latitude climates fuelled an ongoing militarization of Arctic research in Canada that affected scientists, military personnel, civilians, animals, and the environment.

More precisely, this book argues that scientists played a fundamental role in fighting Canada's Cold War. Canadian historians have examined the Cold War through diplomatic, intelligence, military, cultural, and social lenses; scientists and engineers, when discussed at all, are typically viewed as peripheral historical actors and are seldom the focus of studies. Yet the Arctic research performed by scientists in Canada during the Cold War informed how the Canadian state perceived and approached its domestic security and defence issues in the North, while also tying Canada to the scientific and military agendas long woven into the wider North Atlantic Cold War experience of Britain and the United States. Academic scientists and non-government researchers often portrayed themselves as objective and apolitical, yet their work and ideas contributed to decisions made about Canada's national interests in the North, with knock-on effects on northern residents and communities.

Recent international literature on the Cold War shows us the value of examining historical actors who have long been left out of standard studies of post-war government policy-making. As Andra Chastain and Timothy Lorek contend in their work on the Latin American Cold War, moving beyond the spheres of diplomacy and culture is essential if we are to understand how the global conflict manifested itself differently in regions, states, and communities outside the United States and Europe.[22] Uniting military history and the history of science will shed light on the importance of scientists to Canada's Cold War. This is neither a rebuke of superpower diplomacy nor a challenge to the government officials who shaped Canadian foreign and defence policy during a pivotal era in world history. Canada's position and role vis-à-vis the United States, Britain, and the Soviet Union is fundamental to any thorough analysis and understanding of the post-1945 period. An examination of the significant role that Canadian scientists played in the Cold War will tell us a lot about how the Canadian state understood and protected its national interests.

The history of the polar regions is a growing field of interdisciplinary research. Scholars have documented and examined the International Polar Years and the role of global knowledge production in relation to scientific interest in cold. Over the past ten years, the Cold War has received increasing attention as a focal point of studies. Matthias Heymann's research illustrates the centrality of Greenland in Cold War–era efforts to understand and conquer cold climates, while Matthew Farish's extensive body of research on US military approaches to the North American Arctic and cold landscapes provides a detailed perspective on how geographical considerations and concerns affected and influenced prominent officials in Washington.[23] The strong focus on US entanglements with science, technology, and the Arctic during the Cold War leaves several histories open for investigation, however. While a growing body of literature examines the history of ice and snow from the perspective of the Soviet Union, for instance, other Eastern Bloc countries beyond the frame of the Cold War superpowers require increased attention and scrutiny. This applies to historical actors outside the US and the Soviet Union as well, as revealed by Sverker Sörlin's useful transnational examination of Swedish geographer and diplomat Hans Ahlmann, who bridged the East–West divide through glaciological research.[24]

Notwithstanding the critical importance of international and transnational studies to polar history, this book makes no claims to being a full history of military research in northern Canada or the North American Arctic during the Cold War. I focus on scientific activities and technological developments conjured in the minds of Canadian researchers and carried out with the encouragement and sponsorship of government authorities in Ottawa. In other words, *Frontier Science* is a Canadian history and centres Canada and Canadian military research. Scientific ideas, practices, and activities that originated outside Canada certainly affected Canadian approaches to Arctic research during the Cold War, but the core of this study highlights and investigates the role of science within the Canadian defence department and the corollary impact on Canada's post-war military preparedness efforts in the North. The transnational history of military research in northern Canada and the North American Arctic is a topic for fuller, detailed examination elsewhere.

Canada, the Cold War, and Defence Research

Exactly whose interests the state represents has been the focus of a wide and important body of literature on Canada and the Cold War. Among others, Reg Whitaker, Gary Marcuse, Gary Kinsman, Patrizia Gentile, Richard Cavell, Franca Iacovetta, Christabelle Sethna, and Steve Hewitt have exposed and investigated the history of Canada's national (in)security apparatus.[25] State surveillance of suspected communists, the harassment and interrogation of gays and lesbians, immigration control and gatekeeping, targeted oppression of feminists and the women's liberation movement, and other forms of social policing that transpired in Canada during the Cold War era illustrate the depths that state power sank to as well as the priorities of the government actors who determined and "protected" the country's so-called national interests. But scientists, too, informed officials and decision-makers close to the centre of the post-war Canadian state, as evidenced by the concerns, voices, and activities of the men (and select women) associated with Canada's Defence Research Board.

The DRB's history has received increasing attention in the past two decades. Jonathan Turner's 2012 dissertation provided an updated and sweeping institutional history of the entire organization, building on Captain Donald Goodspeed's 1958 official history.[26] Historian Jason Ridler wrote a biography of the DRB's first chair, labelling Omond Solandt – whose position and power in post-war Ottawa should be familiar to every historian of Canada – a "maestro of science."[27] Solandt chaired the DRB between 1947 and 1956, overseeing the emergence and growth of Canada's official peacetime military research program and wielding considerable influence over the direction and distribution of federal research capital allocated to the armed services under the national defence budget. He and other leading scientists who comprised the DRB's core expert pool during the early Cold War years – H.M. Barret, Otto Maass, E.L. Davies, Archie Pennie, and Guilford Reed, just to name a few – received detailed scrutiny in Donald Avery's most recent study of Canada's involvement in biological, chemical, and bacteriological warfare research, *Pathogens for War*.[28]

Avery's work delves into the history of Suffield Experimental Station (SES), one of several research establishments operated by the DRB during the peak years of its mandate in the 1950s and 1960s. Scientists, engineers, and technicians carried out research and associated activities for the DRB at National Defence Headquarters (NDHQ) in Ottawa and in laboratories and field stations across the country. Aside from SES, however, the scientific activities of the DRB's other military research facilities have received little direct attention. Setting aside a valuable collection of internal institutional histories written or edited by former DRB employees, the two clear published exceptions are from Andrew Godefroy and Edward Jones-Imhotep, both of whom examine the history of Canada's Defence Research Telecommunications Establishment (DRTE): Godefroy from the perspective of Canadian space science and Jones-Imhotep from the perspective of radio communications and ionospheric research in northern Canada.[29]

Among the locations chosen by senior DRB authorities for studying and generating scientific and medical knowledge for the armed services was Churchill, Manitoba, a small community on the western shore of Hudson Bay in the northeast corner of the province. Because of its location, Churchill served the US military well as a useful node on the

Figure 0.1. Defence Research Board coat of arms, ca. 1950s–60s. Photograph by Matthew S. Wiseman.

Crimson Route, a series of landing strips jointly planned by military leaders in Ottawa and Washington for ferrying airplanes and materiel from North America to Europe during the Second World War.[30] The Allies never used the transport route during wartime, but the airfield at Churchill would later come to serve as a military foothold in the North. Personnel and equipment from southern Canada and the United States soon expanded the on-site facilities; the result was Fort Churchill, a military garrison for year-round training and research in Arctic warfare. The DRB's northernmost facility in Canada was Defence Research Northern Laboratory (DRNL), a scientific research building constructed as part of the Fort Churchill garrison.

In April 1966, the year after DRNL closed its doors, former superintendent and long-time DRB employee Archie Pennie compiled a large and important collection of personal reflections written by several scientists who worked and performed research at the Fort Churchill laboratory between 1947 and 1965.[31] Pennie's collection is the single most detailed account of DRNL's history, which is told through the experiences and perspectives of the employees who carried out Ottawa's military-sponsored Arctic research program on the ground in northern Manitoba. Except for the work of Margaret Carroll and historian Andrew Iarocci, and one chapter in Goodspeed's official history of the DRB, the institutional story of military activity and scientific research at Fort Churchill and in the nearby subarctic areas of Hudson Bay remains largely untold.[32] This book begins to fill out that story using newly opened military records, government reports, scientific publications, and selected oral interview material, ultimately drawing attention to the motivations, priorities, and implications of the militarization of Arctic research in Canada between 1945 and 1970.

Canada's Long Second World War

War is central to understanding and contextualizing the history and historical actors at the core of this study. As historian Alex Souchen makes clear in his compelling work on the complicated and environmentally harmful impacts of surplus munitions disposal, the end of the Second World War was a harbinger of new issues in Canadian politics and military affairs. Demobilization marked a beginning rather than an end, or a "rebirth" (Souchen's word) that gave new life to war machines and created new problems for the armed services, government officials, civilians, and the environment.[33] In this context, the phrase "legacy of war" is misleading. Reconstruction in post-1945 Canada was less a consequence and more a prolongation of war.

If reconstruction in post-war Canada amounted to a *continuation* of war machines, how else did the Second World War continue into the immediate and early post-war period, and more importantly, how did wartime continuances affect policies and people in Canada? Building on Souchen's call for historians to study extended processes as fundamental to historical analysis, this book encourages scholars of Canada and the Cold War to consider the possibility that other elements of Canada's Second World War experience did not end in September 1945.[34] As the following chapters demonstrate, military leaders, government officials, and senior scientists in Canada understood and approached the early post-war years as a time of continued warfare. The enemy had changed, from Germany and Japan to the Soviet Union and China, from Nazism and imperialism to communism, from total war to total annihilation, but the threat to Canadians and their democratic freedoms remained. Clearly, the circumstances of war were alive in the transition to peace, shaping how Canadians in positions of government power perceived the world they were now defending.

Frontier Science focuses on one wartime continuance in particular: the rise of military science. In Canada after the Second World War, senior defence officials not only continued but also – through targeted funding and resource allocation – expanded the breadth and scope of military-sponsored scientific and medical research performed on Canadian soil. Chemical, biological, and bacteriological warfare research at Suffield in southern Alberta and at Grosse-Île on the St. Lawrence River near Quebec City – two active wartime research facilities – quietly resumed operations in peacetime under the auspices of the newly formed DRB. Operated during the war as part of Canada's contribution to the Allies' secret military testing program, both facilities implicated Canadian scientists and military leaders in controversial trials and experimental research that affected participating soldiers and the environment, as detailed in the valuable work of historians Brandon Davis, Susan Smith, and Amanda McVety.[35]

After the war, senior defence officials in Ottawa relieved the National Research Council (NRC) and its associated laboratories of the responsibility for military research. But DND authorities did not divest the government of its existing commitments to military research, opting instead to transfer ongoing projects to the newly created DRB. Hundreds of scientists, engineers, technicians, and graduate students across the country, in the physical as well as social sciences, applied for DRB grants and gained access to federal capital allocated for military research under the postwar national defence budget; those government monies paid for new and continuing research projects.[36] The experience of war undoubtedly influenced the attitudes and priorities of the government authorities and scientists who formalized and approved the policies for establishing and operating the DRB, Canada's principal source of government capital for military research during the Cold War.

To contextualize and investigate exactly what shaped the constitutions of the privileged and powerful historical actors at the core of this study, it thus becomes useful (and arguably necessary) to adopt the framework of Canada's long Second World War. This framework posits that military leaders and senior officials in DND deliberately prolonged their wartime responsibilities to advance the state's military interests after the war. Several military leaders, government officials, and top defence scientists embraced the concept of military research to advance individual and collective agendas for national security, knowledge acquisition, and scientific progress.

To be clear, Canada's long Second World War is not a comment on the men, women, and children who lived at war between 1939 and 1945. The soldiers, sailors, pilots, nurses, doctors, factory workers, farmers, and supporting citizens who fought, worked, and died overseas or on the home front deserve our utmost respect and gratitude. They sacrificed for the Allied cause, helped to expedite victory, and freed the world of a totalitarian evil. Instead, this framework offers a way of thinking about the purposeful perpetuation of wartime policies and practices for the benefit of specific individuals and institutions. Consider the men who conceptualized, established, and developed the DRB, for instance. In the immediate months and years after the formal surrender of Japan and the cessation of armed conflict, Omond Solandt, Wallace "Wally" Goforth, and other influential decision-makers in the Canadian defence department continued to advise and serve the federal government *as though they were still at war*.

Merely calling the conflict that military and defence officials in Ottawa waged after September 1945 the "Cold War" ignores the imprint of the previous six years. The mandarins of Canada's post-war DND who encountered a new foe in the transition to peace viewed the world in which they lived and worked through a wartime lens. Experience had made clear the value and utility of military science, both for the battlefield and for the laboratory, leaving senior government authorities with the impression that the union of scientist and soldier was mutually beneficial for the armed services and the country. These individuals understood the critical importance of carrying forward Canada's wartime momentum in national research, and they tackled the new military problems of the immediate post-war period with the belief that science was *the* answer to the country's pressing defence issues. The experience of war left an indelible mark on the men who fostered a peacetime military research organization (the DRB) within DND. This underscores the importance of interpreting the events documented in the following chapters through a prism of both the old and the new: Canada's long Second World War.

The principle of functionalism is also useful for understanding the history of military-sponsored Arctic research in Canada during the early decades of the Cold War. Devised by Canadian diplomat Hume Wrong, functionalism stated that Canada's position and voice should reflect its contributions to international security.[37] As a middle power, Canada had little choice but to rely on its Western partners to secure Canadian interests after the Second World War. The term middle power emerged in the immediate post-war years as a way of explaining Canada's international role or status in relation to other countries. "Canadians were of greater consequence than the Panamanians but could not take on the obligations of the Americans, or even the French," former diplomat John Holmes wrote in 1984.[38] Holmes's work referred specifically to Canada's post-war foreign policy and to the pragmatic approach of senior Canadian officials, who viewed multilateral bodies – the UN, the Commonwealth, and NATO – as opportunities to assert Canada's "voice" and wield increased influence in international affairs.[39]

In this book, the term middle power refers specifically to Canada's position relative to Britain and the United States. Selected Canadian scientists informed and influenced Canadian defence policy by defining and emphasizing Canada's unique or special contribution to North Atlantic security. Aware of the centrality of science to security and national defence for military planners in both London and Washington, senior military and defence officials in Ottawa created the DRB in part to strengthen Canada's commitment to the perceived post-war military needs of the country's two closest partners. As historian Isabel Campbell argues when analysing the Canadian brigade deployed to Germany in the early 1950s, in a world dominated by nuclear arms, authorities in Ottawa emphasized non-military means to assert Canada's limited voice internationally and move the needle with respect to the country's policy position among the North Atlantic security partners.[40] The DRB's primary mandate was to provide scientific and technical assistance to the Canadian armed services, but this could only be achieved through international cooperation. To gain and maintain access to US and British resources and expertise, DRB officials supported military research in areas considered distinct or unique to Canada and Canadian expertise. The decision to invest heavily in Arctic research was thus a deliberate and functional approach intended to leverage Canada's northern geography into political, economic, and military capital. In so doing, DND strengthened Ottawa's security position in an evolving Cold War world while maintaining Canada's territorial sovereignty and bolstering the country's defences in the North.

Northern Canada and the Arctic as a "Frontier"

As the unceded territory of sovereign Indigenous nations, the Arctic is a space of vibrant peoples and communities who understand and view this vast region of the world as their homeland.[41] Outsiders have long travelled above the Arctic Circle, intruding on the lives and lived spaces of the region's Indigenous peoples to pursue European and settler-colonial goals of exploration, conquest, and exploitation.[42] The ambivalence and allure of the Arctic, as conceived and remade in the minds of newcomers, has conjured myth-making tales of untouched or pristine wilderness and inspired colonizers to lay claim to Indigenous lands, waters, and resources. Central to the mythology of the Arctic is the concept of the "frontier" – an imaginary geography that has featured prominently in debates about Canada and its northern history.

Frontiers loom large in past and present understandings of North America, largely as regards the imagined geographies of the American West and the northern regions of the continent. The word "frontier" is associated with saloon shoot-outs, male rivalry, range wars, and other Hollywood depictions of life in the so-called Wild West. It speaks to individual freedom, the spirit of adventure, the impulse to explore, the dream of finding gold, and the prospect of a new life. But it also represents an endless and seemingly elusive search. When American settlers reached the Pacific and declared the western frontier closed at the end of the nineteenth century, an identity crisis ensued that sparked a national search for a new frontier. Americans headed north in large numbers during the Klondike–Alaska Gold Rush, helping to cement the Far North as the next great frontier.

American historian Frederick Jackson Turner formulated the frontier thesis in 1893, theorizing that the perceived availability and abundance of unsettled land was the key determinant of national development in the United States.[43] He defined the American frontier as the zone between the populated eastern regions of the country and the supposedly unpopulated and open wilderness to the West, suggesting that the pursuit of "free land" was central to the creation and development of American identity and national unity. Contact with the wilderness, Turner argued, transformed colonists into Americans. Settlers expanding west across the continent encountered opportunities and challenges unsuited to the traditions and institutions of Europe, forcing the eventual collapse of class distinctions and the rise of a unique democratic American society.

Turner's frontier thesis gained popularity among historians in the United States, Canada, and elsewhere during the twentieth century. Most applied the concept within the unilateral framework of national boundaries, exploring the political, economic, and cultural contexts of a chosen country in isolation or comparatively through contrasting analyses with bordering states. In Canada, historians A.R.M. Lower and Frank Underhill considered and debated Turner's framework as a method for exploring and explaining Canadian history.[44] Conceptualizing the Dominion as a frontier allowed scholars to envision and debate how British and French settlers expressed distinct "Canadian" identities. There were clear distinctions in the history and socio-economic development of the Canadian West and American West, however. Settlement and immigration patterns, political and economic laws, and ongoing ties to Europe all distinguished Canada and the settler-Canadian experience from similar nation-building processes in the US, as did the evolving concerns over US encroachment northward and the corresponding protectionist Canadian response(s).

In the context of Canada, the frontier concept had strong associations with the North. As an integral part of the settler/southern-Canadian imagination, linked in so many ways with national symbolism, the North assumed a mythological status akin to the West in American cultural identity.[45] Terms like the "uniquely Canadian frontier" and Canada's "last frontier" fit and encouraged discourses of exploration, fortitude, endurance, and survival, shaping how Canadian settlers communicated and remembered "their" northern identity.[46] The correlation and connection between the waning of the Wild West and the rise of the Far North (Alaska) also drove associations linking the frontier concept with northern Canada and the Arctic, and Euro-Canadian encounters with "pristine wilderness" became apt descriptors for defining and exploring national history as a unified or collective experience.

But the northern frontier differed from the western frontier. Canadian settlers lived in an environment and climate largely perceived as diametrically opposed to the American West, and this led to distinct understandings of Canada's geography and national experience. Although colonization and conquest on both sides of the border drove similar pursuits for land, the perceived hostile environment of the northern frontier conditioned settlers to view the North as a resource-rich area suited to transient exploitation rather than permanent settlement. Canadian settlers assumed what literary theorist Northrop Frye termed a "garrison mentality," frequently viewing their environment as a threat that had to be overcome.[47] The absence of large urban centres and large private corporations, combined with vast, densely settled areas, also meant that the northern frontier represented a space suited primarily to economic and industrial development. In short, as Barbara Giehmann writes, "the northern frontier is more than a western frontier transplanted to the north because it has its own imaginary geography."[48]

Freed from the burden and limitations of the national development debates of old, current scholarship on Canadian history has fostered a very different understanding of the frontier concept and its role in shaping past and present perceptions of northern Canada and the Arctic. The frontier no longer represents a "nationalizing force," to quote historian William Coleman, who, writing in the pages of *American Historical Review* in 1966, tested Turner's thesis as a method for interpreting key developments in the history of science in the United States.[49] Rather, the word frontier is both a cause and a symptom of failed attempts to engage the North using outside knowledge and communication. If perceptions are relative to spoken and written language, as the "weak" hypothesis of linguistic relativity suggests, then how will Canadian settlers ever purport to know and communicate northern geography, climate, resources, traditions, or any other lived realities of the North and its traditional stewards using English or French?[50] In the words of historian and feminist scholar Joan Sangster: "Should we not turn the North, including its cultural construction, over to its own First Nations and peoples for re-visioning as they work through the project of decolonization and self-determination?"[51]

Not surprisingly, then, the word frontier, understood in both conceptual and spatial terms, has declined drastically in use in historical scholarship about northern Canada and the Arctic. The term remains useful for interpreting the very constructs of the North that Sangster brings into focus, as historians Peter Kikkert and P. Whitney Lackenbauer illustrate in their research linking contemporary tropes of frontier masculinity with the "Prospector-Soldiers" of Yukon in the First World War.[52] The term also finds reference in the context of the post-1945 period, albeit – as historical geographer Matthew Farish explores in his work on the "engineered" Arctic that became a so-called frontier for Cold War science in Canada and the US – "both imaginatively and materially."[53]

Whether romantic or utilitarian, common perceptions of what constitutes Canada's northern frontier illustrate the narrow and flawed positionality of outside observers. As historian Janice Cavell points out, contrary to the view that Canadians have assumed a northern identity since Confederation, nineteenth-century settlers disregarded the Arctic as a part of the nation, "and even in the twentieth century they found it difficult to integrate the cold, remote 'second frontier' into their paradigm of national development."[54] Only fairly recently have Canadians embraced the North as a national symbol, perhaps underscoring the difficulty and resistance encountered by literary writers and historians who struggle to incorporate the North into national narratives.[55]

Yet the role of science in shaping understandings of northern nature features predominantly in the study of Canadian history. The authority associated with scientific advances in meteorology, climatology, and terrestrial magnetism supported flourishing settler ideas of expansionism and nationalism, for instance, and the discovery of the north magnetic pole near Hudson Bay represented Canada as a northern country. "These ideas formed the intellectual tap-root of the 'northern myth' seized upon by Canadian imperialists and other nationalists late in the nineteenth century," observes historian Suzanne Zeller.[56] While the debate over Canada's enigmatic, setter-colonial northern identity remains open, the sociocultural role of scientific knowledge has deep-seated and undeniable roots in discourses of northern Canada and the Arctic.

Frontier Science

The title of this book is not meant to suggest or imply that Canada was or is a frontier country. Canadian historians have spilled much ink debunking the myths and cultural baggage associated with the imagined frontier and the traceable, tangible impacts of western and northern development. As scholar Jesse Thistle points out, there never was an American or Canadian frontier, but as a concept, the frontier was a powerful idea that drove settler expansion and justified the forceful invasion of Indigenous land.[57] European settlers, backed by successive interventionist governments, punched west and north, constructing railways, homes, farms, ranches, factories, and mines to spur development projects and protect the interests of "their" country. They colonized the territories and communities of sovereign Indigenous nations, violently displacing the traditional stewards of the land, killing off its animal life, and stripping away its resources.

Having been founded as a mechanism to protect the commercial interests of the Hudson's Bay Company (HBC) in Rupert's Land and to police the so-called frontier territories of the Canadian West, the North-West Mounted Police (NWMP), the forerunner of the RCMP, violated territorial treaties between Indigenous nations and the Crown by forcibly removing Indigenous peoples from their traditional lands during consecutive uprisings in the nineteenth century.[58] Canada's first prime minister, Sir John A. Macdonald, was an architect of policies intended to starve entire communities, "clear" the western plains, and assimilate Indigenous children into Euro-Canadian society through forced familial removal and the horrendous residential school system.[59] The settler-colonial history of North America is one of violence and successive invasions. That history illustrates the power and persistence of the frontier myth, which has repeatedly been co-opted and reimagined by discoverers, explorers, settlers, pioneers, state representatives, police, doctors, scientists, and all those others who have laid claim to Indigenous lands and resources.

Scholars Brenda Macdougall and Brittany Luby, among others, have demonstrated that the memories and recollections of Indigenous elders are particularly important for unearthing and contextualizing the real-life impacts of settler-colonialism in Canada.[60] This is precisely because Indigenous stories represent more than sources of information about the past; they are themselves part of an individual and collective identity that links people together and connects them with the land, water, and resources of their traditional territories.[61] As cultural anthropologist Jeffrey Schiffer explains, long before Europeans arrived, Indigenous societies on Turtle Island practised complex ways of knowing and living – traditional forms of mathematics, science, health, and economic trade – that were "grounded in an understanding of the interconnectedness of all life."[62]

Of course, Indigenous sources and oral histories are also indispensable for understanding the complete picture of Canada's Cold War history. "Stories are the foundation of our resurgence, and it is how we have been so resilient in the face of continued colonial engagement," historian Lianne Leddy writes in her important work on resistance among the Serpent River First Nation to uranium mining at Elliot Lake, Ontario.[63] Extractive industrial work in the so-called Uranium Capital of the World produced environmental contamination of the Serpent River, poisoning a previously safe source of community drinking water and damaging traditional hunting and fishing resources. Anishinaabek saw their role as traditional stewards compromised by their inability to prevent or mitigate the damage caused to their territory, a reality of post-war paternalistic governance in Canada that Leddy calls "Cold War colonialism: community leaders had the appearance of control, but their choices were limited by economics, and by both policy and practice."[64] Leddy uses the word resurgence to foreground Indigenous voices and experiences, and in doing so illustrates the role of traditional knowledge and oral histories in producing scholarship that addresses past and ongoing injustices.

Oral testimony has been especially relevant and beneficial to the telling and decolonizing of the history of Canada's northern regions, where settler mythologies have distorted the actual experiences of Indigenous peoples and newcomers alike. For this very reason, settler-scholars who study and write about northern Canada and the Arctic find it a daunting challenge to compose a historical narrative. As geographer Emilie Cameron writes in an eye-opening book on Samuel Hearne and the 1771 "Massacre at Bloody Falls," stories about the North and its Indigenous peoples, conceived and told by "non-Inuit, non-Indigenous peoples or settlers," have played a central role in enabling and justifying settler-colonial claims to land and resources.[65] Cameron suggests that settlers have a responsibility to assume ownership of narrative composition and produce different ideas/stories about the North, thus addressing the past and ongoing relationship of power and violence to memory. For settler-historians, this means interrogating the sources and methods used to reconstruct past events and inform historical inquiry – a reckoning with the systemic practices that have long shaped and *obscured* the history of northern Canada and the Arctic.

This book does not purport to address or overcome issues of narrative composition specific to the telling of Canadian history. As a White, male scholar who lives and works in southern Ontario, I gazed upon the North from an outside perspective, scouring national repositories in Canada, Britain, and the US to research and write a selective history, informed largely by the very government and institutional records that today are increasingly recognized as biased, flawed, and misleading. I travelled to Churchill, Manitoba, where I spent two weeks researching community records at the public library and discussing local history with the residents, who willingly shared their experiences and recollections of the now-demolished Fort Churchill military base (see the Acknowledgments). That trip was both enriching and enlightening, but it also made it clear to me that I was incapable of researching and writing a history reflective of the lived experiences of the Indigenous peoples and settlers who historically or presently inhabit the parts of northern Canada and the Arctic discussed in this book. Rather, my contribution to the literature would reflect and interpret the experiences and positionality of historical actors from the South – individuals and institutions that sought to define and protect Ottawa's interests in a vast region perceived as inextricably bound to Canadian security and national defence at the beginning of the Cold War.

Science has played a central role in outsiders' approaches to the Arctic. In attempting to understand and overcome the challenges and material agency of this imagined space, where the harsh yet fluid environment represents a resisting force, newcomers to the Arctic – explorers, fur traders, missionaries, whalers, doctors, state representatives, journalists, and scientists and engineers, to name a few – have relied on science in myriad ways.[66] Knowledge in this context refers to Western scientific processes and traditional Indigenous ways of knowing. As Sverker Sörlin, Michael Bravo, Andrew Stuhl, and others have explored, the intimate and practical environmental knowledge of Indigenous communities in the Arctic has made possible many of the movements and actions of outside visitors.[67] Sophisticated technologies such as the kayak and the dogsled serve as proof of how well Indigenous peoples adapted to the harsh conditions of the Arctic, as Shelly Wright explains.[68] Similarly, caribou clothing is more warmer, more versatile, and more effective than any natural or artificial material developed by outsiders in the South, and the igloo is a more practical winter shelter. Newcomers to the North sought to appropriate, alter, and improve on all of these technologies for their own purposes.

Taking inspiration from Edward Jones-Imhotep's original and insightful research into the history of technological failures in the Cold War, the following chapters view the militarization of Arctic research in postwar Canada as reflective of Western science's fallibility. Radio geographies, Jones-Imhotep asserts, represented *objects* – "the products of a convergence of existing mapping practices, interventionist government ideologies, and the detailed measurements of electromagnetic phenomena throughout the Canadian North" – that revealed the limitations of machine-based order even while informing how southerners imagined the North.[69] The technological failures of radio communications in northern Canada shaped how defence scientists understood and defined the natural hostilities and geopolitical vulnerabilities of the nation-state; similarly, the limitations and fallacies of experimental research determined how DRB scientists conceptualized the human challenges of effective and efficient military operations in northern Canada and the Arctic. From this perspective, scientific failures played an important role in shaping contemporary understandings of northern Canada and the Arctic as a multifaceted space of hostile nature, military weakness, and human fragility.

Indeed, the history of Arctic research in Canada during the Cold War is important precisely because it illustrates how science helped assert and protect a specific Canadian identity. That identity was conceived and propagated according to the sociopolitical and liberal-democratic values of several like-minded actors who leveraged the power and authority of scientific expertise to inform policy-level discussions in the defence department. Some scientists gained access to government resources and served as agents of the state, reinforcing deeply entrenched settler-colonial views of Western superiority and dominance over nature, land, peoples, and resources. Others used federal funding to carry out apparently apolitical research or so-called pure science for the advancement of knowledge and the betterment of society.

All scientists acted with good intentions, pursuing the similar cause of a safe and prepared Canada, but individual pursuits also characterized the choices and actions of the scientific experts who conducted research for the federal government during the Cold War. And northern authority was never uniform. As Tina Adcock explains, the contested concept of expertise is a central theme in Arctic history.[70] Outside travellers, explorers, scientists, and government representatives have long competed among one another as well as with local knowledge holders in their efforts to define and claim northern authority according to shifting categories that resist generalization. Stephen Bocking's research suggests that the relationship between knowledge and space is particularly useful for studying human interactions and explaining contested notions of authority in the North.[71] Settler-scientists operating in northern Canada and the Arctic conceived their own Whiteness as part of an imagined boundary for asserting social and cultural superiority. "They defined this boundary not only in terms of methods and theory, but in terms of race, by distinguishing between the objective non-racial 'white' identity of science and the racialized identity of Indigenous knowledge," Bocking observes.[72]

Whatever the individual and local contexts, the shared experience of performing Arctic research for the armed services linked Canadian scientists together in several ways. As Mary Jane Logan McCallum's research demonstrates, medical professionals in twentieth-century Canada defined Indigenous health in terms of isolation.[73] Southern perceptions of the North as an isolated space informed how health authorities justified the expansion of federal services that brought Indigenous peoples under increasing government control and reinforced the assimilative goals of the settler-colonial nation-state. Discourses of isolation are also useful for contextualizing the militarization of Arctic research as a distinctly Canadian frontier initiative. Military science occurred elsewhere in the circumpolar region during the Cold War, of course, but the Arctic research activities, experiments, and projects supported and funded by the DRB served specific Canadian needs.

The findings discussed in this book make no attempt to dismantle settler-colonial binaries that contrast Western and Indigenous knowledge systems. In recognizing the agency and validity of Indigenous traditional knowledge, settler-scholars must be sensitive to past and present attempts at reconciling the real and imagined differences to epistemological approaches. Whereas scholars have historicized traditional knowledge to show how newcomers (mis)understood and (mis)represented the Arctic and its inhabitants, Indigenous peoples have an outlook towards their traditional homelands and ways of knowing/living that differs from Western or Euro-Canadian world views. Inuit knowledge is strong, resilient, and adaptable, regardless of the verbal and written perceptions of newcomers and settlers. As Julie Cruikshank emphasizes when analysing knowledge production in the North, local and traditional ways of knowing are autonomous from Western ontology.[74] Inuit leader Terry Audla, who grew up in Qausuittuq (Resolute Bay), has called for the coexistence of traditional and Western knowledge systems.[75] This requires a new approach to cultural awareness that respects knowledge sharing as an essential, cooperative function of reconciliation. But the responsibility for change rests with newcomers and settlers, whose words and actions must express humility and acknowledge the fallibility of outside knowledge and technology.

Forced relocation, land dispossession, medical treatment without consent, and severe socio-economic disparity represent a small but sobering fraction of the lived experiences and generational trauma thrust upon Indigenous peoples and communities in the name of national progress and settler-colonial domination.[76] As historians Liza Piper and John Sandlos explain, the social and economic drivers of imperialism are root causes of the disease spread and population decline in the North that set the stage for Canada's northern industrial economy.[77] But the history of human activity and cultural interaction in the North also resists clear-cut hierarchical explanations. While the voices and perspectives of the oppressed offer a fuller picture of the past, the actual processes of knowledge production on the ground were historically multilayered and reflect dynamic cultural exchanges that often fractured and contested binary power structures.[78] The agency of Indigenous peoples and communities, moreover, indisputable in past events and present-day human interactions the world over, underscores the importance of local contexts for understanding settler-colonialism in specific contact zones, not the least including the many spatial and cognitive spaces of northern Canada and the Arctic.[79]

Although settler-scholars must be cognizant of the overt and latent colonialisms embedded in the stories collected and retold to inform historical consciousness, where the history and militarization of Arctic research in Canada during the Cold War is concerned, it is my contention that historians have much more to learn about the extent and impact of military and non-military research – scientific activities that, I argue, can be traced to concerns and motivations induced by the Cold War itself. Through access to information requests and rigorous archival research, informed and guided by decolonized research practices and a deepening engagement with a wide array of sources and scholarship, historians can learn more about how the prolongation of wartime science practices affected policies, institutions, communities, and the environment, and ultimately apply those lessons to inform current and future discussions of Canadian history. In short, the colonial archive, although saturated with omissions and biases, still holds important value.

Canada's Cold War Specialization: Arctic Research

Scholars in the US and elsewhere have paid considerable attention to the many implications of military-sponsored research in the Cold War, but only in the past two decades have historians begun to investigate similar themes in the context of post-war Canada.[80] The opening of previously closed records, largely in Canadian, US, and British repositories, coupled with the rise of new and creative research methodologies, has allowed for the collection and use of textual, verbal, and visual sources that offer new and exciting ways to read and interpret the past. The various dimensions of Arctic research in post-1945 Canada collectively demonstrate the long reach of the military-industrial complex and its influence on the health and safety of wildlife, individuals, communities, and resources. This is particularly apparent in the history of land dispossession and (ab)use, especially considering that the Canadian military appropriated and (mis)used Indigenous lands for training during the twentieth century.[81] Negotiations between state representatives and Indigenous leaders often resulted in mutual agreements governing the military's use of traditional lands, but in some cases the federal government acted independently and simply expropriated territory. Training took place from coast to coast to coast, in southern regions of the country, and extended north into subarctic Canada.[82] The armed services tested both men and equipment, performing trials of ground weapons, aircraft, carrier and supply vehicles, and newly refined or upgraded kit, often displacing communities and leaving behind a damaging cultural and ecological imprint.

The state expropriated and used Indigenous lands for military purposes other than training, of course. Environmental awareness, activism, and land claims have brought increasing attention and scrutiny to the widespread harm caused to Indigenous lands and resources by the federal government and led to extended processes of remediation and the removal of hazardous pollutants and contaminants from soil, sediment, and water. The clean-up effort for the Distant Early Warning (DEW) Line, one of three radar strings constructed in the 1950s as part of the North American air defence network, took three years and cost $575 million.[83] Inuit and Inuvialuit worked in partnership with National Defence, Health Canada, Indian and Northern Affairs Canada, and Environment Canada, planning the safe removal of hazardous materials for the sake of the immediate and long-term health of local ecosystems and communities. Elsewhere in the country, both on land and offshore, unexploded ordnance and production residues generated by military activities represent sources of ongoing contamination that continue to threaten food supplies and essential resources.[84] The state's insatiable appetite for land in the name of national security, military expediency, and industrial preparedness impacts Indigenous and settler-communities long after the training and trials end.

Understanding how and why military research affected people and the environment in northern Canada during the Cold War is a challenging but important task. Science mattered to key decision-makers in Ottawa, and the history of Canada's peacetime military research organization reveals the reciprocal relationship between security and knowledge. Senior authorities in the DRB implemented and controlled a wide scientific research program at several facilities across the country, with each designed to investigate military problems and support the research and development needs of the armed services. The DRB's facility in northern Manitoba, Defence Research Northern Laboratory, was established in 1947 as part of the Fort Churchill military base, on the western shore of Hudson Bay, in the northeast corner of the province.

The Fort Churchill garrison stood three and a half miles southeast of the town of Churchill near the mouth of the Churchill River. The US government established a military post there in 1942, when American personnel constructed the Crimson Route airfield for potential transport and evacuation services to and from Europe. Four years later, the Cold War–era military base came into being when the Canadian Army took over control of the wartime post and renamed it Fort Churchill, reviving the name of a former HBC trading post and also distinguishing the garrison from the nearby town.[85] Senior officials in the defence department subsequently designated the site as a joint Canada–US experimental testing station for Arctic warfare.

As a multipurpose research facility, DRNL was an important venue for DND and the Canadian military between 1947 and 1965. The importance of military research performed in or related to northern Canada emerged towards the end of the Second World War, when senior military officials began to assess the strategic importance of the Arctic for post-war security. Fearing the possibility of bomber and rocket attacks on urban and industrial centres near the Great Lakes, military planners in Ottawa looked north to secure vital air routes over the Arctic. Military strategists considered an air attack launched by expeditionary forces lodged on the fringes of the Canadian Arctic potentially more dangerous than an attack launched from the Soviet Union itself. Although the likelihood of an offensive ground incursion in the North was remote, federal authorities funded science in this part of the country with the goal of understanding and overcoming the practical issues associated with northern military operations.

The establishment of a permanent research laboratory at Fort Churchill, where the US military had been stationed since the war, gave the Canadian government a stake in matters considered important to North Atlantic security. As a research space unique to Canada's northern geography and climate, DRNL fulfilled a specific need. Science was important to military planning in both Britain and the US, and senior

Canadian defence officials funded research in the North to support the wider scientific and technical military needs of Ottawa's two closest allies.

Arctic research performed in Canada during the Cold War was not restricted to Fort Churchill, however. DRNL served as a base of operations for scientists and military personnel engaged in various projects related to security and national defence, both within and outside Canada. Scientists often travelled through Churchill to gain access to distant locations in northern Canada and the Arctic, and military funding paid for cold-climate simulation research conducted at government and university laboratories in southern regions of the country. A post-war interest in the defence and development of northern Canada coalesced into strong relations among government departments, military leaders, and scientists. In this effort, DRB authorities pursued the Arctic sciences to address and pronounce Canada's commitment to North Atlantic security and Western defence.

This book highlights the DRB's emergence and growth as a means to contextualize and investigate the social, environmental, and political implications of Arctic research in Canada during the first two decades of the Cold War. The circumstances documented underscore the influence of scientific ideas, the power of federal priorities, and the function of government administration in Canada during a significant period in world affairs. In a Cold War world dominated by security politics and the spectre of nuclear war, Canadian officials relied on scientists to navigate some of the country's most pressing military issues. Scientific research addressed crucial concerns about northern Canada, including long-standing government anxieties over territorial control in the North and specifically the role of science and technology in asserting that control. The value placed on science by leading Canadian officials positioned the DRB at the centre of a broader political agenda to protect and strengthen Canada's security and defence interests in the North.

To maximize the benefits of working with Britain and the US, the military officials and government scientists who created the DRB avoided duplication of effort and pursued research in areas where Canada could make a *unique* or *special* contribution to the collective efforts of the North Atlantic security partnership. Decision-makers in Ottawa leveraged the Canadian North to strengthen Canada's national security and international position vis-à-vis the interests of its partners on either side of the Atlantic. Scientists fulfilled their role by playing on Canadian stereotypes about cold weather and the North. Who knew cold, snow, and ice better than Canadians did, and which country of the three North Atlantic partners had the territory and experience to investigate and understand the impact of a northern climate on military personnel?

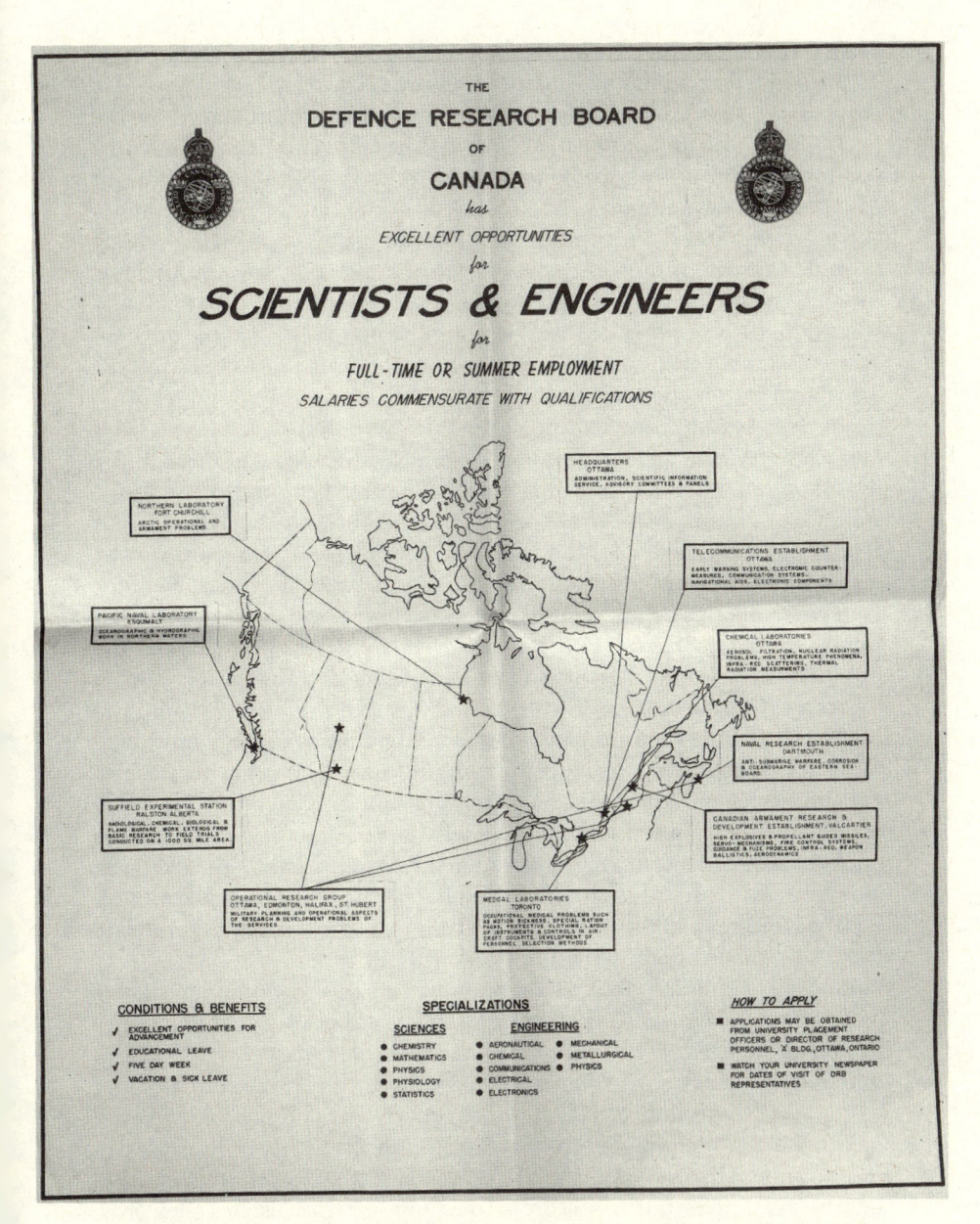

Figure 0.2. Defence Research Board recruitment poster circulated to Canadian universities and academic scientists, ca. 1958–60. University of Saskatchewan, University Archives and Special Collections, J.W.T. Spinks fonds, MG 74, series 10, box 72, file DRB 58–60.

As a case in point, consider Omond Solandt's May 1946 policy brief calling for a federal military research organization (what would become the DRB): "The Canadian Services are regarded as authorities on the problems of life and travel in the Arctic, and it is most important that a consideration of the problems of the Arctic should permeate all our research and development. The results of Arctic research will be one of the most important contributions that we can make to associated powers in return for the results of their Service Research."[86] Backing this statement, Ottawa carved out a niche role for Canada and the Canadian military at the dawn of the Cold War. It would be up to scientists to see these ideas through to fruition and demonstrate that Canadians were indeed world experts on cold and the North.

The frontier trope was central to the Cold War North as conceived and conjured in Ottawa. "Frontiers are of great significance and value in the development, materially and psychologically, of the nation," commented Lester Pearson, the Under-Secretary of State for External Affairs, in a top-secret Cabinet memorandum. "The Arctic frontier promises to be almost as significant in this connection as our western one has been."[87] Pearson wrote this memo in January 1948, a little less than two years after publishing an article in the US journal *Foreign Affairs* that stressed the potential for northern development and the significance of protecting Canada's Arctic sovereignty.[88] He contended that the mineral wealth of the "Land of the Midnight Sun" – silver, gold, radium, copper, uranium – was important to Canada's post-war development, effectively reaffirming the North as a top priority on Ottawa's political agenda.[89] Two years on, Pearson continued to highlight the immediate need to challenge encroachments on Canadian soil and assert territorial sovereignty in the North, underscoring the significance of directing defence and development activities northward to safeguard Canadian interests and cultivate "the importance and potentialities of our 'last frontier.'"[90]

Canadian scientists became fundamental to this vision as articulated by Pearson, which was shared among Ottawa's liberal-democratic elites. Scientists engaged, complicated, and shaped discourses of the North to define and assert Canada's post-war position on the international stage. In so doing, both individually and collectively, they embraced an imagined northern identity, claiming ownership of Arctic research expertise to navigate and overcome the Cold War security challenges threatening their country and way of life. How these individuals and their government sponsors understood and approached Canada's Cold War as a nationally contingent phenomenon underpins the history and analysis covered in this book. While government officials and scientists in Canada shared the values and concerns of many of their counterparts in Britain and the US, the influential historical actors documented in the following pages perceived the challenges of the Cold War through a distinctly Canadian lens. They played to the expectations of their time, leveraging geography and the trope of the "Great White North" to assert national identity and lay (false) claim to Arctic expertise.[91]

More precisely, authority lent a wide group of Canadian scientists access to DND's inner circle, as well as the platform to influence policy-level discussions about the value and role of Arctic research. Ownership of Arctic research expertise granted these scientists access to the financial and physical resources to conduct research for the armed services and several other federal departments; this then reinforced their supposed authority in the eyes of their peers and government sponsors. *Frontier Science* thus attributes the rise and militarization of Arctic research in Cold War Canada to a powerful faction of the country's professional scientific community – a select group of experts and government advisers who believed that novel scientific research was crucial to ensuring national defence and protecting the health and safety of Canadians when it mattered most. The following chapters explore and explain the choices they made, the values they protected, and, most importantly, the consequences of their actions.

Note on Terms and Definitions

Many of the government and military research activities explored and investigated in this book were conducted in subarctic or Arctic Canada, two expansive territorial zones that in terms of topography and ecology are remarkably diverse. From the perspective of leading scientists and military officials in Ottawa during the early Cold War, subarctic Canada referred to the wooded and scrub-covered region below the treeline that encompassed northern Manitoba and Saskatchewan, as well as parts of the Northwest Territories (NWT), the mountains of northern British Columbia, and Yukon. Arctic Canada, on the other hand, referred to the barren region north of the treeline that skirted the north coast of Labrador and stretched northwest across northern Quebec, along the coast of Hudson Bay, to the mouth of the Mackenzie River.[92]

In the following chapters, the "Arctic" refers to the region north of the treeline that includes the Arctic Archipelago as well as the islands and waters situated to the north of the Canadian mainland.[93] The treeline is not as definite a line on the surface of the earth at it is on the map, but rather constitutes a zone or area where the weaker species of trees have disappeared and only the hardy ones remain.[94] In the northern portion of this zone, factors of wind, rainfall, and soil confine even hardy species to the sheltered valleys stretching northward from the main mass of subarctic forest. As the tree cover diminishes, this zone transitions into the Arctic. Churchill is on the treeline, in the northern portion of this zone, on the boundary of subarctic and Arctic Canada.

Between 1946 and 1964, when the Canadian Army held responsibility for Fort Churchill, policy-makers in Ottawa defined the Arctic as the area north of the treeline. Fort Churchill's location on the treeline provided year-round access to a relatively remote subarctic environment suited to northern military science, enabling scientists to study the extremes of climate and terrain encountered by soldiers in northern latitudes. The term "the North" is a more inclusive term and is employed in reference to the regions commonly referred to as "North of 60," including parts of Yukon, the Northwest Territories, and present-day Nunavut. Significant portions of the book concentrate on Churchill and other locations below the sixtieth parallel, referred to generally as subarctic or northern Canada.

Scientists and government officials often referred to research in or about northern Canada as "Arctic research," even though a great deal of this activity took place outside the geographical boundaries of the Arctic. While Canadian scientists conducted government-sponsored research within the Arctic Circle, senior officials painted the North with a broad brush, and science funded by the federal government followed suit. I capitalize the terms "Arctic research" and "Arctic science" for consistency and clarity, but both terms are general references to activities concerning the Canadian Arctic and often refer to activities that took place in other regions of the country. Indeed, that the term *Arctic research* is a general reference with widespread employment in the archival record demonstrates the simple and often flawed perceptions of authorities and scientific experts in Ottawa, who regularly misunderstood and (worse) misrepresented the peoples, animals, lands, waters, and resources of northern Canada and the Arctic.

Important geographic distinctions also exist among the terms "the North," the "Canadian North," and "northern Canada," but I employ all three interchangeably owing to context of the period and the terms and definitions of the governments documents, institutional records, and military files used to research and write this book.[95] According to the 1948 definition used by the Department of Indian Affairs, the NWT constituted the land portion of Canada situated north of the sixtieth parallel, between Hudson Bay in the east and Yukon in the west.[96] By extension, the NWT also included the islands in between the Canadian mainland and the North Pole, including those in Hudson Bay and the Hudson Strait. Under the Northwest Territories Act (Chapter 142, R.S.C.

1927), local administration was the responsibility of a territorial government comprised of a commissioner of the NWT, a deputy commissioner, and five councillors appointed by the Governor General.[97] The council functioned as a legislative body and in an advisory capacity to the mines and resources minister on matters pertaining to the administration of the NWT.

In the post-war context of northern military research, the terms "sovereignty" and "security" are not interchangeable. Internationally, the protection of sovereignty usually refers to state protection of boundaries from foreign interference. Conversely, security commonly describes the methods by which a state protects the well-being of its citizens from a foreign threat.[98] Protection of Arctic sovereignty refers specifically to the protection of Canada's northern boundaries, while the protection of Arctic security refers to the response taken by the federal government to protect the well-being of northern citizens from foreign threats. By large measure, security is the primary focus of this book. Other terms and definitions appear in notes where required to explain the context of the relevant chapter or topic.

1 Fort Churchill and Defence Research Northern Laboratory

In early October 1946, the Canadian Army assumed control of a worn-down encampment near the rock-strewn shores of Hudson Bay in northeastern Manitoba and named the site Fort Churchill to distinguish it from the nearby town of Churchill.[1] Recognizing the importance of the wartime infrastructure constructed on location, Canadian defence officials established a Joint Services Experimental Testing Station (JSES) at the military garrison.[2] There were many reasons behind this decision, including facilities, resources, climate, and terrain. The Hudson Bay terminal provided serviceable rail access, and the Port of Churchill offered deep-sea transportation services through the Hudson Strait from late July to mid-October. As such, Churchill was both Canada's northernmost locale with a year-round rail link, besides having a serviceable harbour that provided access to ocean-going transport during three months of the year. The site also offered a functional aerodrome, a serviceable airfield, and radio and meteorological equipment operated by the Department of Transport. Moreover, Churchill's location allowed the armed services direct access to open territory and bushed areas for a range of subarctic training exercises carried out on various types of terrain.

Officials in the defence department poured resources, time, and personnel into Fort Churchill for several additional reasons. The base itself had offered shelter, amenities, and access to the ground and air transportation services required to facilitate a fully functional northern military station. Arguably, however, climate and terrain were equally important factors. From a meteorological point of view, Churchill offered Arctic-like climate conditions in the winter, including low temperatures and high winds. Located as it was at the junction of the barrens and the treed areas of northern Manitoba, the base was a launch point for the testing of soldiers and equipment over various subarctic terrain and in year-round conditions. The heavy-packed snow near Fort Churchill was relatively easy to cross in the winter, but except for the limited protection and shelter provided by sparce treed areas, the open barren lands provided little protection from sheering winds and cold. The summer months produced warmer weather but a host of new environmental challenges for military training. The layer of earth that sits atop northern Manitoba's permafrost tends to become wet and boggy after the spring melt, producing very spongy walking ground and the perfect breeding conditions for thick swarms of mosquitoes, blackflies, and other irritating biting insects. All conditions, no matter the season, restricted movement of men and supplies, thus creating seemingly endless opportunities for military research and development.

Military interest in the strategic value of the Arctic increased dramatically during the Second World War. Before the December 1941 Japanese attack on Pearl Harbor, the Atlantic, Pacific, and Arctic Oceans isolated the North American continent from external threats. There had long been defences against attacks from the sea; now, the development of the air threat had altered the strategic balance of global security. After Pearl Harbor, North America represented a single geostrategic entity, and continental defence evolved in this context. Signalling the heightened military significance of the North was a combination of wartime events, namely the Japanese invasion of the Aleutian Islands and the capture of Attu and Kiska in June 1942, the establishment of British and Soviet east–west routes for the transport of aircraft, and the initiation of large construction projects by the US. Infrastructure and facilities built to fortify the northern regions of North America included airfields, weather stations, the Alaska Highway, and an oil distribution system named the Canol Project that extended northwest from Edmonton, along the Mackenzie River, to Yukon and Alaska.[3]

Recognizing the growing significance of continental defence and the importance of meaningful bilateral cooperation between Canada and the US, government officials in Ottawa and Washington signed successive cost-sharing agreements during the war. August 1940 saw the creation at Ogdensburg, New York, of the Canada–US Permanent Joint Board on Defense (PJBD); April 1941 saw the signing of the Hyde Park Declaration.[4] The framework of these two agreements laid out the principles and procedures for mobilizing resources in defence of the North American continent. By war's end in September 1945, the US had replaced Britain as Canada's leading partner in foreign investment and trade, and the PJBD had become the principal forum for security negotiations among senior military and defence officials in both North American capitals.

Another global conflict seemed unlikely; even so, the founding of the UN in October 1945 and the promise of post-war peace did not quell international tensions. As government officials in Ottawa and Washington took steps to bolster security and continental defence, state representatives in London and Moscow considered the prospect of rearmament before the hostilities ended. This uneasy atmosphere convinced DND officials of the vital need to develop and fund an active post-war military research program, one that would marshal Canada's scientific brainpower in support of collective mobilization and preparedness.

With the interdependence of Canada and the US clear at the end of the war, officials in both countries expressed concern about the growing ambitions of the Soviet Union and worried about possible foreign activities in the Far North. Across the Canadian Arctic lay air routes crucial to Western Europe. Wartime developments in aviation had opened the northern skies to commercial interests, and at the same time, technology had opened the door to unwelcome and potentially hostile military activity. Recognizing the importance of military cooperation to North American security, authorities at a June 1945 meeting of the PJBD agreed that it was important to test the capabilities and equipment of combined military forces operating in a high-latitude climate in a region viewed as strategically important to continental defence.[5]

As advances in aviation and airborne technology altered the threat against the North American continent, Ottawa and Washington paid increasingly close attention to Soviet activity in the Arctic. US officials proposed to Canada a unified defence plan in May 1946, the Joint Canada–US Basic Security Plan, which was approved by the newly created bilateral Military Cooperation Committee (MCC).[6] The plan declared that the oceans and the sheer size of the Arctic were no longer adequate barriers to protect the northern half of North America against long-range bombers, weapons, or invading armies. This created a policy conundrum for Prime Minister William Lyon Mackenzie King, whose Liberal government needed to mollify a war-weary public while protecting Canadian interests in the transition to post-war peace. Ottawa did both. Demobilization had reduced the size and capability of the armed services, but at the same time, increased cooperation with US military forces provided access to the resources required for post-war defence. As for Fort Churchill, the rising strategic importance of the North afforded Ottawa an opportunity to carve out a niche role for DND in support of North Atlantic security.

Under the 1946 bilateral security plan, military officials in Ottawa agreed to provide one airborne or air-transportable brigade for service in the Arctic. Known as the Mobile Reserve, the brigade comprised three infantry battalions with combat support and service support units. Rebranded as the Mobile Striking Force (MSF) in 1948, this specially trained collection of soldiers and paratroopers functioned as a preventative land element, designed to deter the Soviets from establishing forward operating bases in the Canadian North.[7] At the time, long-range bombers were incapable of round-trip flights over the North Pole, and the MSF theoretically operated to prevent the Soviets from establishing refuelling stations on the fringes of North American territory. In practice, the brigade also promoted Canadian claims to territorial sovereignty by facilitating operational cooperation with the US military – a subtle but important function of Canada's combat forces at the dawn of the Cold War.

To succeed, the MSF required military personnel trained and equipped to land and fight in northern Canada and the Arctic year-round. Ottawa established the Canadian Rangers in 1947, creating a military space for a group of citizen-soldiers from coastal and northern communities.[8] The original concept for the Rangers, writes historian P. Whitney Lackenbauer, "held that civilians, pursuing their everyday work as loggers, trappers, or fishermen, could serve as the military's 'eyes and ears' in remote regions where demographics and geography precluded a more traditional military presence."[9] Equipped with nothing more than obsolete .303 Lee Enfield rifles, 200 rounds of ammunition per year, and armbands, the Rangers relied on local terrain knowledge and traditional methods to fulfil their duties. "With little training and equipment," Lackenbauer continues, "the Rangers could act as guides and scouts, report suspicious activities, and – if the unthinkable came to pass – delay enemies using guerrilla tactics."

While the Rangers took up the important role of supporting the armed services and defending Canadian interests in the North, authorities in the defence department poured considerable resources into training, preparing, and equipping soldiers from southern Canada for northern military service. If a potential invasion in the North required Canadian soldiers, aircraft could theoretically fly and drop the specially trained MSF to form a northern frontline resistance and counter the enemy on the ground or the floating sea ice. Combined with the local support and knowledge of the Canadian Rangers, the MSF functioned as a cost-effective alternative to stationing large standing forces in the North.

Winter Warfare vs. Arctic Warfare

Events during the Second World War cast a spotlight on the unique problem of cold-weather military operations. In the wake of the Russo-Finnish War during the winter of 1939–40 and the German campaigns in the Soviet Union, leading military officers and professional soldiers thought it urgent to consider the challenges of mounting successful military operations in cold regions during the winter months.[10] Canadian military leaders first broached this issue in 1941, developing and publishing a winter warfare training pamphlet titled *Instructions for Winter and Ski Training*. "The many problematic insights that it offered, such as directing soldiers to briskly rub frostbitten limbs with snow to restore circulation, reveal that Canada's 'northern-ness' did not inherently translate into ready-made aptitude for Northern operations," note P. Whitney Lackenbauer and Peter Kikkert in a useful collection of documents detailing the Canadian Army's experience with northern operations between 1945 and 1956.[11]

Responding to concerns about the inability of Canada and the US to defend against a potential enemy attack in the northern regions of the continent, military leaders in wartime Ottawa advocated that Canada specialize in the development and use of equipment and techniques for winter warfare.[12] During the winter of 1941–2, the Canadian Army opened a winter warfare school in Petawawa, Ontario, and conducted various experiments concerning the effects of snow and cold. Researchers also tested power-driven toboggans and employed adapter kits to "arcticize" vehicles for operational use at temperatures as low as −40°C.[13] Additional research was conducted at Shilo, Manitoba, where the army experimented with vehicles and weapons in extreme cold, developed and tested special clothing for dry and wet cold conditions, and performed user trials to assess transportation capabilities over ice and snow. By the end of winter in 1944, Canada's rudimentary understanding of winter warfare had evolved considerably, and the armed services had developed a keener operational awareness of cold-weather techniques and equipment.[14]

Once the Allies realized that winter warfare was unnecessary for achieving victory in Europe or the Pacific, the US Army scaled back its interest in northern operations. Military officials in Canada, by contrast, still saw value in cold-weather training and implemented an independent program to prepare the armed services for tactical winter warfare. A series of exercises performed by the Canadian Army in the winter of 1944–5 – Eskimo in northern Saskatchewan, Polar Bear in northern British Columbia, and Lemming in the area between Churchill and Eskimo Point (Arviat) in the Northwest Territories – attracted the attention of military leaders in Washington and London, who sent observers to document the Canadians' experience.[15] Interestingly, these wartime exercises led some defence planners to assert the plausibility of military operations in the Arctic, setting aside past misconceptions of the impregnable "fireproof house" sitting atop the North American continent.[16]

Military exercises conducted in wartime and early post-war Canada "revealed the critical distinction between *winter* warfare and *Arctic* warfare," although commentators at the time frequently conflated the two concepts.[17] Winter warfare is a significant part of Arctic warfare, of course, but the latter accounts for military engagement under climate and topographical conditions found year-round in an Arctic operational environment. Arctic warfare, in other words, is *all-season* warfare. This realization became increasingly clear to Canadian military officials, soldiers, and scientists beginning in the late 1940s, not surprisingly, in connection with summertime exercises and military training at Fort Churchill.

During the Second World War, as the Canadian Army studied and refined operational techniques in or on the cold and snowy fringes of the Subarctic, military training in northern Canada focused on winter warfare. Some analysts made the mistake of equating winter operations to northern operations, thus betraying their flawed analytical understanding of Arctic warfare. This resulted in a host of problems that the army would have to identify and solve before it could achieve a full operational capability in the North.[18] Participants on military exercises in the Fort Churchill area quickly learned that winter warfare training near major transportation arteries and military bases in Ontario and southern Manitoba was not a substitute for training in the operational conditions encountered in subarctic or Arctic Canada.

Defence Geography and Arctic Canada, Post-1945

To appreciate the full range of challenges associated with military operations conducted at and beyond the treeline, soldiers and their commanding officers required personal experience serving in the North. Testing of men, equipment, manoeuvrability, communication, morale, planning, execution, and all other military considerations was paramount to determining whether the army's capabilities, concepts, and doctrine were adequate for the operational challenges of northern Canada and the Arctic.[19] As will be explored in depth in chapter 4, scientists, engineers, and (especially) those with expertise in operational research participated in field training and devised experiments to optimize the efficiency and effectiveness of Canada's fighting forces. The combination of knowledge and practice produced mixed results. Even so, the directive from senior DND officials was clear: authority over Arctic warfare was necessary for the Canadian armed services.

Colonel Wallace "Wally" Goforth, head of the Canadian Army's Directorate of Staff Duties, who would come to play a key role in the founding and organizing of the Defence Research Board, considered geographical knowledge especially valuable to the armed services. He proposed the creation of a military textbook titled "*Canadian Defence Geography*" in August 1945, and he encouraged officers across the armed services to survey Canadian territory for defence purposes. "Canada offers almost every variety of topographical, climatic and ecological conditions under which wars are fought in any part of the World," Goforth wrote. "The only real exceptions are tropical jungle and desert, yet even these have their close counterparts in Canadian terrain."[20] He thought that a progressive knowledge of military geography was essential for Canadian security and national defence as well as for the effective use of military forces internationally under the direction of the UN Security Council. These ideas would later influence the structure and mandate of Canada's military research program during the Korean conflict of 1950–3.

Goforth was not alone in his assessment of the strategic importance and military value of northern Canada. Matthew Evenden notes that McGill geographer George H.T. Kimble authored *Canadian Military Geography* for DND's Directorate of Military Training in 1949, publishing maps projecting Canada and the North Pole at the centre of the world to educate officers-in-training about the geostrategic air threat facing their country. "Read against the background of the early Cold War," writes Evenden, "the [opening] atlas takes on other meanings: the environment is an instrument of economic and military power; Canada's geography is at once a storehouse of potential and a vulnerable home; geographical knowledge is one element of a defensive and offensive geopolitics."[21]

Such views motivated senior military and defence officials in Ottawa to address Canada's military preparedness in the North after the Second World War. General Charles Foulkes, appointed Chief of the General Staff in late August 1945, and Omond Solandt, the top scientist in the defence department and founding chair of the DRB, prioritized Canadian preparations for Arctic warfare. "Contributions may be made to special advantage in fields which present problems of particular interest to Canada, such as Winter Warfare and also in the fields in which our present establishments and manufacturing potential are particularly suited," stated one of the official recommendations tabled in December 1945 that led to the creation of the DRB.[22] "Our contribution to these problems is likely to be a prerequisite to the free interchange of defence technology between ourselves, the United Kingdom and the United States of America."[23] Cooperation in science was mutually beneficial for post-war North Atlantic security, or such was the determination of the officials who reviewed and ultimately approved the recommendations tabled by Foulkes.[24]

Solandt shared Foulkes's vision for the DRB.[25] Educated in medicine at the University of Toronto, Solandt earned his reputation conducting operational research during the Second World War. He made important wartime connections in the Allied military science community and studied the effects of radiation on the British mission to Japan after the bombing of Hiroshima and Nagasaki. Following the war, he returned to Canada and accepted a newly created position within DND: Director General for Defence Research. Recognizing the increasing geostrategic importance of the Arctic, Solandt focused the DRB's early research efforts on the North. In April 1950 he drafted a five-year plan for Canadian defence research that prioritized Arctic-related projects. "First priority should be given to research work that could only be carried out in Canada and not in other friendly countries," stated the plan tabled before the Cabinet Defence Committee.[26] "This would include work which made use of special features of terrain, climate, etc." In theory, studying the physical environment of subarctic and Arctic Canada would provide military knowledge useful for preparing the armed services to defend the far reaches of the North American continent.

Northern Canada represented a supposedly natural laboratory, which was also ideal for studying men, equipment, and tactics under climate conditions similar to those of the northern Soviet Union. The Arctic sciences would be key to military success should the North Atlantic partners ever need to fight in a cold environment. In this regard, scientific knowledge about the North would reinforce perceptions that Canadians were authorities on the Arctic. As military historian Andrew Godefroy has observed, "understanding how one's adversary fought its wars was a crucial step towards gaining a better understanding of the capability requirements to defeat it on the battlefield – in this case thinking that someday sooner rather than later, that battlefield might be Canada's northern tundra."[27] Conducting scientific research in the North thus doubled as a method for assessing the high-latitude military capabilities of the Soviet Union while reinforcing Canada's special contribution to the post-war North Atlantic security partnership.

Fort Churchill

As it carved out a distinct niche in Arctic warfare, the Canadian military embraced research opportunities specific to the northern regions of the country, thus setting a course for the scientists, engineers, and military personnel who would follow in the first half of the Cold War. Fort Churchill was particularly valuable as a base for army training, trials, testing, and scientific research of military interest under Arctic and subarctic conditions.[28] Canada and the US initiated joint military exercises at the base in the winter of 1946–7, involving approximately 300 Canadian service members and 100 personnel of the US forces. As of July 1947, military officials in Ottawa and Washington had planned further testing for the coming winter involving around 500 Canadians and 300 Americans.[29]

The JSES represented a tangible asset for both militaries, one that offered year-round access to a subarctic location considered ideal for training and testing men and equipment under the climate extremes of northern Manitoba. The base also provided a home for other military units with operational commitments in northern Canada and the Arctic, namely the Royal Canadian Navy (RCN), the Royal Canadian Air Force (RCAF), and the US Air Force (USAF). Fully organized and staffed by Canadian and American personnel, the garrison doubled as a coordinating facility for North American forces in the event they were called upon to carry out active operations against an enemy seeking to establish a foothold in the northern regions of the continent.

In a move to strengthen Canada's military presence in the North, senior officials in the defence department created a short-term construction plan for Fort Churchill, and the army assumed responsibility for repairing the existing facilities. The base did not have running water or an adequate sewage system. Individual huts relied on oil for heat, but delivery of oil by truck from town was inefficient and frustrated personnel on base. New construction alleviated some of these problems beginning in late 1947. Army personnel built permanent married quarters and barrack accommodations for single officers, men, and civilians working on the base.[30] As operations expanded, so did construction to meet the growing military and social needs of the base and its personnel. At the peak, there were as many as 800 civilian workers on site. By the time the Canadian Army withdrew in May 1964, Fort Churchill was an independent community equipped with utilities and amenities that included a school, a library, a theatre, a radio station, two bowling alleys, two chapels, an ice arena, a post office, and an extensive vehicle repair workshop.[31]

Fort Churchill reinforced the growing Canada–US bilateral commitment to security and defence in the North. In early February 1947, William Lyon Mackenzie King and Harry Truman announced an informal agreement for peacetime collaboration in defence of the North American continent. The agreement stipulated unified training, standardized arms, and the shared use of tri-service military facilities on both sides of the border. Within one week of the announcement, high-ranking members of the PJBD visited Fort Churchill to inspect Canada's northern military outpost. "There are certain experiments being carried out there of interest to the whole North American continent," General Andrew McNaughton revealed when he arrived in Winnipeg for a three-day meeting that took place prior to the Churchill visit.[32] "The board

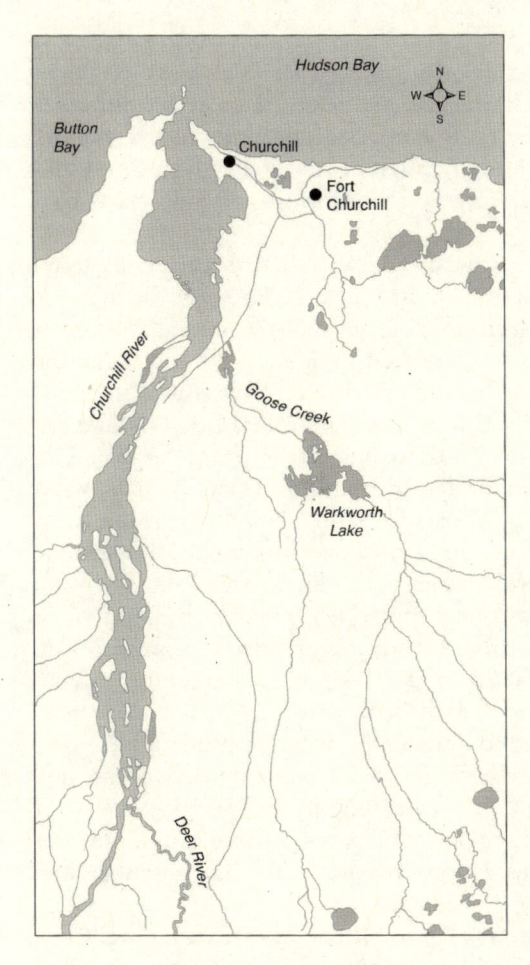

Figure 1.1. Map of the location of the former Fort Churchill military garrison.

is an organization to maintain the good, close, and necessary relations between Canada and the United States in matters relating to the security of the North American continent," he added, before ending with a nod to Mackenzie King: "Permanent means permanent!" Upon their arrival in Churchill, the twelve-member PJBD group toured the JSES with military attachés invited by Canadian officials.[33] Selected members of the press also flew north to Churchill for the two-day visit, although military restrictions limited what they witnessed and heard on base.

Indeed, as military historian Andrew Iarocci explains, Fort Churchill was a closely guarded military garrison with sophisticated administration. Under the Official Secrets Act, the camp and military test areas were declared prohibited places and cut off from unauthorized persons. In place of enclosed fencing, authorities posted signs to ward off potential intruders, and the RCMP investigated all civilians employed at the garrison. Civilian employees received special identity cards that granted access to the camp area, although, as Iarocci observes, given Churchill's relatively small population, "any Soviet interlopers would have been pretty obvious."[34]

The military and defence officials who travelled to Churchill visited the base amid media speculation about Arctic warfare training in northern Canada. Soviet officials in the Kremlin had responded strongly to the joint military agreement announced by Mackenzie King and Truman. Labelling the agreement an "iron fist in a velvet glove," the newspaper of record in the Soviet Union, *Izvestia*, declared that Ottawa had given Washington full control over Canadian territory and the armed services.[35] Canadian officials responded with an immediate denial. "When the Churchill establishment was originally set up last year," Minister of National Defence Brooke Claxton said while announcing the arrangements for the base visit, "it was planned as a joint services' station for the purpose of conducting year-round trails [*sic*] of service equipment in topographical and climatic conditions representative of northern Canada. At no time was it proposed to use Churchill as a base for military exercises or manoeuvres, nor has it been so used."[36] Federal authorities had selected Churchill, the defence minister continued, because it offered not only conditions of extreme cold in winter and moderate heat in summer but also year-round accessibility by rail and air. Intent on dispelling Soviet fears that Fort Churchill was a proving ground for new weapons, Claxton downplayed press reports about increased military activity at the base.

To convince Moscow that Fort Churchill was non-threatening, Canadian officials invited Soviet military leaders to join the tour. Assistant military attaché Major Ivan Pavshoukov attended in place of his superior Lieutenant-Colonel P.I. Domashev, the lead military attaché of the Soviet Embassy in Ottawa. Domashev had visited Churchill a year earlier as an observer of Musk Ox, a three-month Canada–US military exercise in equipment, air supply, and training techniques that began at Churchill during the frigid cold winter of 1946.[37] Pavshoukov remained tight-lipped about the tour, but quelled tensions by reassuring Canadian reporters that military officials in the Soviet Union wanted to avoid any future conflict between East and West.[38]

Bilateral cooperation was imperative to the formation and development of Fort Churchill. In late March 1947, DND accepted a formal agreement with the US War Department that required the Canadian armed services to provide facilities for the US Army at the garrison.[39] In exchange for constructing and renovating new and existing facilities, the Canadian government received $350,000 and a list of directions outlining the supervision of work by the US Army Corps of Engineers. The following month, members of Canada's Inter-Service Committee on Winter Warfare drafted a memorandum outlining a proposed long-term plan for the a joint Canada–US experimental and training station at Fort Churchill. The plan outlined the need for "facilities to develop the art of warfare in the Canadian Arctic through study, experimentation, and training."[40] This framework satisfied two important requirements of Canada's post-war defence policy in the North, namely the training of personnel and the development and testing of military equipment designed to improve the operational capabilities of the armed services in northern Canada and the Arctic. More precisely, the plan outlined Fort Churchill's geographical significance as a locale ideal for studying northern Canada and the effects of Arctic conditions on personnel, supplies, equipment, and logistics.

From a Canadian perspective, Fort Churchill also fit the wider postwar vision for North Atlantic security cooperation in the Arctic. Military officials from Canada, Britain, and the US agreed that preparation for Arctic warfare, supported by proactive research and development, was important for North Atlantic security. Military and civilian agencies from all three countries conducted trials, research, and training under so-called Arctic conditions in the Churchill area. "The general object is the acquisition of knowledge and the development of tactical doctrine and equipment to enable men to live and fight in the arctic," stated a joint organizational mandate written in April 1947.[41] Liaison officers in Ottawa, London, and Washington coordinated joint military exercises and collaborative research projects among the various services and government scientists using Fort Churchill. Officials in Ottawa thus leveraged Canada's northern geography and the subarctic region of Churchill, Manitoba, culling resources, personnel, and brainpower to advance the Arctic military research capacity of the North Atlantic security partners during the early stages of the Cold War.

Arctic Research: Canada's Unique Contribution

When the Soviet Union acquired its own nuclear bomb in 1949 and developed long-range aircraft capable of delivering warheads over the North Pole and down into North American airspace, the northern regions of Canada became a focal point for continental defence. Geography placed Canadian territory directly in between the superpowers, and this spurred military strategists in Ottawa and Washington to champion defence systems designed to detect bombers and protect against a potential nuclear attack. The three strings of radar constructed across the North in 1950s – the Pinetree Line, the Mid-Canada Line, and the Distant Early Warning (DEW) Line – had been funded jointly by Canada and the US to enable detection and advance warning of incoming Soviet bombers.[42] Ottawa also joined Washington in funding a satellite-detection system after the Soviet Union successfully launched Sputnik into orbit in October 1957. That system, which employed instruments specifically designed to identify orbiting space capsules and track satellites, included heavy Baker-Nunn cameras weighing more than three tons apiece. These were operated at military bases in Prince Albert, Saskatchewan, and Cold Lake, Alberta.[43]

Sandwiched in between the superpowers, Canada occupied an inferior position in the Cold War struggle for techno-military supremacy that played out between Washington and Moscow in the Arctic. Ottawa faced the dual challenge of ensuring national defence while maintaining territorial sovereignty, and this led US officials to show restraint when dealing with their Canadian counterparts, at least in public exchanges. In February 1963, for instance, President John F. Kennedy chastised his defense secretary Robert McNamara for bluntly telling Congress that supplying nuclear-tipped Bomarc-B missiles to Canada would draw Soviet fire away from the US in the event of an attack. Canadian officials debated nuclear acquisition extensively, and some even favoured the prospect despite the risks outlined by McNamara. Career public servant Warren Langford co-authored a policy paper in November 1962, mere days after the harrowing two-week Cuban Missile Crisis, in which he argued that Canada's refusal to arm its NORAD squadrons with nuclear warheads was illogical and irresponsible for both Canadian security and continental defence.

Langford wrote his paper while attending a special training program for "cold warriors" at National Defence College in Kingston, Ontario. The program involved visiting defence installations and military establishments in Canada and other parts of the world. Langford travelled with camera in hand, capturing photographic evidence of the physical infrastructure that had been constructed to protect Canada and Canadians from the devastating effects of a superpower conflict. His photographs "reflect both the dream of world travel and the nightmare of nuclear confrontation," explain the authors of a book about Langford's Cold War–era tourism.[44] His stops included Fort Churchill, where he toured the joint Canada–US cold-weather testing facility. Langford photographed the buildings of the Churchill Research Range, a launch pad and experimental facility near Fort Churchill (discussed in chapter 5), where he saw with his own eyes the installations and equipment needed for on-site testing of ballistic missile guidance systems and other elements of the high-latitude rocketry program. Historian John O'Brian describes the 53-metre-high Aerobee rocket launch tower depicted in a Langford slide from January 1963 "as an alien presence in the Arctic environment. It is a fantasy outpost of ice and snow, the tower silhouetted against the sky."[45]

Canadians avoided the nuclear nightmare that Langford warned of in his writing. But fifteen years after his visit to Fort Churchill, a malfunctioning Soviet spy satellite – Cosmos 954 – broke up while re-entering the earth's atmosphere, scattering radioactive debris over northern Canada near Fort Resolution in the Great Slave Lake area. In late January 1978, shortly after that incident, large papier-mâché lobsters swarmed the set of the television comedy show *Saturday Night Live*, with the performers quipping that giant mutant-like crustaceans from northern Canada were invading the US.[46] In actuality, the Cosmos 954 accident posed a severe threat to the Indigenous communities and resources in the affected areas. Moscow at first denied that it bore any responsibility for the clean-up, until a Canadian search under the code name Operation Morning Light located fallen fragments and debris along a 600-kilometre path from Great Bear Lake to Baker Lake. Only then did the Soviet Union admit responsibility for the accident and provide compensation.[47]

The Cold War threat against the northern reaches of the continent and US concerns about the Soviet Union influenced the direction and scope of research activities undertaken by scientists, engineers, military personnel, and technicians in Canada's DRB. In December 1949, the US Research and Development Board (RDB) produced and circulated a "Strategic Guidance" plan outlining the need to prepare for the possibility of war with the Soviet Union.[48] It stated that according to US intelligence, Soviet warfare capabilities would match and potentially surpass those of the US by 1955.[49] "The probable enemy will have atomic bombs and biological, chemical, and possibly radiological weapons, and will use them ruthlessly unless deterred by fear or overwhelming retaliation." Adequate preparation, the plan declared, meant understanding Soviet capabilities in the air, on the ground, and at sea. "[The enemy] will, in addition, use every type of cold-war and guerrilla technique, including propaganda, sabotage, espionage and subversion ... [and] will be able to operate vigorously in cold weather and under usual conditions in the Arctic."[50] Washington responded to all this by strengthening its commitment to military research; it also circulated its contents with the intention of urging Ottawa and London to follow suit. In February 1950, Canada's Directorate of Scientific Intelligence, a division of the DRB, received a copy of the plan.

This was not a subtle reference to Arctic warfare. The plan deliberately outlined "Geographic Problems" to emphasize the importance of environmental military research, citing the value of both hot- and cold-weather warfare studies. A potential war with the Soviet Union, the plan explained, "will probably extend to every type of climate and terrain, and success will depend in large part upon the ability to operate anywhere."[51] The types of operations requiring special emphasis included land and air operations in all-weather climates, namely temperate desert zones and the polar regions. Canadian authorities in the defence department, although concerned over the increasing military and strategic threat in the North, viewed the plan as an opportunity to reaffirm Canada's role in preparations for continental defence and North Atlantic security. Military officials in Ottawa perceived the Arctic as a tangible asset and leveraged Canada's northern geography into political and economic capital.

Senior DRB authorities, including Solandt, viewed scientific intelligence – the practice of analysing scientific information for forecasting the weapons and warfare potential of enemy countries – as a core DRB function and an important contribution that defence scientists could make to Canadian security and national defence.[52] "From the Scientific Intelligence point of view, the more we know about Canada, the more we know about Russia," stated a secret DRB report produced in November 1949. "Canada is the best laboratory available to the Western Allies where full scale experiments can be carried out under 'Russian' conditions."[53] Climate, terrain, geology, flora, fauna, ecology, radio communications, living conditions, and logistics were only some of the geographical and environmental features/challenges shared by northern Canada and the Soviet Union, and senior DRB officials argued that the Canadian armed services required scientific and technical knowledge of these geophysical characteristics and operational impediments to fully understand what was necessary for military preparedness in the North.

If understanding the rapidly evolving post-war strategic threat to North America depended on accurate information about the warfare potential of the Soviet Union and its satellite states, then scientific intelligence offered a practical (and arguably necessary) method for fighting the Cold War. As historian Brandon Webb demonstrates in his research on governmental oversight of communist visitors to Canada in the late 1950s and early 1960s, cross-cultural exchanges between scientists from Canada and the Soviet Union provided officials in Ottawa with useful information about the latest developments in science and technology in the Eastern Bloc.[54] In both civilian and government circles, science could and did function in service to the state, generating threat-assessment knowledge of direct importance to post-war Canadian defence.

While US and Canadian security interests largely coincided, officials in Ottawa also supported non-strategic research. Hugh Keenleyside, commissioner of the Northwest Territories between 1947 and 1950, agreed with Lester Pearson that Canada should support northern resource development and scientific research over strategy and politics.[55] Keenleyside was a high-ranking civil servant who held significant influence in northern affairs. He had received an informal education about northern lands, peoples, and resources from polar explorer Vilhjalmur Stefansson, Arctic geographer Trevor Lloyd, and Danish Canadian botanist Erling Porsild.[56] Captivated by the social and economic possibilities of Canada's northern frontier, he used his position in government to promote the spread of "industrial civilization" northward.[57] Military and strategic considerations in the North were lower on his agenda than the work of scientists, explorers, administrators, educators, doctors, and social workers, among other outside visitors from the South.

Although ambivalent towards northern defence, Keenleyside supported the DRB as a modern scientific research organization. He chaired the DRB's Arctic Research Advisory Committee. He also participated in the creation and subsequent activities of the Arctic Institute of North America (AINA) and helped develop strong ties between that organization and the DRB. Created in 1944 and originally based at McGill University, AINA directed tax-exempt government support for Arctic-related research and education. Much like the Scott Polar Research Institute (SPRI) at the University of Cambridge, AINA facilitated Arctic research through grant funding and fostered networks that connected students, faculty, and government researchers. It also "profited from government contracts (often from military sources) that matched the expertise of academic researchers to the needs of governments," explain Lize-Marié van der Watt, Peder Roberts, and Julia Lajus, whose research demonstrates that Graham Rowley and other researchers connected to AINA "legitimized Arctic research as an academic pursuit." In this way, academics with ties to AINA championed the geopolitical importance of the Arctic for the purpose of carving out careers in polar research.[58]

AINA emerged alongside other Arctic research institutes that came to the fore both during and after the Second World War. Responding to the increased military and strategic value of Arctic environmental knowledge, SPRI assisted the British Admiralty during the war. Polar historian Shelagh Grant notes that the US military formed the Arctic, Desert and Tropic Information Center in 1942.[59] After the war, Norway's Ministry of Industry formed the Norwegian Polar Institute in 1948, and the Arctic Research Institute in the Soviet Union expanded its activities and economic investment in the development of the Northern Sea Route.[60] AINA did not derive directly from military roots, but its proponents welcomed and encouraged collaboration among academia, government, and the armed services, establishing close connections with DRB scientists as a means to access federal research capital and other military resources to expand Arctic research in Canada.

The DRB's senior leadership consistently pointed to the civilian benefits of military research as a means to gain and maintain federal support, and this persuaded officials in Ottawa. Like AINA, Keenleyside saw value in the DRB as a federal and financial resource for the North. The DRB's research-first mandate fit his vision of northern modernization particularly well. "The awakening general interest in the Arctic was in part the result of political and defence considerations that marked the period of the Cold War," he wrote in his memoirs. "But additional recognition of its importance," he continued, "came also from a new appreciation of the economic possibilities of that region."[61] From Keenleyside's perspective, Canadians who lived and worked in the southern regions of the country had a growing social responsibility to northern Canada and the people living there. He embraced science and defence research to support his vision of resource and social development in the North – a vision that appealed to Rowley and others, who worked to establish a strong and cooperative institutional culture between AINA and the DRB.

Authorities in the DRB supported the armed services by allocating monies from the defence budget to enable novel scientific research. Although the DRB funded such notable projects as the Black Brant series of sounding rockets and the two Alouette satellites, the actual creation or development of weapons, vehicles, and related military equipment was not a primary focus of the organization's research mandate.[62] The Canadian defence budget was simply too small for large-scale military development of that sort. Canada could not match the military resources available in Britain or the US, and this led senior Canadian officials in Ottawa to design the DRB as a research-first organization capable of pursuing military projects where Canada could make a special or unique contribution to the North Atlantic security partnership. Cold-weather science and Arctic research thrived under this guiding principle of Canada's early post-war defence policy. As concern about the Soviet threat grew stronger, research teams, administrators, and soldiers pushed north to study and occupy northern Canada and the Canadian Arctic. Collectively, on behalf of the federal government, these individuals worked to defend the North against Soviet aggression while promoting territorial sovereignty amid increasing encroachment from the US.

While the air threat to North America dominated strategic considerations in Ottawa during the early post-war years, senior military and defence officials remained cognizant of the role and value of ground forces in the Arctic. To counter a potential Soviet attack, Canadian military leaders turned to science to help prepare soldiers for the cold-weather battlefield. They deemed cold-climate training important for indoctrination and soldier preparation in the North, considerations underscored by the need to prepare a defence along the shortest and most direct route over the North Pole. As Canadian soldiers learned how to survive and use their weapons and equipment under northern conditions, the DRB's scientists and engineers provided scientific and technical assistance to meet their needs in the high-latitude environment of subarctic and Arctic Canada.

Military Cooperation at Fort Churchill

Fort Churchill, the northernmost military post in central Canada, had year-round access to transportation infrastructure and facilities. The Hudson Bay rail line provided service between Winnipeg and Churchill, with a regular freight transportation service and three weekly passenger trains. Logistic support of the garrison was thus relatively economical. Air transportation was also available from the runway and airport facilities operated by DND. The RCAF and the Military Air Transport Service of the USAF provided air transportation for government and military personnel.[63] Canadian Pacific Air Lines operated passenger and air express service to Fort Churchill twice weekly in the mid-1950s.[64] The airfield also accommodated non-scheduled air services when required, such as transport for scientific research expeditions and mercy flights to and from locations in the eastern Arctic. In addition to all that, sea transportation through the Hudson Strait to the Port of Churchill was possible from late June through mid-October.

Construction of the Fort Churchill garrison was complete by the summer of 1952. At maximum capacity, the base accommodated 1,850 personnel of all ranks.[65] That included around 1,000 permanent occupants, representing both the administrative staff and those permanently engaged in research, experiential trials, and military training. The remaining 850 personnel were members of participating agencies who visited the base or worked there temporarily. The Canadian Army was the single largest segment of the base population, accounting for around 68 per cent at full capacity. US Army personnel formed the second largest group at 18 per cent, while personnel of the RCAF, DRB, Department of Transport, and the RCN accounted for the remaining 14 per cent.

As stipulated by agreement between the armies of Canada and the US, control of the Fort Churchill garrison was vested with a Canadian commandant (post commander) designated by the Canadian Army.[66] The commander's flag was a green triangular pennant with a single polar bear at its centre.[67] As a joint Canada–US and inter-service military base, the garrison accommodated all branches of the Canadian armed services as well as visiting personnel and research teams. The US Army's 1st Arctic Test Detachment was stationed at the base, as were two officers of the UK's Army Liaison Staff, who observed and studied Arctic winter warfare for the British military.[68] Various civilian agencies also used Fort Churchill, often conducting cold-weather research and field trials near the base or travelling farther north to conduct experimental work in the eastern Canadian Arctic. Command of Fort Churchill was the responsibility of Headquarters, Prairie Command, except for matters affecting more than one agency, in which case oversight belonged to the Fort Churchill Co-ordinating Committee in Ottawa.

DND officials intended Fort Churchill to function as an experimental site and training station for the study and development of Arctic warfare techniques in the Canadian armed services, the DRB, and the US forces. Ottawa's policy focus at the base was two-pronged: the individual and collective training of service personnel in northern Canada and the Arctic, and the development and testing of military equipment for operational use in subarctic and Arctic regions.[69] The agencies operating at the garrison implemented this policy by carrying out short- and long-term research projects, user trials of equipment, training and tactical exercises, and on-site Arctic indoctrination courses. Each branch of the Canadian armed services had specific responsibilities at Fort Churchill. The Canadian Army was tasked with studying the environmental conditions of subarctic Canada and the eastern Canadian Arctic region of Hudson Bay, as well as the effect of Canada's high-latitude climate on personnel, supplies, equipment, and logistical problems; to those ends, it conducted exercises and trials to determine the suitability of military items for use in northern Canada and the Arctic.[70] Routine exercises and experimental trials doubled as a method for developing Arctic warfare techniques and training personnel for future battlefields.

Although Fort Churchill was under the army's overall command, security regulations, and administrative control, Canada's navy and air force had their own autonomous operational roles. Command and maintenance of the airfield, hangar control tower, and associated equipment and facilities was the responsibility of RCAF Unit, Fort Churchill.[71] Besides conducting its own experimental projects and operational training, the RCAF provided aircraft and crews to carry out the air responsibilities of the Canadian and US forces stationed at the base. The RCAF also provided air search-and-rescue capabilities and controlled and maintained facilities for the safe landing, parking, and securing of commercial aircraft. Outside the Churchill area, the local RCAF unit serviced and maintained a landing strip at Coral Harbour on Southampton Island and provided services for the occasional air transport of government officials, military personnel, and scientists. For its part, the RCN was largely a separate and self-administered unit at Fort Churchill.[72] Represented by HMC Naval Radio Station and housed in a building on the outskirts of the camp, the navy performed limited experimental work and was the least active of the three Canadian services stationed at the base.

During the 1950s, the Department of Transport operated three divisions with independent sections at Fort Churchill.[73] The Aviation Radio Division operated a radio range station five miles south of the main camp, with a remote unit located in the RCAF Operations Building. The station provided aircraft with a radio aid to air navigation, air-to-ground and ground-to-air radio communication services, and the latest weather and forecasting information. The station also provided radio navigational aids and ship-to-shore communication services for vessels entering and leaving the Churchill harbour. The other two functions of the federal transport department were a Meteorological Section, whose personnel coordinated with the Winnipeg District Aviation Forecast Office to prepare forecast maps and briefings for aircrews flying out of Fort Churchill, and a Radio Field Intensity Section, whose personnel correlated transmission signals to observe and measure ionospheric conditions.[74] During the winter months, the meteorological staff also participated in special lectures carried out in conjunction with Fort Churchill's Arctic indoctrination course.

The primary purpose of the US Army at Fort Churchill was to conduct engineering tests for various types of army materiel and equipment under Arctic-like conditions. The US Army's 1st Arctic Test Detachment (7099th Area Service Unit), organized under the Commanding General, Military District of Washington, provided logistic and administrative support for technical service test teams sent north to perform research and trials.[75] The unit was organized into several sections, including supply, finance, ordnance, operation, passenger and freight transportation, signal, special service, and administrative, in addition to a headquarters company. A motor pool housed and serviced all vehicles, and a hobby shop provided leisure supplies to personnel at the base.

Officials in the US Department of Defense funded military research at Fort Churchill under the belief that military preparations for Arctic warfare required intimate, first-hand knowledge of northern geography and climate. "In view of the increasing probability that a future war will involve extensive military operations in Arctic territory, FORT CHURCHILL provides a valuable testing ground for both equipment and personnel," stated a US Army report written in the early 1950s. "Continued activity by both the United States Army and the Canadian Forces at this Station represents progress toward strengthening national defense on a front hitherto neglected and now of prominent significance – the Arctic."[76] All US personnel ordered to Fort Churchill to conduct or observe tests were attached to the 1st Arctic Test Detachment and undertook duties as assigned by a specific technical service test team. While the unit's commanding officer exercised full jurisdiction over all US soldiers and civilian personnel stationed at the base, each test team worked under the supervision and direction of the chief of the technical service concerned. Most tests were conducted over the course of a single winter season, from November through March. Some continued through subsequent winter seasons and could extend into the summer months, enabling data collection under the extremes of both cold and mild conditions.

Fort Churchill housed all branches of the Canadian Army, including the Army Service Corps and the Army Medical Corps, which fielded the largest numbers on base. The Provost Corps cooperated with the local RCMP detachment, helping maintain law and order. The garrison headquarters functioned like an area headquarters in both size and composition, and the leadership organization reflected army standards and practices. Under the base commander, a colonel, was a lieutenant-colonel who served as deputy commander and general staff officer, and two staff clerks. After the Churchill Research Range was constructed in the early 1960s, the army added a technical staff officer to monitor safety during rocket research and live firings.

In the spirit of cooperation and mutual exchange, senior military and defence officials in Ottawa encouraged US participation at Fort Churchill. Several branches of the US military undertook experimental work on and near the base, including the Chemical Corps, the Medical Department, the Corps of Engineers, the Signal Corps, the Ordnance Department, and the Transportation Corps.[77] American personnel had access to garrison living quarters, messes, hospital facilities, and medical supplies, as well as to standard fuels and lubricants. The USAF also stood guard at Fort Churchill, where a refuelling squadron of Strategic Air Command (SAC) prepared to help carry the next war into enemy territory within a matter of minutes. Officials in Washington withdrew SAC at the end of June 1963, after advances in technology and radar equipment led to modifications in aerial strategy.[78]

Defence Research Northern Laboratory (DRNL)

Fort Churchill enabled DRB authorities to expand Canada's Arctic research program. With officials in both DND and the Department of Mines and Resources keen to maintain northern air bases and photographic mapping in the North, a permanent scientific research laboratory integrated the federal departments operating in the Hudson Bay region of northern Canada. For authorities in the DRB, interdepartmental cooperation meant a sustained budget and the possibility of expansion. Fort Churchill was a suitable locale because the base allowed for laboratory work and military testing in an environment that offered limitless ice in winter and swarms of mosquitoes, biting insects, and spongy tundra in the summer. Archie Pennie, one-time superintendent of Defence Research Northern Laboratory, wrote that "[Churchill was] virtually the crossroads to the Arctic."[79]

Scientists employed or contracted by DND first conducted research at Fort Churchill during the winter of 1946–7, experimenting with different types of clothing and petroleum products intended for winter use in northern Canada and the Arctic. The following summer, entomologists from the Department of Agriculture commenced experimental studies of insect control and environmental management of operational terrain (see chapter 3). In April 1947, the DRB was formally established; in the fall of that year, the first defence scientists arrived at Fort Churchill and commenced semi-permanent laboratory research in temporary huts.[80]

Establishing a permanent research laboratory at Fort Churchill was an arduous task for the scientists involved. James Croal was DRNL's first employee, although the laboratory itself did not exist when he arrived at Churchill in late August 1947.[81] An RCN veteran, Croal arrived at Churchill with a BB X-ray diamond drill and commenced a drilling program to record the temperature of the frozen ground and study permafrost in the area.[82] He worked alone on an old searchlight platform made of skids before joining Captain Bill Crumlin of the US Army Corps of Engineers.[83] Crumlin also expressed an interest in permafrost research, and the two made daily trips to various drill sites near the Fort Churchill base.

By the end of 1947, the DRB had a permanent staff of four scientists and one clerk at Churchill.[84] As research activities increased on site, the Canadian Army provided two small huts for the storage of scientific equipment, warmed by unreliable oil heaters borrowed from the US Army. In the harsh winters of northern Manitoba, scientific work in the two huts was extremely difficult. The small DRB staff maintained an effective permafrost program, nonetheless, and initiated a series of research studies to measure human performance under cold duress.

In June 1948, A.C. Jones took his post as DRNL's first superintendent.[85] Upon his arrival, Jones made plans to transport a hut from the Rideau Military Hospital in Ottawa.[86] Delivered to Churchill and rebuilt on base, the hut provided accommodation for the DRB scientists already on site and for the additional staff who arrived with Jones. The temporary facility served the DRNL team until construction of the laboratory was complete. Work on building a new one-storey laboratory facility began that summer; the first occupants moved in several months later, in February 1949, after its initial wing was completed. Work on an additional two-storey wing commenced shortly thereafter, creating a larger space for scientific instruments and laboratory research.

While Jones assumed full responsibility for DRNL, the Arctic Section of DRB Headquarters in Ottawa maintained coordinating responsibilities for federal scientific research projects and activities conducted at Fort Churchill and other locations in northern Canada. Scientists working on contract for the DRB and all other agencies of the Canadian, British, and US governments travelled to Churchill under arrangements made in Ottawa, where DRB authorities coordinated travel arrangements with the RCAF.[87] As the principal users of DRNL, scientists employed or contracted by the DRB conducted basic research into military problems particular to living, working, and fighting in northern conditions. DRB staff also used the facilities to house visiting scientists operating out of Fort Churchill on short-term or seasonal projects. In this way, the federal government provided a permanent facility that enabled and promoted both national and international Arctic research projects on Canadian soil.

Fort Churchill provided the DRNL team with advanced research facilities and equipment. The location also enabled access to a northern testing environment suitable for advanced military training and research. Scientific projects conducted on and near the base focused on the acquisition of new knowledge and the development of techniques and equipment that would enable men to live and fight under the various conditions found year-round in northern Canada and the Arctic.[88] Specific projects included work on environmental protection of the human body, nutritional medical problems, the design and development of Arctic clothing and equipment, the performance of mechanical parts and supplies in severe cold, and entomological research involving mosquitoes and biting insects in warm weather.[89] The official function of DRNL was twofold:

1. [To] undertake such research on problems of Arctic warfare as can most effectively be carried out in the Churchill area and in areas directly accessible from Churchill. The problems include those of clothing, equipment and shelter, pest control and protective

measures, acclimatization and nutrition, measurement of environmental factors, trafficability and properties of terrain, use of fuels and lubricants, and the behaviour of materials under Arctic conditions.
2. [To] provide laboratory facilities, transportation, equipment, clothing, food, supplies, and communication of scientific workers, test team and observers from other stations of the [Defence Research] Board or its associated agencies, to enable them to carry out research in the Arctic, using the Defence Research Northern Laboratory as a base for their operations.[90]

The DRB's scientific and administrative headquarters provided all resources and funding for DRNL. In late November 1948, the total number of personnel employed at the laboratory increased to fifteen: two professional scientists, six non-professionals, and seven casual staff. The approved operating budget for the first full year was $315,195.[91]

Early in 1951, authorities at Fort Churchill received approval to develop plans for a new wing to augment the two newly constructed one- and two-storey buildings that constituted DRNL.[92] Completed and occupied in December 1952, the adjoining wing housed DRNL's administrative functions, a library, and a special cold-room laboratory. One month later, on 25 January 1953, senior defence officials celebrated the newly completed DRNL at an opening ceremony.[93] By that time, the permanent staff had increased in size considerably; it now included scientists and engineers with expertise in such fields as physics, chemistry, biology, physiology, and psychology.

Between 1947 and 1953, DRB scientists and engineers at Fort Churchill studied cosmic rays, petroleum products, permafrost, rations and food, biting insects, and the physiological responses of men to cold and subarctic conditions; it also interpreted northern terrain from aerial photographs. Work after 1954 expanded to include investigations of heat-transfer equipment, wheeled and track vehicles, and the physiological and psychological effects of operational stress in northern environments, as well as biological studies on insect vectors and operational studies on military problems particular to northern Canada.[94] DRNL was a permanent facility for seasonal and long-term research projects, but the laboratory also acted as an Arctic research space for visiting scientists, engineers, and research technicians from other DRB establishments, government agencies, and university institutions. The Canadian scientific staff routinely aided the various US Army test teams stationed at Fort Churchill; for instance, they cooperated on several military research projects and advanced the JSES's Arctic warfare program.

Figure 1.2. Defence Research Northern Laboratory, ca. 1955–64. Churchill Public Library, Churchill Public Archives, box 992.5, oversize (O/S) 1, document 8/8.

Military Research at DRNL

Canadian military officials understood the need for a capable presence in the North and for its soldiers to be able to live and fight in conditions of extreme cold and an inhospitable environment. To that end, the burgeoning DRB joined the armed services to help military leaders and combat personnel identify and solve the myriad service problems encountered in high latitudes. DRNL emerged under this mandate; its purpose was to conduct research to support its only client: the Canadian military. The research laboratory was relatively small, but its staff included a versatile and competent collection of scientists, engineers, and technicians with various disciplinary backgrounds and educational expertise. Its superintendents included biologist K.C. Fisher, chemist D.B.W. Robinson, and physical chemist Archie Pennie.[95] Under the direction of top government scientists, most of whom were military veterans, DRNL boasted a staff with research specialties in biology, botany, zoology, entomology, astrophysics, mechanical engineering, and other fields; all of these specialties were directly or indirectly linked to the various research needs of the armed services.

DRNL was a hub for all scientific and military work carried out at Fort Churchill in the early post-war years. Between 1947 and 1964, the laboratory functioned as a multipurpose facility for research, training, and education on winter and summer issues specific to subarctic and Arctic Canada. Canadian, US, and British forces trained and developed methods and equipment specifically designed to function under the extremes of climate and terrain encountered in northern latitudes, and scientific research at Churchill became important to Canadian military interests in the North.

According to Pennie, DRNL's one-time superintendent, "joint military and scientific work was required to define the problems which faced man in the Arctic environment – his clothing, feeding, tactical deployment, navigation, re-supply and a host of associated problems."[96] A native of Banff, Scotland, Pennie trained in Canada during the Second World War as an RAF pilot. He returned to Canada in 1948 and joined the Canadian Armament Research and Development Establishment at Valcartier, Quebec, where he directed the facility's work in applied chemistry. Together with James Ross of McGill University, Pennie helped establish factories for the manufacture of RDX, an organic compound used widely as a high explosive in military applications.[97] Appointed DRNL's superintendent in September 1954, he moved to Churchill and directed the DRB's military science program in northern Canada.

Scientists, engineers, and research technicians employed at or with DRNL attempted to study and solve the many operational problems the armed services encountered while living, training, and possibly fighting in northern Canada and the Arctic. From the fear and fatigue associated with performing military duties in the dead of winter on the subarctic barrens to the requirement to develop proper fuels and lubricants for military vehicles and equipment, the armed services encountered many practical problems that opened a wide range of new scientific research opportunities. Authorities in the DRB viewed Canada's military presence at Fort Churchill as an opportunity to foster a scientific and technical staff with qualifications and skills in diverse and overlapping fields. "It is quite possible that within one single day during field exercises in the Churchill area a soldier may encounter a situation which calls for advice and assistance from scientists in all these varied fields of work," Pennie wrote in a 1956 article for *Canadian Army Journal*.[98]

This diversity of work made DRNL unique. In the DRB's other scientific research facilities, the trained staff generally shared expertise in closely related fields; permanent and contract employees of DRNL arrived at Churchill with diverse expertise. The DRB's staff at Suffield Experimental Station in southern Alberta consisted primarily of chemists and biologists, while engineers and physicists at Valcartier studied the design and performance of guided missiles.[99] In contrast, DRNL's scientific and technical staff shared no particular relationship, but they worked closely on the many and varied problems presented by high-latitude military operations – an approach to science that reflected and characterized the DRB's wider Arctic research program. Instead of funding a single primary defence project in northern Canada and the Arctic, the DRB supported several small research studies that added up to a large whole. "This project can be defined as the Effect of Arctic Environment on the Performance of Personnel and Materials in the Field," wrote Pennie, succinctly describing the genesis of Canada's early Cold War military research program in the North. "In other words, the main interest is what happens to men, materials and equipment when exposed to Arctic weather and surroundings, both in summer and in winter."[100]

Scientific research at Churchill was not restricted to military training, however. Winter scientific work at Churchill concentrated on permafrost, problems of clothing and equipment, fuels and lubricants, and nutrition. During the summer months, scientists turned their attention toward entomological research, devising solutions to improve military operations in areas with dense mosquito and blackfly populations. Research at DRNL constituted a variety of scientific disciplines, because problems "dealing with any phase of military operations in the Arctic had to have an input from the operational, the human resources, the biochemical [and] as well as from the engineering side." [101] While Fort Churchill facilitated the scientific research needs of the Canadian armed services, the federal government extended its reach in northern Canada by providing financial and technical support for related research elsewhere.

Arctic research at Fort Churchill expanded quickly in the early 1950s. DRNL had to build a new and larger adjoining wing in 1952 to provide more laboratory space for the permanent working staff. That same year, the DRB's northern laboratory underwent an organizational restructuring to reflect its growing research program; it emerged separated into four distinct but cooperative sections: an Administration Section, a Physical Science Section, a Biological Research Section, and an Operational Research Section.[102] The scientific staff at DRNL participated in various army exercises and training programs, performing field trials and observational studies of soldier aptitude, morale, military leadership, and related aspects of operational behaviour in the unique environmental conditions of the immediate area. They also worked on the low-temperature storage properties of fuels and lubricants and carried out co-operative research projects with visiting scientists from the Department of Northern Affairs, the Department of Agriculture, the National

Research Council, and other branches of the federal government. One collaborative project, undertaken in cooperation with the Office of the Surgeon General of the US Army, saw DRNL staff study sanitation problems in the North – proactive research deemed important to equipping and preparing military forces for temporary and semi-permanent operations in high latitudes and polar regions.[103]

Although the DRB was the fourth branch of the Canadian military, scientists at DRNL, like their colleagues in Canada's other military research facilities, also carried out research projects for the benefit of military forces in Britain and the US. "Omond Solandt always stressed that, in the field of defence science, Canada should concentrate on the areas in which it was best suited to succeed as a result of particular scientific skills, climate and geographical position," Pennie wrote in 1994. "This was particularly true when it came to considering research in the Arctic and Northern latitudes."[104] In the field of operational research, for instance, DRNL's scientific staff evaluated Arctic clothing, equipment, and general stores developed for army units by engineers in all three countries. Operational research scientists also observed and evaluated Arctic indoctrination and training methods, ground navigation and survey techniques, and addressed nutrition and related medical concerns, besides performing statistical analysis on user trials of weapons, vehicles, radios, and ration packs (see chapter 4).

In addition to their own work at Churchill, scientists at DRNL contributed to an expansive effort to photograph and map Arctic Canada from the air. The RCAF began operating out of Churchill in October 1946, providing aircraft and crews to facilitate air requirements for the Canadian and US forces stationed on site. Search-and-rescue operations were another important function of the RCAF at Fort Churchill. Pilot records regularly referenced mercy flights carried out on behalf of individual citizens and communities.[105] RCAF rescue units stood ready and equipped with specially fitted aircraft and highly trained aircrews. Researchers at DRNL frequently questioned survivors who received medical treatment at Fort Churchill. According to former laboratory employee Mace Coffey, DRNL scientists spoke with survivors to collect information that was vital to the modification and improvement of Arctic survival kits issued by the army and air force.[106] In this regard, Fort Churchill served as a gateway to the North, providing government scientists and military personnel with access to locations like Baker Lake and Coral Harbour. As early as 1947, the RCAF Canso made the Churchill base an important stop on its annual magnetic research flight into the Arctic for the Division of Terrestrial Magnetism of the Dominion Observatory.[107] By 1951–2, the RCAF was operating polar navigation training flights out of Fort Churchill and

Air Transport Command was using the airfield as a major staging stop for the increasing northern traffic.

Scientific research at Fort Churchill expanded rapidly during the early 1950s, requiring the expansion of the existing facilities. In January 1953, Solandt travelled north from Ottawa to take part in a ceremony marking the opening of DRNL's new wing.[108] The new building provided additional administrative offices as well as extra laboratory space to accommodate a further expansion of the DRB's Arctic research program. Senior Canadian officials supported and promoted intergovernmental work at DRNL as a means to attract visiting scientists and researchers to Fort Churchill. In February 1954, for instance, military authorities stationed at the base organized an Arctic warfare demonstration for twenty-six military attachés and advisers appointed to Canada. The display of military power included over-snow vehicles, radio equipment designed for northern latitudes, specially constructed snow houses, and a 280 millimetre "atomic cannon." US officials supervised testing of the atomic cannon under cold-weather conditions with conventional shells prior to the demonstration, in part because atomic weapons were neither fired nor stored in Canada. "Although tests have not involved firing of atomic warheads, due to danger and costs," stated one report, "it is known that shells of the conventional type have been fired including at least one 600-pounder, and from data gathered, the actual range of an atomic shell is known."[109] During the two-day visit, Colonel H.A. Millen, commandant at Fort Churchill, facilitated a guided tour of the military base and nearby facilities. The attachés and advisers also observed simulated attack scenarios, designed as a display of soldiers and equipment operating under realistic Arctic warfare conditions.

Authorities in the DRB spoke highly of Fort Churchill in order to maintain federal financial support for the ongoing and expensive scientific projects carried out there. After visiting DRNL on a review assignment in December 1954, E.F. Schmidlin of the DRB's Arctic Section wrote a report emphasizing the importance of the facility for Canadian security and national defence. "DRNL is unique in defence research," he wrote. "It is the only station of its kind in the tripartite family. There is much to be done yet, before warlike operations can be successfully prosecuted in the high latitudes. Because of this I think we are justified in providing adequate facilities, both scientific and domestic. Unless we do, we cannot expect the maximum results for the effort expended."[110]

The Canada–US–UK tripartite emphasis that Schmidlin placed on DRNL reflected wider attitudes within the DRB. As a principal user of Fort Churchill, the DRB conducted basic research concerned with military problems peculiar to living, working, and fighting in the Arctic; but it also used the facilities to house visiting scientists. DND opened the garrison to military and civilian agencies in Britain and the US, which used DRNL to conduct trials, research, and training under realistic northern conditions. This was exactly how Canadian defence officials envisioned Fort Churchill when they were developing a policy framework for the DRB. As Solandt wrote in a policy brief about Canadian defence research in May 1946, "Canada is particularly interested in extending cooperation for technical and operational testing, under Canadian climatic and topographical conditions, of equipments [sic] developed by the UK or the US, which are suitable for employment by UK-Canada-US Forces."[111]

Scientific activity at DRNL was particularly notable in 1955–6, when the facility hosted personnel of the US Environmental Health Laboratory, who travelled to Churchill to study general sanitation problems in the North.[112] Besides serving as a facility for research teams and individual scientists and engineers, DRNL enabled information and intelligence exchange among the North Atlantic partners. Laboratory researchers compiled and analysed statistical data on clothing and equipment from various trials conducted at Churchill by inter-service personnel. In cooperation with the armed services, researchers at DRNL shared relevant information by lecturing about Arctic indoctrination to visiting military and civilian groups from the three tripartite countries. Within a decade, as the Cold War turned hot in the early 1950s, DND had established a permanent scientific research facility at Fort Churchill, where it devoted significant resources and personnel to claim and demonstrate Canadian authority over Arctic warfare.

Conclusion

First occupied by Canadian and US forces during the Second World War, Fort Churchill became important to the continental defence of North America during the early Cold War. Military and defence officials in Ottawa recognized that soldiers and scientists new to the North required training and confidence in order to operate effectively and safely in the high-latitude climate of northern Canada and the Arctic.[113] As Fort Churchill grew increasingly important for post-war security and defence in the North, the location served as a base for tripartite-funded science and Arctic research. Scientists from Britain and the US travelled to Churchill and conducted Arctic research at the DRB's Defence Research Northern Laboratory. In addition, scores of scientists, engineers, and medical practitioners from Canadian universities and various branches of the federal government used DRNL to conduct Arctic research from the late 1940s through the mid-1960s, signifying the peculiar attraction of northern Canada to outside interests.[114]

Fort Churchill was a tangible representation of Canada's unique contribution to the North Atlantic security partnership. Senior officials in Ottawa leveraged Canada's northern geography and military resources to strengthen Canada's position relative to Britain and the US. Science was a key component, and the DRB was Canada's single largest patron of Arctic science during the early years of the Cold War. But as the following two chapters will demonstrate, authorities in the DRB also financed and pursued a range of scientific activities that had only loose connections to the immediate needs of post-war Canadian defence. Cold War concerns about the high-latitude, cold-weather military capabilities of the Soviet Union spurred a government response, accelerating the pace and intensity of military research in northern Canada and the Arctic. Scientists, engineers, and technicians funded or employed by the DRB made meaningful and lasting contributions to northern research; however, the frontier sciences funded by the federal government often lacked responsible and adequate oversight. Investigating the circumstances and attitudes that enabled Cold War–era research in the North sheds light on the implications of military science in Canada during the first two decades of the post-war period, demonstrating the reach and influence of the government power structures that underpinned the work of scientists sponsored and supported by the Canadian armed services.

2 Acclimatization, Cold Tolerance, and Biochemical Experimentation

In December 1950, after completing five field trips to the Canadian Arctic, medical scientist G. Malcolm Brown of Queen's University wrote a progress report for the Defence Research Board that described a series of biochemical studies conducted on Inuit and White test subjects.[1] Designed and carried out as a long-term study of the effects of cold on the human body, the research aimed to determine how much exposure to cold was required to achieve acclimatization. Using funding from the DRB, the National Research Council, the Department of National Health and Welfare, and the Arctic Institute of North America, Brown oversaw the administration of medical treatment services and biochemical work over an eight-year period.[2] Four trips took researchers to Southampton Island (Shugliaq) in northern Hudson Bay and a fifth to both Southampton Island and Igloolik, in the northeast corner of present-day Nunavut. Brown and his team returned to Southampton Island for a last trip in 1954. While on location over the eight years, researchers took samples of blood, urine, skin, and liver from "Eskimo test subjects" and transported the specimens to southern Ontario for independent and comparative biochemical analyses with samples taken from White university students.[3] Records indicate that Brown and his colleagues worked on no fewer than 288 Inuit, including sixteen children ranging in age between one and ten.[4] Although the exact number remains unclear, at least sixty-seven of the 288 were men used to study cold-weather acclimatization.[5]

In the absence of a full physiological description of acclimatization to cold, scientists conducted the research to obtain information considered useful for preparing government and military personnel to work in northern Canada and the Arctic. The theory of acclimatization offered a potential solution to the problem of cold tolerance. Inuit represented acclimatized experimental subjects, while "85 male, healthy medical students" from Queen's University represented the so-called control group of unacclimatized White subjects.[6] Researchers subjected both groups to similar studies, but it seems that Brown and his team of scientists obtained organ samples from Inuit alone. During otherwise routine medical examinations, liver samples were extracted from at least ten adult Inuit by means of needle biopsies.[7] Brown ultimately deemed the studies inconclusive because testing failed to detect definitive evidence that cold acclimatization existed. The circumstances and attitudes that gave rise to the research are indicative of the complex health and social consequences of military-sponsored medical science performed in northern Canada and the Arctic during the early Cold War.

Operating under deeply entrenched race-based understandings of human biology, authorities in the DRB funded Brown's medical research in order to devise a method for selecting soldiers for northern service. Theoretically, studying Inuit biology served as a scientific approach to determining "Eskimo-like" cold-weather-fighting physiological traits in predominantly White soldiers being considered for active duty in northern Canada and the Arctic. In the process of conceiving and conducting the research study, the scientists and government sponsors involved supported a colonial agenda by imposing medical and scientific practices on Inuit communities, asserting claim over the Inuk body, and subjecting Inuit to biochemical experimentation.[8] Unlike wider government initiatives aimed at Inuit health in the early post-war period, though, assimilation was secondary to the primary goal of biological appropriation. In the context of the acclimatization research examined in this chapter, I define biological appropriation as the attempted use of Inuit biology for non-Inuit purposes.

Researchers were not attempting to extract and then transmit Inuit blood or tissue; rather, they were using comparative sampling of human specimens from Inuit and White test subjects to pursue a scientific understanding of a physiological response to cold tolerance. Under the auspices of the Canadian government, comparative biochemical testing was intended primarily to isolate the vascular characteristics that enable the human body to acclimatize to severe cold. Brown and his team believed that this type of information could be useful for devising a process for the "physiological screening of persons considered for service in the far north."[9] Scientists thought that if they could identify the vascular characteristics of cold-weather acclimatization, it might be possible to "provide some guides as to the best method of rapidly acclimatizing a group of men, and of selecting those likely to adapt quickly and completely."[10]

For Brown and his government sponsors, the Inuk body was a new and exciting field of research. Outside the small presence of the RCMP, federal authorities had shown little concern for the North and the people living there before the Second World War.[11] Ottawa had commissioned an Eastern Arctic Patrol in 1935 to investigate the clinical and metabolic health of northern Indigenous populations; a year later, the responsibility for Inuit health care passed to the new Department of Mines and Resources.[12] The provision of Inuit health services changed again in 1945. Under the direction of the Department of National Health and Welfare, the federal government again commissioned vessels to patrol the eastern Arctic and provide medical and evacuation services to Inuit.[13] Northern patrols promoted Canadian sovereignty as well, but the concern over a possible Cold War conflict arising between the US and the Soviet Union drove decision-makers in Ottawa to ponder the changing role and significance of Canada's northern regions.[14] As the North became increasingly important to post-war security, Brown and other government-sponsored scientists studied Inuit to learn about the effects of cold on the human body and to help prepare settler-soldiers for service in a region perceived in Ottawa as cold and unforgiving.

The attempted appropriation of Inuit biology for non-Inuit purposes offers a perspective for understanding colonial attitudes in post-war Canada. In contrast to wider settler-colonial perceptions of the Indigenous body, scientists involved in the DRB's cold-weather acclimatization research viewed the Inuk's ability to live and work in the cold as an "enviable" physiological and biological trait. If science could unlock the "Eskimo" secret to cold-weather survival, the "functional capacity of white men in the Arctic" might increase, thereby improving the ability of Canada's government and military personnel to work in and defend the North.[15] In this instance, scientists and government officials from southern Canada viewed Inuit as having superior, rather than sickly and inferior, bodily traits.[16] That the scientists conceived the research in terms of an imagined racial dissimilarity suggests that they operated under deeply entrenched colonialist positions, nonetheless. The biologized Inuk body, according to the interpretation of the scientists involved, was better suited than the White body to live and work in the harsh climate of northern Canada and the Arctic.

This socio-medical perception of the Inuk body is important to interpreting the health and social consequences of military-sponsored science performed in Canada during the early Cold War. The studies conducted by Brown and his research team did not originate in a military or strategic agenda; even so, DRB authorities assumed financial responsibility for the research on the advice of Arctic policy-makers, who pointed to the potential value of the intended results. Based on the assumption that the environment determines biological traits, acclimatization scientists in post-war Canada aimed to determine biological differences between

Inuit and White bodies as part of a wider international scientific interest in defining the vascular characteristics of cold tolerance. Originally financed by the NRC, authorities in the DRB absorbed Brown's cold-weather acclimatization research with the intention of applying the scientific findings to military service. Canada's northern security did not depend on the successful appropriation of Inuit biology, but acclimatization research found and maintained federal support as a possible means to protect the lives of newcomers in the North during an era of intensifying Arctic activity.

Research Ethics and "Environmental Protection"

The context in which the cold acclimatization research took place is extremely murky. At the time of the research, medical research standards did not require scientists to obtain written consent to conduct human research on the people they studied. Physicians with the Canadian Medical Association first adopted a Code of Ethics in 1868, borrowing standards, practices, and language from the American Medical Association.[17] Nearly a century later, the Nuremberg Code set a base international standard for medical ethics in 1947, but the Canadian government did not implement formal ethical guidelines for medical research performed on human subjects until 1980.[18] Contemporary standards stated that voluntary consent was mandatory for clinical research, which meant that all test subjects needed to agree to participate without coercion and that the researchers were responsible for ensuring that each person understood the risks involved in the research.

Sources indicate that Brown and his colleagues used an "excellent native interpreter" to communicate with Inuit involved in the research, but the details of the verbal contract, if it did exist, went unrecorded and untold.[19] According to the progress report referred to at the outset of this chapter, "the only method of selection used in these surveys was to take family groups as a whole as they became available."[20] Brown expanded this explanation in 1954, writing: "[The] selection of experimental subjects from among Eskimos who continue to live in their traditional dwellings and who still gain their livelihood by hunting and trapping has the merit that such Eskimos as these have an ability to live and work in the cold that permits, on the basis of performance, their acceptance as acclimatized individuals."[21]

Brown's explanation reflected the findings of his research, and the 1954 publication date is significant. When the research began in 1947, Brown and his team pursued innate biological factors to explain the vascular characteristics of cold-weather acclimatization in the human body.

By 1954, the findings of their research pointed instead to environment, culture, and diet as explanations for Inuit cold tolerance. This shift mirrored wider scientific changes with regard to the perceived value of Indigenous knowledge in the North. As Stephen Bocking has argued, the development of post-war research laboratories meant that "scientists no longer needed to live among Indigenous people, learn their techniques for travel and survival, or indeed, have any contact with them" by the mid-1950s.[22] The years between 1947 and 1954 are thus the focus of this chapter. During this period, Brown and his team perceived cold-weather acclimatization in strict biological terms and senior officials in Ottawa funded the research to appropriate the superficial cold-fighting traits of Inuit in the North.

The socio-medical perception of the Inuk body, based on biology rather than cold performance, came to bear on Inuit as developments in science and technology promoted the geostrategic significance of the Arctic. Much like the bureaucrats and doctors responsible for carrying out the assimilative agenda of the Canadian state in the early post-war years, the scientists who engaged in cold acclimatization research exploited contemporary circumstances to achieve distinctly southern goals. As historian Ian Mosby explains in his research on the mistreatment and abuse of malnourished children in residential schools, shifting attitudes in Canada supported race-based medical testing and experimental human science involving Indigenous peoples.[23]

Predicated on the exploitation of Inuit bodies and communities, acclimatization research also occurred within the edifice of colonial science. But a discourse of Indigenous dependency is noticeably absent in records pertaining to the work of Brown's research team.[24] Although the scientists who conducted the experimental research leveraged colonialist attitudes about Indigenous health and welfare to gain access to Inuit communities, cold-weather research on human beings derived specifically from a scientific and medical agenda that sought to describe the superficial connections between the physical body and the natural world. The research was not military or strategic in origin, but Canadian scientists and government officials supported and perpetuated cold-weather acclimatization research in response to distinct concerns about northern security and a desire to pursue science as a solution to military problems.[25]

The perceived need to identify acclimatized bodies for service in the North highlights the importance of White settlement and the grip of the colonial project, especially when considering Canada's unique security and defence issues in the post-war North.[26] While senior officials in Ottawa saw little intrinsic value in funding or facilitating a widespread defence of northern Canada and the Arctic during the early Cold War, geography dictated a concerted response.[27] If Ottawa was not prepared to defend the northern reaches of the continent, Washington might take charge to protect its own interests in the North. Responding to concerns about Soviet and US encroachment on Canadian territory, military and defence officials in Ottawa abandoned any thought of an isolationist post-war security posture.[28]

In addition to bilateral security arrangements and the continental air defence radar network, Ottawa pursued inexpensive options to bolster Canadian security in the North. The concept of the Canadian Rangers and the Mobile Striking Force developed accordingly in the early post-war period (see chapter 1). The experimental research program led by Malcolm Brown, otherwise known as the "Queen's University Arctic Expedition," supplemented this plan. As a northern research initiative, the cold-weather acclimatization study served a specific scientific need that stretched the DRB's military work in northern Canada beyond the confines of Fort Churchill and the western shore of Hudson Bay. Brown's biochemical work on cold acclimatization contributed to a large body of multipurpose scientific research, designed to support simultaneously Canada's independent and collaborative military needs associated with defence and security in the North.

The connection between Brown's medical acclimatization research and the early Cold War military needs of the Canadian armed services is further evident in the DRB's Environmental Protection Program. A 1954 report, published eight years into the program, defined the field of environmental protection as research on "the protection of the serviceman and his equipment against the adverse physical effects of his environment."[29] The program employed various scientific disciplines across multiple branches of the federal government, resulting in cooperation among several departments and the armed services.[30] Coordination of the program was the responsibility of the DRB's Environmental Protection Section. Administrators in the DRB also facilitated field tests carried out in northern Canada and the Arctic by government-sponsored scientists from Britain and the US. In this manner, scientific cooperation in the field of environmental protection brought increased awareness to the exciting new research possibilities of fieldwork in the natural elements of northern Canada, which simultaneously served and reinforced the multilateral potential for military science projects in a safe and accessible cold climate.

Scientific investigations into climatic adaptation and the effects of cold on the human body were a central focus of the environmental protection research funded by Ottawa between 1947 and 1954. At the height of

Brown's work in 1951, scientific study on "physiological stress produced in men by cold" topped the list of fourteen research activities performed at DRNL.[31] Scientists conducted blood and urine sampling on service personnel to determine quantitatively the physiological and biochemical responses to cold in men, with the aim of determining "the degree to which adaptation to cold occurs ... [and] the best methods of bringing about adaptation."[32] Thus, the comparative sampling conducted on Inuit and White test subjects by Brown and his research team did not occur in isolation. Scientists and military leaders in the Canadian defence department showed significant interest in cold-weather acclimatization research, facilitating and funding multiple studies at different locations in the North during the first decade of the post-war period.

Malcolm Brown and Arctic Research

Long-entrenched cultural beliefs shaped how Brown and his team of scientists conceived and carried out the cold-weather acclimatization study. Operating from a privileged position, the researchers and their government sponsors exploited contemporary understandings of Indigenous health to gain access to Inuit communities and test subjects. Powerful individuals close to the heart of the Canadian defence department accepted and contributed to the "discourse of isolation" that historian Mary Jane McCallum identifies as key to interpreting settler-colonial views in medical literature about Indigenous peoples in the North.[33] Brown and his team treated Inuit for nutritional, metabolic, and respiratory diseases, but at the same time, administering medical services in northern Canada and the Arctic gave them access to Inuit bodies for purposes unrelated to Inuit health and welfare. The researchers theorized that "Eskimos" – long isolated from the outside world – were racially pure and biologically adapted to cold. From this flawed point of view, the so-called medical experts posited mistakenly that the Inuk body was the ideal test subject for studying the vascular characteristics of cold tolerance. Nevertheless, while Brown's team co-opted discourses of isolation and assistance to access Inuit communities and research subjects, they were able to extend their acclimatization research study as long as they did because it reflected a distinct federal research agenda that pursued experimental science as a solution to a specific military problem.

Brown graduated in medicine from Queen's University in 1938 and obtained a PhD from Oxford in 1940. Following a three-year research term, he served in the Canadian Army Medical Corps between 1943 and 1946. During two of those years, he was on loan to the Royal Army Medical Corps as a physiologist with the Malaria Research Unit.

Discharged with the rank of major after serving in Britain, Italy, and Northwest Europe, Brown held various academic and professional positions following the war.[34] His appointments included an associate professorship with the Faculty of Medicine at Queen's University in 1946 and a membership on the DRB's Panel on Arctic Medical Research in 1947. He simultaneously held a position with the federal health and welfare department and maintained scientific advisory roles in Ottawa until the 1970s.

Brown's wartime experiences shaped his post-war research as well as his views about cold-weather acclimatization.[35] As a malaria expert during the war, he had witnessed unparalleled death, and this spurred him to pursue and promote the blood sciences as a means to prevent unnecessary death in peacetime. Although his work was independent of the Canadian defence department at the end of the war, his research became particularly important for scientists who conducted Arctic research for the armed services. Shortly after the DRB was founded in April 1947, Brown found himself advising policy-makers responsible for overseeing science and medical research in northern Canada and the Arctic. His work thus influenced how senior officials and military leaders in the Canadian government understood abstract theories of acclimatization and the potential research value of studying Inuit from physiological and biological perspectives.

The war had a profound impact on medical science in Canada. Many scientists turned their attention away from basic research towards narrow practical problems; at the same time, the emergency requirements of the wartime period forced academic researchers and their counterparts in Canadian industry to collaborate closely. The mass production of penicillin was one famous result. Yet the cooperative atmosphere of applied research that accelerated developments in some fields had left others neglected, with the result that Canada's medical science community emerged from the war with mixed capabilities. Nevertheless, as medical historian Alison Li explains, the war "provided an opportunity for proponents of research to demonstrate the utility of medical science and to demand a great increase in funding."[36] One result was that after the war, the NRC's wartime committees, at the urging of prominent scientists like James Collip and Wilder Penfield, expanded the NRC's permanent institutional structure for funding and coordinating medical research in Canada. Collip became first director of an autonomous Division of Medical Research in 1946, managing a $200,000 budget – almost a fourfold increase over the original 1938 budget – that was earmarked exclusively for extramural research performed outside the NRC's central laboratories.[37]

Canada's post-war military research program benefited from this increased investment in the medical sciences. In 1948, for instance, the Privy Council commissioned a national survey of medical research; it was conducted by the NRC's Chester Stewart, the DRB's Morley Whillans, and Ralph MacAulay of the Department of National Health and Welfare.[38] The committee found that 955 persons were engaged in medical research at universities, including full- and part-time researchers as well as graduate students and technicians. The report also documented research conducted in hospitals and institutions separate from medical schools, in provincial and federal health laboratories across the country, and in the research establishments owned and operated by the defence department. Canada's growing body of post-war scientific literature reflects this trend as well: *Canadian Journal of Medical Sciences* and *Canadian Journal of Biochemistry and Physiology* were founded around this time, illustrating the growth and expansion of medical science in Canada during the 1950s.

The first Queen's University Arctic Expedition to Southampton Island took place during the summer of 1947. Brown led a team of researchers, who included medical professor R.G. Sinclair and biochemist Bruce Cronk.[39] Together they travelled north, aiming to study a range of Inuit health issues, including low fertility rates in women and the possibility of immunity to artery issues and cancer. Wrote one intrigued journalist from Toronto:

> The expedition hopes to discover alas whether there is any basic physical difference between the Eskimo and the white man which stands in the way of the latter adapting himself to the Arctic environment in which the Eskimo thrives. If there is such a difference then the white man may well have to forego his dreams of some day developing the mineral wealth believed to lie under the Arctic ice fields; he may have to add new fears to his mental picture of a third world war fought on an Arctic front; and he will have to write off the Arctic regions as possible elbow room for the white race when the pressure of population becomes acute.[40]

Students and faculty at Queen's University followed the exploits of Brown's team of Arctic travellers with great interest. "With most Queen's students sitting out the summer in the sweltering heat of Southern Canada, a group of Meds men had a better time of it," boasted an article printed in *Queen's Journal* in October 1948. "The group – five in number – spent the summer amid the barren and rocky wasteland of Southampton Island in Hudson Bay ... [rubbing] noses with Eskimos."[41] The so-called Queen's Expedition travelled north from Kingston to continue the research initiated under Brown's direction the year prior. On his second trip north, Brown returned to Kingston within ten days with human specimens for diagnosis and biochemical analysis in the university's laboratories. The other members of the research party spent eight weeks studying the physical and nutritional health of Inuit at Coral Harbour; this included extensive observation of traditional dietary customs and habits. Cronk led the research in Brown's absence. The researchers included Fred de Sinner, who joined the group from Switzerland, and fifth-year medical students James Gibbons and John Green.

The same news article credited Dennis Jordan of Toronto, a double graduate of Queen's in arts and medicine, with inspiring Brown's research. Jordan had made several trips to northern Canada as a medical officer with the federal government's Eastern Arctic Patrol. He did so aboard the RMS *Nascopie*, the maligned HBC supply ship that operated between 1911 and 1947, when it sank after striking an unmarked reef near Cape Dorset.[42] Jordan reported to officials in Ottawa that a "vast field of study for Medical men" awaited in the Canadian Arctic, and Brown, with federal and institutional support, followed up on Jordan's lobbying with a preliminary survey: the first research trip of the Queen's Expedition.[43] While on location, the researchers observed and recorded what they believed were serious nutritional deficiencies and other health-related problems among Inuit. Determined to make important contributions to medicine and pathology, Brown returned to Coral Harbour the following year with the intention of expanding the survey.

Military and government support was crucial to the Queen's Expedition. With RCAF assistance, Brown's second-year party transported more than 3,300 pounds of scientific equipment to Southampton Island, including a portable X-ray machine. They also borrowed a significant amount of Canadian army gear to combat adverse weather; this enabled them to stay on location for several weeks.[44] Throughout the study, the Northwest Territories and Yukon were administered by the federal government, with the Northwest Territories Council serving as the territories' governing body; at the same time, the Bureau of the Northwest Territories and Yukon Affairs of the Department of Mines and Resources managed executive functions. The Advisory Committee on Northern Development advised Cabinet on northern matters and ensured consistency in policy where the interests of the various government departments were concerned. The Arctic Research Advisory Committee was under the auspices of the DRB, but it included representatives from various government agencies and coordinated scientific research activities in the North. Scientists who aimed to conduct research in the Northwest Territories required licences issued by the Bureau of the Northwest

Territories and Yukon Affairs. Brown's research conformed to federal standards, according to the minutes of the first meeting of the DRB's Panel on Arctic Medical Research.[45]

Support for cold acclimatization research took hold in Ottawa once the members of the Arctic Research Advisory Committee established a section designed to expand Arctic medical research in Canada. The first meeting of the newly created Arctic Medical Research panel, a division of the larger advisory committee, was held on 16 December 1948.[46] Members of that panel drew attention to advances in Arctic medical research and suggested areas of potential research interest for the Canadian military. Indeed, the terms of reference for the panel stipulated that its members were to review and report on the progress of Arctic medical research projects of real and potential interest to both the DRB and the armed services.[47] The work was confidential, and all members were bound to an "oath of secrecy ... sworn before a Justice of Peace or Commissioner for Affidavits."[48]

Brown was one of the six original members of the panel, and he served as chair between 1952 and 1954. In this role, he reported directly to both the Defence Medical Research Advisory Committee and the DRB's Arctic Research Advisory Committee. Under the oath of secrecy, Brown, as a DRB member, had full security clearance to discuss and write about cold acclimatization research.[49] His authority to communicate research was autonomous, confined only by the limits of the Official Secrets Act, Chapter 49 of the Revised Statutes of Canada. Under that act, it was "an offence to communicate to any person, except under lawful authority, information which might be useful to a foreign power – or to fail to take reasonable care of, or to endanger the safety of such information in one's possession or control."[50] These restrictions did not prevent Brown from publishing extensively about cold acclimatization, which speaks to both the popularity of the science and the results of the work. When the research failed to isolate the vascular characteristics of cold acclimatization in Inuit, Brown had no secrets to protect.

The studies administered by Brown and his colleagues were a direct, non-military extension of the DRB's wider cold acclimatization program, which studied the physiological adaptation to cold of personnel operating in the Canadian North.[51] About 40 per cent of the total amount granted by the DRB for Arctic research in the late 1940s went to medical projects that supported "basic studies of the Eskimo and experiments on nutritional problems, on physiological and other responses to cold, and on conditions resulting from exposure to cold."[52] These studies provided scientists an opportunity to conduct fieldwork in a natural cold environment and the defence department an opportunity to assess a theatre considered vital to Canadian security and national defence.

Acclimatization Research on Inuit

Arctic research in the early Cold War period was a highly cooperative venture. By the time the DRB began funding northern research in 1947, the federal government was already supporting Arctic science projects through the NRC. In addition to participating in several scientific investigations, the DRB served as coordinator for the transportation of scientific personnel and the organization of Arctic research.[53] At the request of NRC administrators, the DRB accepted five medical research projects in 1948.[54] All five concerned human performance and physical response under various stressful environmental conditions. The five projects included an investigation into cold acclimatization by Louis-Paul Dugal of Laval University as well as Brown's study, titled "Clinical and Biochemical Studies on the Eskimo." Both projects were originally financed by an NRC research grant, but following examination "as to their suitability for support by the Defence Research Board, and with the permission of the applicants, the Defence Research Board assumed responsibility" for them,[55] at which point the DRB absorbed both projects into the military research agenda of the defence department.

The first acclimatization project financed by the DRB commenced at Fort Churchill in December 1947.[56] Dugal collaborated with R.E. Johnson of the US War Department Medical Nutrition Laboratory to "prove that ascorbic acid is necessary for acclimatization to cold."[57] Dugal's previous work had indicated the importance of ascorbic acid for the acclimatization of animals to cold, and based on that, the DRB decided to fund his Arctic research. His project aimed to determine whether ascorbic acid increased cold acclimatization in the human body – scientific knowledge that conformed to the military's wider defence initiatives in the North. Under the supervision of the DRB's Guy Marier, Dugal's team at Churchill experimented on a group of thirty-six volunteer soldiers engaged in winter warfare training. The soldiers underwent a two-day physical examination prior to and following the test program. Examinations included blood samples, urinalysis, chest X-rays, and dental inspections. After being divided into three groups, the soldiers took vitamin C pills daily without knowing the dosage. Group A received placebo pills containing zero vitamin C; group B, a 300 mg dosage; group C, a 1,000 mg dosage. In the end, the trials were inconclusive because of a "shortage of accommodation and other administrative difficulties" on location during the winter of 1947–8.[58] Nevertheless, the scientists involved believed that the research had produced valuable information for further study.

Figure 2.1. A group of medical scientists and Inuuk guides in front of a temporary clinic hut near Coral Harbour, 1948. "Arctic Health and Diet Studies Carried on by Queen's Expedition," *Globe and Mail*, 16 September 1948, 15. Public Domain.

Brown's research entailed collaborating with the US War Department; except for that, his research followed the same trajectory as Dugal's. Brown's experience with peripheral blood vessels and diseases made him ideally suited to research cold acclimatization for the Canadian armed services. He first flew to Southampton Island in the summer of 1947, along with the three other members of his team, in RCAF planes, by way of Winnipeg and Churchill. Abandoned wartime huts at Coral Harbour, an Inuit community on the south coast of Southampton Island, served as a temporary medical clinic for examining Inuit and administering tests.

It is reported that during the first trip in 1947, Brown's team brought some 80 per cent of the Indigenous population of Southampton Island by boat to the clinic for medical examination and testing.[59] The examinations revealed that respiratory tract infections and tuberculosis were the primary causes of illness and death among the local population. Researchers also reported enlarged livers in one third of the patients examined. "Specimens of liver obtained from two subjects showed that the enlargement was due to the presence of large amounts of fat, and further work is being done on this problem which is of considerable interest," stated one research report.[60] In response, Brown and his team carried out further nutritional intervention experiments that today draw obvious and grim connections with other federal studies of malnourished Indigenous people in Canada during the same period.[61]

Clinical research performed on Southampton Island extended beyond Inuit health. In a comparative physiological study, researchers used water immersion to analyse cold tolerance in Inuit and White test subjects. Brown and his colleagues studied Inuit on location during June and July, and the so-called White control group during October and November, in comparable outdoor temperatures in Kingston. Although described as acute or short-term exposure tests, researchers immersed the hand and forearm of each test subject in water for a duration of one to two hours at temperatures between 5°C and 45°C.[62] Archival records and open scientific publications provide the results of these immersion tests.[63]

Researchers measured rectal temperature at the conclusion of each immersion and compared the data with measurements taken prior to the test. Inuit tolerated the coldest water temperature for nearly one hour longer than the White subjects; however, the coldest conditions were so severe that all persons subjected to the experiment experienced a drop in core body temperature. A separate test measured hand and forearm blood flow as well as skin, subcutaneous tissue, muscle, and rectal temperatures while subjects rested in a room at 20°C.[64] Researchers also determined skin, tissue, and rectal temperature after each subject's hand and forearm was clothed with cotton wool for thirty minutes. The results of both experiments indicated that Inuit maintained greater blood flow through the extremities than the White students. Researchers attributed the difference to hormonal thyroid activity, postulating that greater metabolic heat production in the Inuk body resulted in increased thyroid secretion, thus enabling the vascular system to maintain a higher level of heat distribution to the extremities.

Brown and his group of researchers from Queen's University posited that hyperthyroidism might be a physiological explanation for cold acclimatization in Inuit. The wider medical community was intrigued by this but also cautious. "[Brown] finds that the liver in the Eskimo is markedly enlarged by clinical standards, and plainly palpable," stated Alan Burton, the founder of the Department of Biophysics at the University of Western Ontario. "Yet liver biopsies obtained from a number of very cooperative natives, have shown no microscopic abnormality whatever."[65] Burton considered the findings inconclusive; he thought that only a seasonal change could show the existence of acclimatization. Brown and his colleagues continued the research, investigating further the links between hormonal thyroid secretion, blood circulation, and cold tolerance in the human body.

Acclimatization, Cold Tolerance, and the DRB

Brown's acclimatization research aligned with the military's long-held interest in the effects of cold on the human body. The DRB organized a conference on that topic in 1948. Scientists, engineers, and research technicians from Canada, Britain, and the US convened in Toronto for that conference, where they presented and discussed the latest research on acclimatization, tolerance, hypothermia, frostbite, and other cold-related medical problems. Concerned that much of the wartime work on cold would remain buried in classified military reports and unpublished in the open literature, the Canadian representatives at the meeting agreed to produce a reference text on the topic. Subsequently, the DRB contracted Alan Burton and his colleague Otto Edholm to research and write *Man in a Cold Environment: Physiological and Psychological Effects of Exposure to Low Temperatures*.[66] First published in 1955, the book served as a detailed yet accessible scientific reader on the principles of thermal insulation, vascular and metabolic reactions to cold, acclimatization, hypothermia and resuscitation, and local cold injury. The authors discussed experiments performed on animals and human subjects, in both military and civilian settings, and concluded with a brief section about problems for future research.

Acclimatization and cold-related research sponsored by the DRB involved other scientists besides Brown and Dugal. Numerous scientists undertook related projects at university and government laboratories – research that often highlighted Arctic experiments funded by the federal government. "The lemming is a low-slung problematic little beastie that lives in the tundra around this northern Manitoba military base," wrote columnist Lex Schrag in a May 1952 reference to acclimatization research conducted at Fort Churchill and DRNL.[67] Schrag had interviewed biochemist J.S. Barlow, who conducted low-temperature metabolism research for the DRB's wider acclimatization program. Barlow's research was an extension of laboratory studies conducted at the University of Toronto by physiologist Edward A. Sellers, who had shaved and cooled tropical white rats to near-freezing temperatures as a test of acclimatization. The cold-exposed rats showed an increased metabolic reaction, which was boosted even higher with hormone injections. Intrigued by Sellers's findings, Barlow applied the same method to study lemmings at DRNL.[68] Explained Shrag: "If the hormones help the lemmings weather the loss of their coats, will stimulation of the production of the same hormones benefit the soldier or airman who has to serve under Arctic conditions? That's the more important question. The answer, obviously, is still hidden in future research."

While Barlow studied the effects of hormones on lemmings, his colleagues at the DRB's northern research laboratory conducted research closely linked with Brown's acclimatization study. During the winter of 1952–3, scientists at DRNL conducted a series of experiments and observational studies comparing the physical and psychological performance of Indigenous soldiers with that of soldiers new to the North. One study of physical aptitude observed a group of Indigenous soldiers from the Northwest Territories while they performed manual tasks alongside a group of White soldiers. "It was noted by means of timed tests, as well as observations," stated a confidential research report, "that their manual performance was decidedly superior to that of white troops in the field although much of their ability would appear to be attributable to their organization." As an explanation for the performance discrepancy, the researchers "hypothesized that existence of a cold skill might be responsible and that it might be possible to train white subjects in this skill, possibly under warm conditions."[69] To test this theory, the scientists subjected Indigenous and White soldiers to various tests of manual dexterity inside DRNL's cold room, with the aim of investigating acclimatization under controlled experimental conditions.

Scientists at DRNL investigated the racialized theory of an Indigenous "cold skill" by comparing the decline in manual dexterity across a range of individual tests. This included tests of tactile two-point discrimination, kinaesthetic sensitivity, strength of finger grip, and decline in joint flexibility. Researchers also observed soldiers for pain associated with chilling, appearance of frostbite, and changes in skin temperature. According to one research report, "native troops showed less decline in manual dexterity than white troops under similar conditions in the cold ... The hypothesis of the cold skill was therefore weakened and [it] appears that the superior performance of the native troops was due to physiological rather than purely psychological factors for the most part."[70] The research did not stop there, however. While in the cold room, scientists subjected the Indigenous soldiers to verbal interviews about working in the cold. Indigenous men withstood greater pain, worked longer under cold duress, and performed outdoor tasks in an organized and efficient manner. Researchers sought a clear explanation for this behaviour, scientific or not. "The main point made was that these Native troops have been working in the cold since early childhood, are used to working barehanded in ice, slushy, water setting traps and their only 'secret' was that they did plan their work to a certain extent," concluded the DRB scientists. "It was our observation that the personal pride of these people in their ability to work in the cold played a part in their superior performance."[71]

Owing to the superior performance of the Indigenous soldiers in northern Canada and the Arctic, military officials considered employing a special unit in connection with the Mobile Striking Force. Research reports indicate that scientists at DRNL studied a selected group of ten Indigenous soldiers stationed at Fort Churchill.[72] The soldiers were steadily observed over the course of one calendar year; this included individual and group performance evaluations under both summer and winter conditions. Scientists observed the soldiers in training, compiled case histories, and administered various intelligence and personality tests. They also took physiological measurements, including basal metabolic rate, reflex and reaction time, and manual dexterity in conjunction with the laboratory cold room tests. The year-long study was intended to determine the suitability and reliability of using Indigenous soldiers with the MSF, assessments considered valuable for developing selection criteria for northern military service.

Scientific research at DRNL also included disturbing biochemical studies involving human blood samples. In August 1954, a mobile unit of the Red Cross visited the Fort Churchill military hospital to obtain blood donations for Manitoba's blood bank. The unit gathered 357 donations from military personnel and northern residents, including Indigenous and White civilians. Scientists at DRNL then "salvaged the minute residue of blood left in the tubes after a blood donation ... to determine what provides the Eskimo's extremely high resistance to cold," explained an article printed in the *Winnipeg Free Press*.[73] Medical scientist J.S. LeBlanc, DRNL's lead project researcher, told a reporter who travelled with the blood clinic that "the Eskimo is able to handle with his bare hands metals such as guns and ammunition in temperatures as low as 60 degrees below zero with no danger of frost damage."[74] Unlike Brown, who focused on biology, LeBlanc grounded his race-based theories about Inuit cold tolerance solely in nutritional science. Convinced that the traditional Inuit diet affected cold tolerance, he examined blood samples to find a correlation between nutrition and human efficiency in extreme cold. LeBlanc theorized that blood samples obtained from Inuit would yield scientific clues for ways to achieve a high degree of cold tolerance in soldiers new to the North.[75] Neither he nor the Red Cross obtained donor consent to use the acquired blood for experimental purposes.

The Fort Churchill hospital operated by DND provided comprehensive medical care to a wide group of patients. Military personnel and their dependents from the Canadian and US services stationed at the base received care at the hospital, as did civilian employees and their dependents, residents of the Churchill area, and transients who arrived at or passed through Churchill by land, sea, or air. By reason of Churchill's strategic location and accessibility in northern Manitoba, the hospital also served as a community medical centre and dental clinic for people residing north of the 57th latitude as well as for Indigenous and settler patients from the District of Keewatin.

There were no family doctors on site, but the facilities and staff offered excellent medical care to patients who had access to the services provided. Eileen Jacob of Victoria, British Columbia, who moved to Churchill and started a family with her military husband, Jack, in the early 1950s, gave birth to two of her three children at the garrison hospital. Army doctors assisted in both births, and Jacob recalls the hospital staff treating White and Indigenous civilians, some of whom the RCAF flew in from isolated areas farther north, along with service members. "One time the cast of the Tommy Hunter Show visited patients at the hospital," recalls Jacob, who has fond memories of military and family life at Churchill.[76]

Others encountered different challenges in their attempts to access medical treatment at Fort Churchill. Although the hospital provided fair and equitable services to military and civilian patients, personal and familial wealth represented a barrier to treatment for some. DND assumed responsibility for all military personnel stationed at or visiting the base, ensuring that soldiers and their dependents, by reason of their terms of employment, received free care and treatment. This policy applied to all personnel regardless of service and rank. Civilian patients had equal access to the hospital but had to pay a portion of the treatment costs if they wanted or needed care. Medical and surgical services to in- and outpatients followed the rates paid by the Public Service Surgical-Medical Insurance Plan; in other words, DND used federal civil service standards to determine the costs of civilian treatment. Insurance through the Manitoba Hospital Commission or Territorial Hospital Insurance Services covered some of the treatment costs for civilian patients, but the net result was often a revenue surplus for DND between the mid-1950s and the early 1960s.[77]

As evidenced by DRNL's close integration at Fort Churchill, the garrison hospital also provided defence scientists with access to research facilities and experimental subjects. The physiological studies that LeBlanc referenced during his interview were conducted concurrently with the research involving Inuit blood. In a series of experiments designed

Acclimatization, Cold Tolerance, and Biochemical Experimentation 85

to investigate the use of military equipment under northern conditions, DRNL scientists observed White soldiers and Inuit handling metal objects in sub-zero temperatures. Mace Coffey, DRNL's head of operational research, carried out the research with biophysicist Charles Eagan. The two scientists studied a selected group of research subjects while on location in December 1954, hoping to gain valuable information for the Canadian armed services. Ten White soldiers and Inuuk men participated, individually exposing bare skin to a metal object for as long as they could bear the pain.[78] "Experiments have shown that an Eskimo can hold for several seconds a piece of metal that has lain indefinitely in the cold without having the flesh of his hand frozen," stated one report.[79] "A white man's flesh freezes instantly." Eagan called the experiments partly biophysical and partly biochemical; reportedly, he said in an interview that "tests showed the Eskimo could hold for several seconds a metal object while the temperature hovered between 35 and 40 below, with a wind blowing from 35 to 40 miles an hour."[80]

It is no coincidence that Coffey and Eagan conducted the experiment in late 1954. After recorded incidences of frostbite spiked among the multinational fighting forces in Korea, scientists in Canada and the US devoted considerable attention to studying the effects of cold and wind on the human body. Army doctors in the US military used experimental drugs during the Korean conflict to control spasms produced by frostbite and to repair damaged nerves and tissues in treated soldiers.[81] In Canada, scientific intelligence officers in the DRB paid special attention to frostbite and hyperthermia research conducted by scientists in the Soviet Union, collecting and translating any medical literature or press reports on the topic and supplying the information to Brown and his colleagues on the Arctic Medical Research panel.[82] Experimental work at Fort Churchill conformed and contributed to this larger research agenda and scientific fixation with the medical treatment of military-related cold injuries, work firmly grounded in the rapidly evolving military context of the early Cold War.

Scientific interest in the metabolic responses of Inuit also conformed to a wider pattern among North American researchers, as medical historian Maureen Lux has demonstrated in her work on the history of Indian hospitals in Canada.[83] DRB scientists expressed particular interest in obtaining and testing the blood of Inuit patients treated at such locations as the Charles Camsell Hospital in Edmonton, as did their US counterparts. Researchers from south of the border used blood samples taken from hospitalized Inuit patients for a race-based study of different ethnic groups conducted in the mid-1950s. This kind of experimentation differed from standard clinical trials in that the patients from whom researchers obtained blood did not consent. As Lux explains: "Given the very unequal power relationships between medical staff and patients, coupled with often perplexing language barriers, and the widespread perception that many Aboriginal people would not willingly accept treatment, patient consent for treatment was often simply taken for granted."[84] Settler-colonial medical practices of this sort underscore the gravity of the experimental work performed on Inuit blood at DRNL, where Canadian and US researchers had immediate access to the facilities and patients of Fort Churchill's military hospital.

Acclimatization Research in International Context

Brown and his colleagues were part of an international cohort of scientists engaged in climate-related medical research. In February 1954, for instance, fifteen scientists from Canada and the US travelled to DRNL for a five-day conference about medical problems in severe cold climates.[85] Besides Brown, the Canadian group included Alan Burton and James Stevenson of the University of Western Ontario, Edward Sellers of the University of Toronto, and Louis-Paul Dugal and Edward Page of Laval University. Jointly sponsored by the Josiah Macy Foundation for Medical Research and the DRB, the closed-invitation event included tours of the Fort Churchill military base, the neighbouring townsite, and a field trip on the snowy barrens of northern Manitoba. The attendees discussed medical research relating to frostbite, hypothermia, and acclimatization to cold, among other subjects.

While DND authorized the scientists, engineers, and research technicians who visited and performed experimental work at Fort Churchill, DRNL scientists openly shared information vetted and pre-approved by DRB's public relations officials. In fact, the DRB took a leading role in pursuing and promoting interdisciplinary research into the effects of cold on the human body. During the 1955 annual meeting of the Canadian Physiological Society, held at the University of Western Ontario, several DRB-sponsored scientists presented research findings on the hazards of cold for soldiers, sailors, and pilots. More than 300 physiologists, pharmacologists, biologists, and other medical scientists attended the meeting, representing both civilian and military research institutions in Canada and abroad. Among the DRB attendees were J.S. LeBlanc and Charles Eagan, who discussed the effects of cold on fingers and on manual dexterity.[86] These two researchers explained that the lubricating fluid between the finger joints thickened when cold – like oil in a combustion engine – and that this effect offered a partial physiological explanation for stiff fingers in cold conditions.

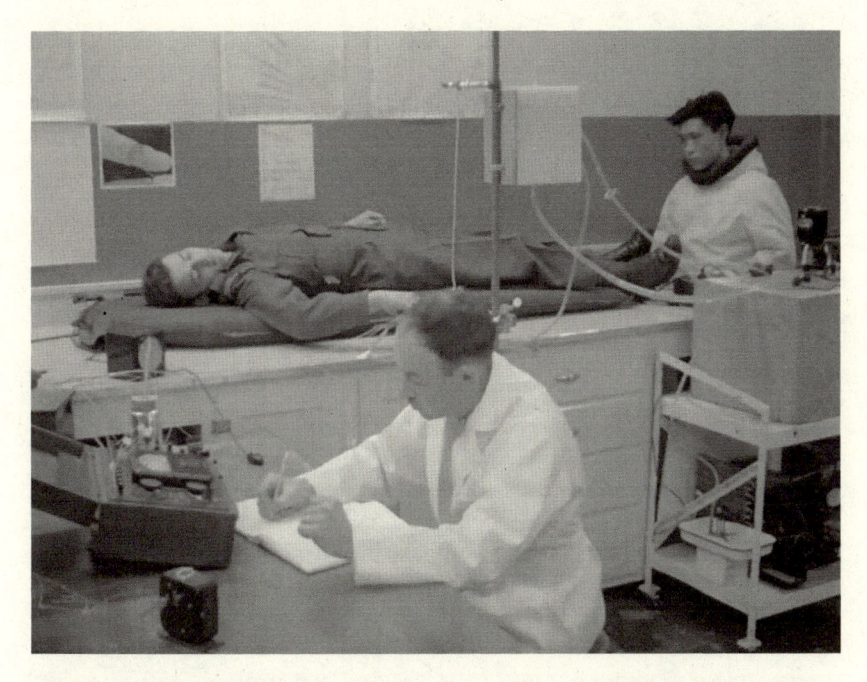

Figure 2.2. A White soldier and an Inuk man undergo scientific tests inside Defence Research Northern Laboratory, 1954. *Vigil in the North*, National Film Board of Canada, 1954. Public Domain.

Comparative testing performed on Inuit and White research subjects at DRNL had supposedly confirmed this finding. Eagan developed a device called a continuous flow calorimeter while conducting cold-room experiments with Burton at Western.[87] He relocated to Fort Churchill in February 1954 and joined the DRB's staff at DRNL, where he used his device to observe and record heat loss in the fingers of White soldiers and Inuit. Each subject placed one finger inside the device, thus exposing an extremity of the body to varying degrees of cold.[88] Eagan observed the physiological responses of each test subject, theorizing that small tissue pressures inhibit "cold vasodilation" – blood flow and temperature regulation in the fingers. Scientists contended that this research had practical applications, in that it explained the coldness of tight gloves and hand gear.

Acclimatization and cold tolerance research was conducted outside Canada as well. As Vanessa Heggie's research on the history of exploration and extreme physiology shows, the work of scientists who studied human adaptation and acclimatization to tropical, high-altitude, and Arctic environments often overlapped and intersected with research on

Indigenous bodies. Theory and consensus were different matters altogether, though. Physiologists Laurence Irving and Per Fredrik Scholander, for instance, used fieldwork and studies of living organisms in the Arctic to argue that cold-weather adaptation in Inuit derived from experience and skill, not biology. "Time and again," Heggie explains, "the Eskimo, and other cold-region dwellers, were specially excluded from discussions of biological adaptation, and while in many cases their exclusion reinforced a hierarchy that put tropical peoples below those of temperate regions, by midcentury it was also the source of criticisms of environmentally deterministic theories."[89]

Irving, Scholander, Burton, and others pursued non-biological explanations of acclimatization and cold tolerance. Race-based theories of biological adaptation in the Indigenous body were not exclusive to Brown and the Queen's Expedition, though. Military considerations led US officials to sponsor studies of cold-weather science and human adaptation in the Arctic. As historical geographer Matthew Farish documents, extensive militarization during and after the Second World War turned Alaska into a veritable laboratory of Cold War science.[90] Researchers at the Arctic Aeromedical Laboratory in Fairbanks conducted an extensive acclimatization and cold-survival research program that included experimentation on Indigenous Alaskans. Testing a hypothetical connection between hyperthyroidism and cold tolerance, scientists used a radioactive medical tracer to measure thyroid activity in 120 subjects, including "19 Caucasians, 84 Eskimos, and 17 Indians."[91]

In 1938, American scientists J.J. Livingood and T. Seaborg announced that they had discovered a longer-living radioisotope of iodine, iodine-131. Researchers Joseph Hamilton and Mayo Solley quickly put iodine isotopes to use in medical experiments, administering radioiodine orally to patients. The experiments showed an uptake of more than ten times as much radioiodine in patients with overactive thyroids, and this laid the groundwork for the widespread use of iodine-131 in the treatment of hyperthyroidism and for related biological experiments that attempted to develop therapies and medical treatments with radioisotopes.[92] The administration of radioactive iodine was "one of many methods deployed to understand the physiology of the (singular) Eskimo as a gateway to military success in the North," Farish explains.[93] Ethical questions regarding how the participants were selected and the associated medical risks of the research led to a public inquiry in the 1990s. After hearing testimony from medical scientists and persons directly involved in the study, the committee leading the inquiry published a report that described in detail what it viewed as a gross disregard for human life.[94]

The use of tracers predates the atomic age. Scientists employed naturally occurring radioelements in the 1920s and both artificial and stable radioisotopes in the 1930s; the development of nuclear reactors during wartime then enabled the mass production of radioisotopes and increased the scale and use of radiotracers in the Cold War. By the early 1950s, as historian Angela Creager explains, the US Atomic Energy Commission had vastly increased the overall consumption of radioisotopes at home and abroad while Canada and Britain were selling radioisotopes in conjunction with their atomic energy programs.[95] The Soviet Union supplied radioisotopes to its own institutions and satellite states, thus contributing to the global circulation of radioactive products and a corollary explosion of biological and medical research with radiotracers.

Notwithstanding certain and obvious similarities, no direct evidence links the experiments in Alaska with the acclimatization research conducted in Canada. The experiments in the US were performed after the Canadian researchers returned from their final trip to Southampton Island in 1954, and Brown's personal correspondence indicates that neither he nor any member of his research team had any direct involvement with the US scientists who conducted the research in Alaska. Brown did, however, read extensively on the experimental use of radioactive iodine in thyroid treatment.[96] On 2 November 1954, he wrote to Keith Wightman of the University of Toronto's Banting Institute, inquiring about the practice of administering therapeutic doses of radio-phosphorus.[97] In his reply, Wightman confirmed that he had treated cancer patients with doses of a radioactive isotope of phosphorus and radioactive iodine. The correspondence ended with the reply, and there is no evidence that Brown inquired with the intention of administering radio-phosphorus in his own practice.

Brown was careful to distance his research from similar US studies. When George Mann of Harvard University sought permission to co-publish results about acclimatization research in 1955, Brown declined. "Despite any estimates," Brown wrote to Mann, "I don't believe anyone really knows what is the average fat intake in the Eskimos and it is a mistake to say that the results of our carefully done but necessarily restricted dietary experiment provide such a figure."[98] Mann specifically wanted to co-publish the results of a study that examined the relationship between diet and serum lipid in a group of 161 Inuit of various ages, but Brown declined because his results were derived from separate research on the biological characteristics of cold acclimatization. Nonetheless, the correspondence in Brown's personal papers reflects a fascination with race and science. His team may have conducted research in relative isolation, but the acclimatization studies involving Inuit conformed to a wider pattern of experimental science in Canada during the early Cold War.

From Biochemical Science to Cultural Appropriation

Perhaps most visibly in suggestions for future projects submitted to the DRB's Arctic Medical Research panel, military science in Canada embraced colonial perceptions of the Inuk body. Light reflection from snow made military operations difficult on Arctic terrain, and some senior officials thought that science could provide a useful solution to the problem of "snow blindness." One concept suggested "a study on the special senses of the Eskimo, especially eye function ... [T]he Eskimo is 'racially pure' and has had high ultra-violet exposure for generations."[99] This hypothesis proposed research into the biological functions of the Inuk eye, which, based on rudimentary scientific observation, seemed capable of resisting snow blindness. In other words, researchers thought that studying the eye function of Inuit could prove beneficial for developing protective gear to assist the vision of White soldiers serving in northern Canada and the Arctic. Another idea submitted for consideration was "a study of the adaptability of the Eskimo to unfamiliar tasks," because officers in the armed services expressed frustration about Indigenous soldiers' discipline and knowledge of military regiment. Although both suggestions seem to have gone unexplored scientifically, Canadian scientists in the DRB pondered and discussed a range of possibilities for Inuit test subjects.

Malcolm Brown made a final request for grant monies from the Defence Research Board in mid-January 1954. "Considerable discussions arose regarding acclimatization [and] Brown indicated his reasons for believing that his work did constitute a study of acclimatization itself and not racial differences," stated the minutes of the panel's tenth meeting.[100] The specifics of Brown's reasoning went unrecorded, but the panel approved his funding request. He used the funds to make another trip north, which proved to be his last. Brown's research ended in 1955 without clinical evidence showing the existence of cold acclimatization.

Federal funding for cold-weather research continued after Brown concluded his work, although grant funding from the DRB went increasingly to non-human cold studies concerning weather, terrain, oceanography, and other environmental research areas. Defence scientists continued to support the armed services through collaborative projects in the North, and the military continued to train soldiers for Arctic warfare, in both winter and summer conditions. In the process, the armed services turned to cultural appropriation of Inuit. Military records refer to the value of

Indigenous shelter and living techniques, and DRB engineers fabricated special northern attire from Inuit clothing.[101] In 1947, for instance, the DRB's Wilbert Cowie initiated a research project to develop Arctic clothing for the armed services. Impressed by the "comfort of the Eskimo garb," he brought a caribou skin parka and wolverine fur back to Ottawa after a research trip to Churchill. With chemical engineer Arthur Blouin, Cowie tested the fur for oils and waxes and conducted examinations to determine its tensile strength. Five years later, the researchers produced eleven variations of a nylon fur suit. "The men gave their [Inuk] guide a suit to try on and had difficulty getting it back," read one press release. "It is superior to caribou and wolverine because it won't tear, it will wash and dry quickly and natives want it because one who has smelled wet caribou skins will walk 100 miles to get something else."[102] After testing in the North, the so-called survival suits underwent operational trials at the RCAF's Survival Training School in Edmonton.

Authorities in the DRB proudly advertised the synthetic survival suits as a scientific achievement, one that showed both civilian and military promise.[103] The organization applied for nylon fur patents in several countries and signed a contract with a British manufacturer. In Canada, researchers involved with the project touted the wide benefits of nylon fur. Information about the project became public three years before Cowie and Blouin released the eleven suit varieties. For a long time, one of the military's greatest problems was keeping soldiers and pilots warm in the frigid conditions of the Arctic, reported Gerald Waring over a CBC radio broadcast in June 1949:

> The Eskimo keeps warm by wearing caribou skin clothing; there aren't enough caribou in Canada to outfit the army and the air force. Cowie and Blouin solved that problem by developing a synthetic caribou fur made out of nylon, and at the same time they opened the door to what may become a tremendous new industry: synthetic furs. A friend of mine, Flight Lieutenant Scott Alexander of the air force, wore a nylon fur suit up in the Arctic for two weeks in March. He was testing it, living in igloos, and traveling by dog team across the barrens in 50-below weather. When the Eskimos first saw his nylon fur suit, they couldn't believe that the White man had made it. They thought it must be fur like their own suits. He had to turn his pocket inside out to show them the woven fabric, and if you can fool an Eskimo on furs, you can fool anyone.[104]

Similar reports appeared when Cowie and Blouin announced the successful production of the eleven suits in 1952. The mink, beaver, and muskrat manufactured by DRB chemists and engineers could be any texture and dyed any colour, reported a featured article in *Reader's Digest*: "*Is it warm?* Warmer than caribou – until now the warmest fur in the Arctic – say the researchers. *Does it look real?* Real enough to fool the Eskimos who thought the DRB experimental suits were white sealskin and tried to buy them."[105] In their efforts to keep service members warm in northern Canada and the Arctic, Cowie and Blouin received accolades for stumbling "on the answer to every girl's dream of a fur coat."[106] The DRB celebrated the military and civilian applications of nylon, in part because advancements in science and engineering gave all Canadians access to a symbolic piece of the rugged and mythic North.

For the armed services, the development of winter clothing demonstrated cultural *difference* through science and engineering. The military created its own forms of winter clothing, but Inuit attire was essential to the research and development process. Science had failed to appropriate the perceived cold-fighting biological traits of the Inuk body, but traditional knowledge remained valuable for Canadian military personnel and scientists alike. Replicating and "improving" Inuit clothing was appropriation by other means, yet another manifestation of the colonial authority credited to Western science by newcomers in the North.

The Canadian military poured resources into the design and development of Arctic medical supplies as well, further demonstrating the perceived value of science and engineering for overcoming the environmental hazards of northern climate and terrain. Cowie and Blouin developed a medical tourniquet for use by Canadian and US forces, for instance. At the request of army officials, the two chemical engineers worked with physicist R.A.F. Carruthers and chemist E.M. McPherson on the development of a rubber tourniquet designed for self-application in operational circumstances. The device applied pressure to a wounded limb, limiting but not stopping the flow of blood. Officers in the Royal Canadian Army Medical Corps tested it in the field under northern conditions and claimed that it held up at 40 degrees below zero.[107] Fabric tourniquets were susceptible to fungus and other infectious bacteria, especially in hot climates or cold and wet winter conditions, whereas the rubber tourniquet developed at the DRB's chemical laboratories served as a versatile replacement suitable for military use.

The push to develop new products for civilian and military use in Canada during this period reflects the wider technological changes of the mid–Cold War. During the 1950s and 1960s, rapid laboratory advances enabled US chemical giant DuPont, British Imperial Chemical Industries, and other large industrial firms to create and distribute a wide variety of improved synthetic polymers. Before the widespread manufacture and use of plastics in the 1960s, polymers and other synthetic products represented technological advances. They caused little or no anxiety about the environmental consequences.[108] As historians J.R. McNeill and Peter Engelke point out in their work on the acceleration of climate change since 1945, the technical mastery of rubber and plastics also represented social progress, and the relative affordability of synthetics fuelled world production of them until it reached some 6 million tons by the start of the 1960s.

Press reports about the DRB's northern military clothing and medical equipment played into preconceived notions about the Arctic among observers in southern Canada. Descriptions of long and arduous sledding expeditions near Cambridge Bay and other seemingly remote locations on the Arctic islands north of the Canadian mainland presented military personnel, scientists, and engineers as virile risk-takers. Flight Lieutenant Scott Alexander and Blouin reportedly tested various suits in "villainous weather," trusting themselves to the engineered synthetics.[109] Backed by the benefits of science and engineering, these men had the intellectual capacity and physical stamina necessary to overcome a climate and terrain perceived as harsh and unforgiving. One report from January 1953 credited Alexander, then squadron leader and chief of the RCAF's survival school, with suggesting that the DRB pattern winter suits after Inuit clothing. The same report indicated that researchers issued the nylon suits to Inuit as a test. "They wanted to keep the suits and get supplies of the cloth," recounted one RCAF pilot. "Their only complaint was that it was too warm."[110] The noise of the suits apparently drew the ire of soldiers, who expressed concerns about sneaking up on a potential enemy. "Yet the Eskimos would gladly trade their caribou for this nylon fur," one member of the RCMP claimed, "and they don't think it would be noisy. It isn't."[111] Removed from the full context of interaction in the North, such claims reinforced the notion that Western science was superior to Indigenous traditional knowledge. At the same time, however, the theory of acclimatization remained firmly imbedded in the minds of outside observers from southern Canada.

Conclusion

When read more than sixty-five years after its conclusion, the acclimatization research on Inuit and White test subjects represents a disturbing and complex symbol of Canadian science in the early Cold War. The research assumed that human testing might produce civilian as well as military applications, and ethical issues concerning the use of human subjects did not deeply penetrate the scientific or medical discourse. The studies contributed to a popular and growing field of environmental scientific inquiry, and acclimatization research was not highly restricted or classified. Acclaimed scientists received support from state and academic institutions to conduct the research and publish the findings in reputable scholarly journals, illustrating the militarization of medical science in Canada as well as increased integration between the armed services and civilian scientists in the early post-war period.[112] When the studies failed to yield practical results, Canadian researchers moved on and the experimental work gradually faded from relevance. Yet the survival of medical papers, unpublished reports, and military records makes it possible to investigate the purpose of the research and contextualize the studies in relation to the perceived scientific intent.

Records indicate that Malcolm Brown and his team of researchers did not operate with the primary aim of helping Inuit reduce any perceived strain they placed on the Canadian state. Although acclimatization scientists helped introduce the welfare state to the North by providing medical treatment services to Inuit, they did so while pursuing an unrelated goal. Interest in the Inuk body, rather than a desire to reform Inuit health care, was the primary motivation for the scientists engaged in acclimatization research. In this, they differed from those who were responsible for carrying out the government's extensive Inuit relocation program during the same period.[113] The superficial problem of Inuit dependency merely opened the door to a different form of settler-colonialism, one in which the biologized Inuk body served a distinct scientific agenda rooted in Cold War concerns about Canadian security and national defence in northern Canada and the Arctic.

The perception that the Inuk body was acclimatized to cold was a constructed idea. The scientists engaged in the research pursued an unattainable goal that had been conceived and was thereafter perpetuated by racialized perceptions of northern Indigenous peoples. Sources refer to the so-called dependency of the "Eskimo," but most of those references are outside records pertaining to cold acclimatization. Where cold-weather research was concerned, military research reports and medical publications largely avoided discussion of the "Eskimo problem." Those engaged in acclimatization research biologized Inuit in a process that advanced southern interests first and those of the colonized second. The southern interests pursued a dual-purpose agenda that was distinct from cultural assimilation.

Available sources have little to say about the personal convictions of the scientists who researched cold-acclimatization. Published medical papers describe Inuit and White human test subjects as material objects. Brown's personal papers are much the same. Correspondence remains between Brown and his colleagues, but written records apparently make no reference to interactions between researcher and subject. Brown was heavily invested in Arctic research and medical activities, nonetheless.

He enjoyed his work and valued his chance to contribute to the Canadian medical profession through government initiatives. He died in 1977 and was inducted posthumously into the Canadian Medical Hall of Fame in 2000. His participation in acclimatization research should be viewed as part of a wider intersection of complex circumstances and events. His personal records provide a window of clarity without match, but he was one among many who were engaged in the medical treatment services and acclimatization research activities involving Inuit. The many who jointly conceived, supported, and contributed to the research and experimental work also deserve attention and scrutiny.[114]

Imbued with visions of dominance and superiority, cold-acclimatization research in Canada provided an opportunity for medical science to serve multiple agendas. The research intrinsically posited the possibility of biological appropriation, contributing another disturbing layer to the colonial (mis)treatment of Indigenous people in Canada. The nordicity of Inuit was absolute in the eyes of the Arctic scientists, and the Inuk body became eminently well suited to meet the needs of the Canadian military and defence department when reduced strictly to a biological function. As Brown wrote in 1954, "[because] of their performance in the cold it seemed safe to assume that [Eskimos] were acclimatized, though in the beginning uncertainty had to be admitted." [115] Military and defence officials endeavoured to exploit Brown's assumption, but cold acclimatization extended beyond the control of the state. Adaptation to the cold Canadian North remained elusive, and by the end of the research, a dejected Brown could only conclude that "the degree of acclimatization seen in the Eskimo is not really important for any purposes but theirs."[116] He published a final research paper four years later. "To be an Eskimo is not necessarily to be acclimatized to cold," he and his co-authors admitted, "but it is our conclusion, for the reasons given, that our selected Eskimo subjects were so acclimatized."[117]

While available records describe a complex set of circumstances, the perceived acclimatization of Inuit served a specific scientific pursuit. Originally conceived as part of a medical effort to investigate cold tolerance, acclimatization research offered a potential solution to military problems. Scientists' failure to define the vascular characteristics of cold acclimatization is thus irrelevant when assessing the impact of their research. The imperative point is that they had pursued an abstract biological variation between the Inuk and White body. An idea conceived by medical science took on an agenda distinct from and unrelated to its original purpose. Perpetuated by a calculated response to a specific military problem of the early Cold War period, acclimatization research represents a troubling but important intersection between the colonial state and the Canadian defence establishment.

3 Entomology, Insect Control, and Biological Warfare

The publication of Rachel Carson's *Silent Spring* in 1962 spurred an environmental awakening in the US and changed public discourse about the widespread impacts of chemical production. Carson, a marine biologist who studied at Johns Hopkins University and worked for the US Bureau of Fisheries in the 1930s, ignited a controversy over chemical residues in food and the environment that put the insecticide dichloro-diphenyl-trichloroethane (DDT) at the centre of a growing public conversation about the dangers of synthetic chemicals. She argued that the Second World War had helped create an ideological belief in technological development. Like many other wartime innovations, DDT was a product of economic ties between the US military and private industry. It was viewed as a miracle chemical that came with clear civilian benefits and relatively little risk to people. As historian Edmund Russell explains, agricultural experts in the US were calling for "total war against man's insect enemies, with the avowed object of total extermination instead of mere 'control.'"[1] From this perspective, DDT and similar pesticides appeared safe, affordable, and necessary in the human battle for survival against insects. *Silent Spring* contended otherwise, however. "Carson did not just expose the risks posed by pesticides," writes environmental historian Nancy Langston, "she also challenged a worldview that insisted science could and should master nature."[2]

Prior to the environmental awakening that Carson helped inspire among activists in the US and elsewhere, DDT was the insecticide of choice for many civilian and military users in North America. Indeed, the first decade of the Cold War saw the mass spread of wartime chemical agents across newly industrialized, domestic landscapes in both Canada and the US. As historian Kate Brown describes, one of the more harmful environmental consequences of the Second World War was the vast increase both in chemical production and in the residual by-products of newly developed substances. The US chemical plants that produced ammonia for explosives also used it to make DDT for combating mosquitoes in tropical climates where Allied soldiers were fighting, and after the war, these same companies used mass marketing campaigns to sell wartime chemical surplus for everyday domestic use.[3] It seemed that insecticides offered a viable solution to North America's bug problem in both military and civilian settings.

Scientists, soldiers, and government workers posted to northern Manitoba learned quickly that wet, boggy muskeg is a perfect breeding ground for biting insects. Swarmed by clouds of mosquitoes, blackflies, and deer flies so thick as to blot out the sun, newcomers realized that extreme cold and drifting snow were not the only natural environmental challenges posed by the North. Summer conditions introduced different but equally challenging problems for military operations. In July 1947, drawn there by the prevalence of northern insects, a joint Canada–US party dubbed Operation Bugbait camped in a swampy forest near Goose Creek, a waterway connecting the Churchill River and Warkworth Lake in the Hudson Bay drainage basin immediately south of Churchill. The members included scientists from the Bureau of Entomology and Plant Quarantine, on behalf of the US Army Committee for Insect and Rodent Control. Those scientists worked alongside Canadian entomologist C.R. Twinn and his team assembled from the divisions of entomology, botany, and plant pathology of Canada's Department of Agriculture.[4] The party camped for three days at Goose Creek, where it surveyed biting flies and tested chemicals for mitigating them.[5] They tested the effectiveness of shelters, sprays, clothing, and repellents and released various concentrations of DDT into both air and water.[6] These experiments marked the first in a series of entomological research projects carried out across northern Canada over the next eight years – experiments that demonstrate the entangled histories of ecological science and military affairs in northern Canada during the first decade of the Cold War.

Between 1947 and 1955, military-sponsored scientists from Canada and the US carried out entomological research experiments, seeking ways to eradicate biting insects from selected locations in the North. In theory, if biting insects were eradicated then soldiers would be able to train more effectively and efficiently; furthermore, controlling insects in isolated northern areas could provide the Canadian and US forces a distinct advantage in the event of a Soviet land attack. "Insects proved a big menace," journalist Cyril Bassett commented in February 1952. "Mosquitoes up there, said the boys of the RCAF, were as big as fighter planes and in their way just as deadly."[7] The first experiments were conducted at Fort Churchill. Using military equipment and aircraft, scientists sprayed DDT and other chemical insecticides over large areas of open terrain near the Churchill River. The results led to further insect-control studies at various other locations in northern Canada, where entomologists experimented with radioactive isotopes to study insects as part of biological warfare research. Senior military and defence officials in Ottawa and Washington expressed their concerns about a possible vector-borne attack against North America, and this prompted a series of diverse and long-running entomological research experiments that began with eradication and eventually extended to a sinister area of the Cold War sciences.

The entomological research conducted near Churchill in the summer of 1947 kick-started a long series of experiments in insect eradication and repellency.[8] "One to five acre tests of aerosols, insecticide smoke generators, and hand sprayers to protect the camp site from biting flies using DDT, 666, and pyrethrum indicated that under high population pressure only partial and temporary relief can be obtained by such measures," wrote Twinn in his report.[9] The Canadian and US entomologists released various chemicals into both air and water in various concentrations, exposing mosquito and blackfly larvae to DDT for periods of fifteen minutes and longer. "No evidence was seen of injury to fish," Twinn claimed in his progress report. The party evaluated eighteen different repellents in field tests against subarctic and Arctic species of Aedes mosquitoes and blackflies, marking the start of a series of investigations into insect control methods for the North. Twinn travelled to Washington three months later and discussed continuing entomological trials at Churchill the following summer. The US representatives pressed for air spray trials early in the mosquito season to prevent the emergence of insect larvae.[10] Twinn lacked the authority to make any firm commitment in that regard but returned to Ottawa convinced that US assistance could further entomology and insect-control efforts in northern Canada.

Entomological research at Churchill involved extensive use of military equipment and experimental chemicals. During the summers of 1947 and 1948, Canadian and US scientists used ground equipment to spray approximately 250 experimental plots of land ranging in size from 1,000 square feet to two acres with various dosages and formulations of DDT and other insecticides.[11] With the assistance of the RCAF and specially equipped C-47 Dakota aircraft, the entomologists sprayed DDT preparations on an additional thirteen plots as large as two and a half square miles. Generally, the results indicated the superiority of DDT as a mosquito insecticide.

To determine the effectiveness of aerial spraying for insect control in limited areas, the DRB and the RCAF expanded the experimental research program. Spraying occurred at various RCAF stations during the spring and summer of 1949, including Whitehorse and Watson Lake in Yukon, Fort Nelson and Fort St. John in British Columbia, Rockcliffe in Ontario, and Goose Bay in Labrador.[12] Dakota aircraft fitted with tanks sprayed a 4 per cent DDT–fuel oil solution over a total of seventy-one square miles, releasing insecticide by gravity flow through a calibrated vertical emission pipe below the fuselage. Entomologists involved in the research recorded an estimated dosage of one quarter-pound of DDT per acre – enough insecticide to kill 91 per cent of larvae in sprayed areas and control the mosquito population at each RCAF station for several weeks.

The long history of insecticide use in Canada suggests that DND officials supported aerial spraying under the assumption that the practice was safe, affordable, and effective. Jennifer Bonnell's research on insecticides and honeybee poisoning in the Great Lakes Region during the late nineteenth century, however, illustrates the deep roots of agricultural industrialization in North America and its associated environmental consequences.[13] In the 1880s and 1890s, concerned beekeepers in Ontario pressed for prudent insecticide use through legislation, education, and advocacy. Bonnell likens those late-century beekeepers to "early advocates of environmental protection," a descriptor worth applying to Canadian researchers outside the federal government, who, as will be seen later in this chapter, eventually pushed back against aerial spraying and voiced their concerns about the use of chemical insecticides in Canada during the mid-1960s.

But the roots of Canada's aerial spray war against forest insects were deep and strong. As Mark Kuhlberg's work illustrates, the interwar years saw concerted attempts to battle and control spruce budworm in Nova Scotia, kill and eliminate hemlock looper in the Muskoka region of Ontario, and combat various forest insects in Vancouver's Stanley Park and the British Columbia Lower Mainland. Kuhlberg writes that in using aircraft to dust infested woodlands with chemicals that were toxic to pests, "environmental preservationists and recreationists frequently resorted to the rhetoric of war in an effort to portray their endeavours in terms that resonated with a population that had all too recently emerged from the world's most gruesome conflict."[14] In their war against bugs, which was waged on both sides of the Canada–US border, entomologists aimed to protect natural woodlands for both business and pleasure. But nature resisted human attempts at control and manipulation, exercising its own agency and creating new challenges for the evolving chemical industry.

The military's efforts to study and overcome biting insects in northern Canada during the early Cold War were strikingly similar to the aerial spraying campaigns of the interwar period. However, the goal this time was not preservation – it was complete eradication. Entomologists worked in lockstep with the armed services to eliminate insects, treating nature as a military impediment. From this perspective, northern Canada did not represent an environmental playground. Climate, terrain, and environmental conditions were variables in the rapidly evolving equation of security and national defence, and the overlapping sciences of entomology and chemistry seemed to offer a viable solution to the problem of warm-weather military preparedness in the North.

The Canadian Army and Insect Control

It would be an understatement to say that mosquitoes and blackflies hampered military exercises and training in the summer months around Fort Churchill. Head nets and repellents offered only limited protection, and exercises that involved long marches over boggy terrain or canoe trips in marshy water educated soldiers and officers in the reality of insect life in the North. Canadian military officials expressed frustration about the insect-control efforts undertaken by Twinn and his team of entomologists. In early January 1948, the Army's Director General of Medical Services, Brigadier William Coke, sent a memorandum to DRB chair Omond Solandt proposing research into the irritating substances injected by mosquitoes and blackflies during the biting process. Coke maintained that repeated small blood loss or systemic toxicity altered the morale of individual soldiers exposed to numerous bites. "In all individuals and particularly in those who react most violently to the bites," he wrote, "there is a decrease of morale and unwillingness to venture into the most heavily infested areas."[15]

Coke criticized the protective clothing and repellents used by soldiers in northern Manitoba and called for better insect-control methods to be found than the conventional ones that had been developed by the entomologists in the agriculture department. To emphasize his point, he referenced the apparently low tolerance of northern newcomers: "Natives of the area, or individuals who have long lived in the area, appear to be less effected than new-comers to the area." Only after repeated exposure to bites over several years did newcomers to the North develop "a state of 'immunity' in which they react less violently," he wrote.[16] As a solution, he proposed a controlled scientific study during the summer to either develop a screening process for northern service based on insect tolerance or a method for artificial immunization.

Coke's memorandum spurred the DRB to act. Within two months, various panels and committees had convened to discuss insect-control methods in the North and devise a solution for the armed services. Coke received a formal response on 8 March. Writing on Solandt's behalf, DRB vice-director General E. Llewelyn Davies agreed that the discomfort and irritation caused by biting insects posed a serious problem for soldiers' morale and effectiveness. Davies was a British scientist with extensive experience in defence research. He served as head of medical research in chemical warfare for the British at Porton Down before becoming superintendent of the Suffield Experimental Station in southern Alberta in 1941.[17]

As a prominent member of the DRB, Davies championed biological and chemical warfare research to support the Canadian armed services in peacetime. In his letter to Coke, he acknowledged the deficiencies of conventional control methods in locations far from insecticide supply and proposed that DRB scientists undertake a "preliminary psychological investigation of the reaction of troops who have not previously served in areas where biting insects are a problem." Davies also appealed to Coke's second suggestion, proposing that insect saliva be isolated and analysed for subsequent "immunization trials with unexposed subjects."[18]

The first of Coke's two suggestions received immediate attention. Ten days after Davies sent his response, D.C. Williams of the University of Manitoba submitted a plan to the DRB's Biological Research Division for a study of the psychological effects of insects on army personnel. Previous studies had shown that the complete eradication of pesky insects from northern Canada was impossible, Williams wrote; he then proposed a study of "human adjustment" to determine what differences exist in a person's capacity to tolerate insect attack and whether developing psychological or physiological immunity was possible.[19] Williams's proposal was no more than a formality. In the original memorandum to Solandt, Coke expressed his frustration with the Department of Agriculture by advising the hire of a university scientist working outside government. Williams received an invitation shortly after, agreed to participate, and began drafting a formal research proposal. His research received approval from officials in the Canadian Army in April 1948, and he undertook preparations for Fort Churchill.

Ideas about masculine strength and resistance to pain strongly informed Williams's research. He studied the psychological effects of biting insects on army personnel by observing and interviewing soldiers new to the North. He sought information about individual performance, morale, efficiency, and related personality qualities associated with northern service. He showed a particular interest in attitudinal adjustment, even attempting to define the personality qualities "associated with variations in sensitivity and reaction." He theorized that information about sensitivity in soldiers would lead to an understanding of the "possible effectiveness of systematic orientation or indoctrination in modifying attitudes toward exposure to insect pests."[20] In other words, Williams wanted to screen soldiers for pain threshold and devise training methods to toughen new recruits.

The Northern Insect Survey

Obtaining reliable data about insect populations and dispersal patterns in northern Canada proved challenging for Canadian defence scientists. While Williams studied soldiers, the DRB and the Department of Agriculture launched a widespread study called the Northern Insect Survey. Under the direction of Department of Agriculture scientist T.N. Freeman, survey teams boarded RCAF planes and flew to several bases in northern Canada. Those teams generally consisted of one entomologist, one botanist, and a university student who served as a research assistant. Between June and August 1948, researchers surveyed biting insects near the Alaska Highway from Dawson Creek, British Columbia, to Snag, Yukon. They also conducted surveys at Reindeer Station and Richards Island on the Mackenzie River, Sawmill Bay near Great Bear Lake, Coral Harbour on Southampton Island, Frobisher Bay (Iqaluit) on Baffin Island, Ungava Bay and Fort Chimo (Kuujjuaq) in northern Quebec, and Goose Bay in Labrador.[21] The map of the expanded survey conducted in the summer of 1949 shows locations from coast to coast to coast, including Gander, Newfoundland, Moose Factory on James Bay, Resolute on Cornwallis Island, and Whitehorse, Yukon.

Although the survey was coordinated by Freeman and the agriculture department's science service, the study as a whole took place because of the DRB's focus on a particular area of military research called environmental protection. The name given this area of research is extremely misleading. "Environmental protection" was not about protecting the environment; rather, it encompassed all scientific research activities undertaken to protect *man* from the geographical, topographical, and climatological stresses of his operational environment. The DRB placed special emphasis on human protection against adverse factors in sub-arctic and Arctic environments, performing so-called environmental protection research to understand the stresses caused by "exposure to the forces of nature." The goal was to improve military clothing, shelter, equipment, and repellents and develop methods for overcoming environmental challenges.[22]

Faculty and graduate students from several universities participated under the supervision and control of federal scientists. "Shock troops against the north's greatest summer enemy, eight field parties are moving

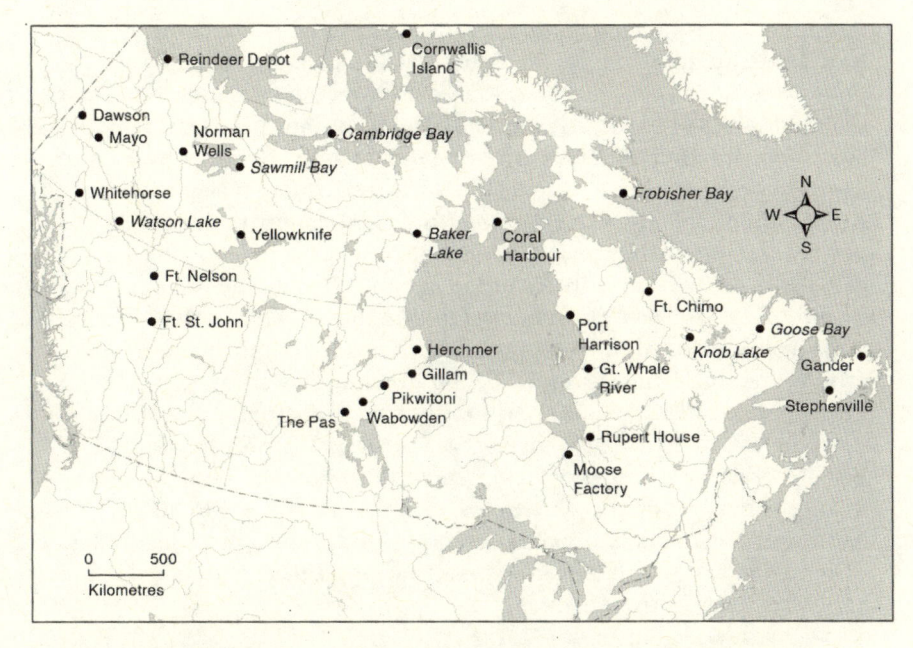

Figure 3.1. Map of Northern Insect Survey bases, 1949. Canada, Department of National Defence, *Annual Report on the Progress of Environmental Protection Research, Report No. DR. 25* (Ottawa: Defence Research Board, 1950), iii. Map reproduced from original.

into Tundra-land as Canada's first line of defence against bugs," read a press release printed in Brandon, Manitoba's local newspaper.[23] Each party of two or three researchers collected insects, observed breeding and biting habits, compiled meteorological data, and tested the effectiveness of various repellents and control methods. During one experiment, researchers sprayed snow near Churchill in an attempt to kill insect larvae before the spring thaw. A primary goal of Operation Insect was to create an entomological map of northern Canada with detailed information about biting insects at the various research locations. The survey was marginally successful from a military perspective. Federal entomologists obtained large quantities of data, but the information recorded derailed the possibility of complete insect control in the North.

The insect-control program was conducted in southern Canada as well. In the spring of 1950, RCAF pilots in a specially equipped Dakota aircraft carried out low-level spraying operations along the flood marshes of the Ottawa River. Using aerosol sprays attached to the aircraft's exhaust system, the pilots initiated an experimental spray program, jointly conceived and facilitated by scientists in the DRB and the agriculture department.[24] Flying at an altitude of 150 feet and carrying a 2,000-pound load of insecticide, the Dakota sprayed a solution of DDT and fuel oil in wide strips across areas with heavy mosquito and larvae populations. One gallon of the mixture covered an acre of land or water, and the insecticide remained active for several weeks upon release. Ground surveys conducted after the spraying indicated fatality rates of about 84 per cent among targeted insects and over 90 per cent for the larvae. C.R. Twinn coordinated the entire program, while entomologist A.W.A. Brown of the University of Western Ontario directed the spraying operations in the field. After declaring war on mosquitoes near the national capital, the RCAF also sprayed DDT concentrations at military bases in Fort St. John and Fort Nelson, British Columbia, Watson Lake and Whitehorse in Yukon, Norman Wells in the Northwest Territories, and Goose Bay in Labrador. Entomological research of this sort, although not confined geographically, was conducted at the request of the RCAF, largely because senior military officials wanted to understand the effectiveness of aerial spraying for controlling insects where it mattered most for the armed services: in the North.

Insect Vectors and Biological Warfare

DRB records illustrate the extent to which military concerns about an unconventional weapons attack against Canada affected how the country's top defence scientists approached entomological research involving northern insects. At a meeting of the DRB's Biological Warfare Research panel in late January 1949, Guilford Reed spoke about current research efforts in the field of biological warfare (BW).[25] Reed had been trained in bacteriology at Harvard University and was one of Canada's foremost microbiologists during and after the Second World War. His research focused on such areas as tuberculosis, gas gangrene, and the various effects of toxoids and vaccines. He was a prominent member of the DRB's BW advisory committee, as historian Donald Avery notes in his work on North American biodefence.[26] Before Reed retired in the mid-1950s, he carried out research for the DRB and advised senior military and defence officials about possible BW threats against Canada.

Reed stressed that scientific work involving biological agents focused almost exclusively on the dissemination of air-borne infections. He referenced research on ground contamination and the use of fly baits but emphasized research on the use of biting flies as vectors. In epidemiology, a disease vector is any living organism that carries and transmits an infectious pathogen between humans or from animals to humans.[27]

Bloodsucking insects are the most common disease vectors. Mosquitoes, ticks, flies, and other carrier insects ingest disease-producing microorganisms during a blood meal and later transmit the disease by injecting a new host.

Reed's brief on vector-borne infections convinced DRB authorities of the severity of the biological warfare threat against Canada and the US. Scientist D. Mitchell expressed his support and argued that information about the distribution of insect pests found in the larger populated areas of likely enemy countries would be useful for assessing the existing vector-borne threat against Canada.[28] Over the following weeks, DRB members involved in biological warfare research, entomological research, Arctic research, and scientific intelligence agreed on a plan to investigate scientists conducting insect vector research in the Soviet Union. The goal was twofold: Canadian scientists wanted a record of all Soviet entomologists engaged in BW research, as well as a list of all geographical locations in the Soviet Union that had climatic and geological characteristics similar to those of the Canadian North. In theory, understanding shared geography was a method for determining insects common to both Canada and the Soviet Union. To carry out this mandate, senior DRB authorities allocated federal funds to expand entomological research studies in northern Canada.

The outbreak of the Korean conflict in 1950 had intensified interest in Canada's biological warfare program, and the DRB's network of research laboratories expanded to meet the evolving needs of Canadian biodefence.[29] A primary military lesson of Korea was that too many enlisted personnel among the multinational forces lacked essential combat skills, physical conditioning, and battlefield morale. As historian Brain Linn observes, tactical or small-scale atomic war represented a new challenge and demanded a higher standard of combat readiness.[30] The US Army's post-Korea Basic Combat Training was thereby adjusted to include long, gruelling marches, increased physical exercise, and detailed instruction in fieldcraft, equipment maintenance, and marksmanship. New recruits were from now on to understand that nuclear bombs and other newly developed weapons had become part of the American fire-support arsenal, and they were to be trained to respond to evolving weaponry.

Canada's biodefence program evolved in this context during the early Cold War. Scientists like Reed supported the armed services by studying the latest developments in advanced military weaponry. Because Reed was a prominent bacteriologist with extensive experience in war-related vector research, the DRB asked him to compile a report about the major scientific and technical developments in the field of biowarfare.[31] Based on open and unclassified sources, Reed concluded that Canadian biodefence measures were insufficient against a potential Soviet bioweapons attack. He recommended that the DRB increase its efforts in the field and train specialists to develop Canada's capacity for detecting and immunizing against biological agents.

Reed received the resource boost he requested in April 1953, becoming the full-time superintendent of the newly established Defence Research Kingston Laboratory (DRKL). Operating in response to the perceived Soviet threat, the DRKL reflected the DRB's diverse and evolving research agenda. Using science to defend against military threats during the early Cold War meant developing a wide and sophisticated research program supported by a diverse collection of highly qualified scientists, engineers, technicians, and administrators, who would conduct and oversee fieldwork and laboratory research across the country. DRKL joined the DRB's growing network of research facilities tasked with improving Canada's scientific and technical resources for security and national defence. Reed's work advanced the country's overall capacity in the field of BW research.

Radioactive Mosquitoes

In June 1949, Dale W. Jenkins of the US War Department's Medical Division applied for radioactive materials to the Isotopes Branch of the NRC's Atomic Energy Project at Chalk River, Ontario, an atomic pile and nuclear research facility on the Ottawa River 180 kilometres northwest of the Canadian capital.[32] Isotopes form when a chemical element becomes radioactive in an atomic pile. While university laboratories fitted with high-voltage equipment had the technical capability to produce isotopes for experimental purposes, government scientists relied on Chalk River as their primary source of radioactive materials.[33] Jenkins submitted the application to obtain isotopes for use in a biting insect survey and research program, to be jointly conducted by the DRB and the Department of Agriculture.

Canadian and US scientists had developed the program to learn about mosquito migration, particularly as it related to the protection of military personnel on summer duty in northern Canada and the Arctic. Jenkins's application stated that he was ordering the isotopes in order to produce "radioactive mosquitoes" for dispersal studies.[34] He submitted the application through the DRB, whose administrative personnel then facilitated the procurement process. Funds for the radioactive materials came out of the research budget of the RCAF, which had allocated funding specifically for air spray trials.

Jenkins conducted his experimental entomological research study east of Hudson Bay in northern Quebec. During a two-week testing period in

July 1949, he reared about one million radioactive mosquitoes in the vicinity of the Great Whale River. He dissolved radioactive phosphorous (P-32) in distilled water, reared mosquitoes in the solution, then used Geiger-Müller counters to track the dispersal pattern of the tagged population. He alone handled the radioactive material, with clothing and equipment brought from the US. Jenkins took rudimentary steps to protect the health and wellness of the local population and environment. After the experiment, he disposed of unused radioactive waste in "a deep trench located at a considerable distance from human habitation, away from any water source." He did not record the exact location of the trench, opting instead to bury the material "marked 'Dangerous' together with the date of disposal, in both the English and Eskimo languages."[35] His records provide no indication of the amount of radioactive waste generated and buried, nor does his correspondence with DRB staff suggest any concern about the residual effects of the disposed radioactive waste.

The method of control for the experimental mosquito population is equally alarming. Curiously, Jenkins's research description said nothing about the dispersal of the one million mosquitoes reared in the radioactive solution. His experiment left him convinced that radioactive tagging was useful for entomological research, regardless. "Highly radioactive larvae, pupae, and adults were produced," he wrote in an article describing his method for tagging mosquitoes with radiophosphorus. "The average radioactivity of the adults was 1.44 milliroentgens/hour or ca. 4300 cts./min. This method for producing radioactively marked mosquitoes in nature can be recommended for dispersal, flight range, migration, predation, and other ecological studies."[36] Jenkins's research convinced the DRB that it should fund his research, and he returned to northern Canada the following summer, where he expanded his experimental work at Fort Churchill.

Jenkins was certainly not the only entomologist to experiment with radioisotopes during the early Cold War. One of the first published accounts of radioactive entomological research appeared in August 1949 when scientists John Bugher and Marjorie Taylor published a preliminary report about the potential use of radioisotopes in insect population studies.[37] Their report suggested that the release and subsequent recapture of labelled or marked mosquitos could serve for an estimate of the total mosquito population of a particular area. They also indicated the potential use of radioisotope marking for the study of age distribution and longevity among groups of mosquitoes reared separately. Also known as tagging, the process of labelling or marking mosquitoes allowed entomologists to track migration and dispersal patterns over biweekly and monthly testing periods.

By the 1950s, with the Cold War in full swing, speculation about biological weapons and chemical warfare was increasing. In 1951, the newspaper of the Soviet Navy, *Krasnyi Flot*, alleged that US scientists had tested BW weapons against Inuit in the Canadian Arctic during an experiment conducted two years earlier that caused an "epidemic of plague" among the persons exposed to the biological agents.[38] This allegation is highly suspect. Military propaganda was common in the heightened political atmosphere of the early Cold War, and although government scientists in both Canada and the US researched the effects of chemical simulants, there is no evidence to suggest that a live agent experiment involving human research subjects was ever conducted in northern Canada or the Canadian Arctic.[39]

Nonetheless, entomological research involving radioactive marking expanded during a period of growing intrigue in the peaceful applications of atomic energy. On 5 March 1953, the president of Atomic Energy of Canada Limited, C.J. Mackenzie, told a special House of Commons committee about exciting new applications for radioactive isotopes.[40] To validate federal funding for the nuclear facility at Chalk River, he discussed the use of cobalt-60 in cancer treatment. He emphasized the importance of peaceful atomic research and later reiterated the civilian benefits of atomic research to the press.[41] Lauding the work of government scientists who had tagged and studied pine weevils and mosquitoes with radioactive isotopes, Mackenzie announced plans for a large-scale tagging of freshwater fish to determine migration and spawning habits at selected locations in Canada.

Although radioactive marking sounds dangerous and irrational to non-specialists, the basic scientific principles of radiation decay help explain the entomological research projects undertaken in northern Canada. The danger level of radioactive material varies according to rate of decay. Each chemical element exists in the form of two or more isotopes, which have the same electrical charge on the atomic nucleus but a different atomic mass. Certain isotopes, such as radium and thorium, are naturally unstable. The phenomenon of radioactivity occurs when the nuclei of unstable isotopes undergo spontaneous disintegration. Writing for the World Health Organization in 1956, malariologist Leonard J. Bruce-Chwatt described radioactivity as "the emission of elementary particles or the release of electromagnetic energy in the form of gamma rays, or both."[42] His article described the latest application of radioactive isotopes in mosquito control research, providing a clear explanation of the contemporary scientific understanding. "The speed of the disintegration of a given nucleus or the rate of decay of a radioisotope," wrote Bruce-Chwatt, "is usually expressed in terms of half-life, namely,

Figure 3.2. Entomologists perform a radioactive experiment with mosquito larvae near Fort Churchill, 1952. Library and Archives Canada / National Film Board fonds / e011175869.

the period at the end of which there remains only half of the radioactivity initially present."

Half-life represents the time required for one half of a radioactive substance introduced into a living organism or ecosystem to undergo elimination by natural processes.[43] The half-life of radioactive phosphorus is 14.3 days, which means that a source containing the P-32 isotope retains half its radioactivity after 14.3 days, one quarter after 28.6 days, and so on. Entomologists thought that compared to other isotopes, P-32 was relatively safe and practical for marking mosquitoes because any radiation decayed relatively quickly.[44] This allowed Jenkins and other entomologists to tag biting insects without concern for spreading dangerous traces of radioactivity to the environment and the local population.

Biological Warfare and Western Security

US concerns about Soviet intentions influenced the entomological research funded by the Canadian defence department. In late August 1949, medical liaison officer A.A. James of the Canadian Joint Staff in Washington wrote to the DRB's Director of Scientific Intelligence A.J.G. Langley to inform him of the US position regarding biological warfare research involving insect vectors. James told Langley that H.I. Cole, the executive director of the US Research and Development Board's committee on biological warfare, believed that studies of vector-borne insects were "really in the field allotted to Canada by common consent of the U.S., Great Britain and Canada."[45] James added that the US General Staff thought scientists in the Soviet Union had conducted microbiological research on the offensive uses of toxins or viruses. "The Committee on B.W. feel that the Russian offensive B.W. potential is fairly high and surmise that their present objective is sabotage though fifth columns," wrote James. Cole went as far as to suggest that Soviet forces could initiate a sinister biological warfare incident, "even to the point of starting an epidemic or killing off by toxin some of their people and using it as a casus belli" (i.e., an act to provoke or justify war).

Convinced that insect vectors posed a serious threat to the North Atlantic partners and Western security, James recommended that the DRB study chemical insect attractants and repellents. His recommendations quickly assumed a serious tone. One week after his first correspondence with Langley, James sent a second letter claiming that the Soviet Union was ill-equipped to manage a vector-borne attack. He had spoken with a US expert on biological and medical intelligence who allegedly agreed that the "Russians are in a poor way to sustain extensive biological attack due to their shortage of decently trained technical and medical personnel."[46]

That same month, DRB scientific intelligence officers produced a brief about the possible use of insects for biological warfare in northern Canada and the US. That brief stated: "Considerable study has, of course, been made in the dissemination of B.W. agents via insects, rodents, etc. in specials localities, i.e. the propagation of lice-borne typhus, flea-borne bubonic plague, tick-borne encephalitis, etc., but it appears that further studies of such possibilities are required, particularly in areas north of Lat. 45°N."[47] Gaining reliable information about insect population and dispersal patterns in the North proved extremely challenging for

Canadian and US scientists; however, entomological research involving radioactive mosquitoes offered a method for estimating the biological warfare threat against Canada, Britain, and the US.

Intelligence records from November 1949 suggest that Canadian defence scientists had sinister plans for using radioactive insects in the event of Arctic warfare. "It should be emphasized that neither Canada nor the USSR lie entirely within the Arctic or sub-Arctic," stated one assessment. "Large areas of both lie within the cool temperate region where, again, close similarities exist. This will certainly be a factor of high importance in dealing, for example, with project (b) listed in Appendix D ... Report on B.W. possibilities using insects as vectors."[48] Theoretically, creating radioactive mosquitoes to learn about dispersal patterns would allow scientists to study the defensive and offensive aspects of biological warfare involving insect vectors. On behalf of the DRB, entomologists sprayed chemical insecticides not only to eliminate mosquitoes from the North but also to identify DDT-resistant insects for possible use as vectors of biological warfare agents.

Insect Repellency Research

When DDT became available during the latter stages of the Second World War, the armed services and particularly the Royal Canadian Navy used the chemical compound as a dusting powder to delouse personnel.[49] Canada obtained DDT from the US during the war; after the war, the DRB opened a pilot plant in Ottawa to meet the increasing postwar demand. Defence Research Chemical Laboratory (DRCL), another DRB facility, began producing DDT in early 1948. Scientists there studied the defensive aspects of chemical and biological warfare; this included developing protective equipment and clothing to protect military personnel in northern Canada and the Arctic. The laboratory produced just enough DDT to meet the military's requirement for dusting powder and liquid spray for insect control. DRCL supplied the armed services until the commercial production of DDT met the demand, at which point the pilot plant switched to producing gammexane (benzene hexachloride), a newer insecticide that showed promise as a chemical for combating DDT-resistant insects.

In the late 1940s and throughout the 1950s, Canadian scientists used DDT extensively on fur farms in southern Ontario. In the 1947 annual report of the Ontario Veterinary College, A.A. Kingscote, head of the Department of Parasitology at the University of Guelph, argued that DDT was safe when applied under normal precautions.[50] Dust or spray DDT 1068 (lethanes and pyrethrum) proved effective for reducing flies and fleas on farms, Kingscote's research indicated. Combined with a single spray of commercial DDT kerosene, one application of pyrethrum resulted in the complete eradication of parasites from a barn of animals. This led to further DDT studies and the creation of new insecticides, which researchers applied and tested rigorously on open farmland, in barnyards, and in forested areas throughout southern and northern Ontario.

Senior defence officials touted the civilian applications of military research to explain and justify the DRB's work in northern Canada and the Arctic. Besides conducting entomological research and insect control efforts, scientists and engineers studied fuels and lubricants for military vehicles and equipment. Researchers at Fort Churchill tested snowmobiles, tanks, and aircraft in cold conditions, refined oil dilution techniques, and developed new synthetic materials for use at very low temperatures. Similar research took place during the summer months: mechanical engineers tested and refined heavy military equipment and vehicles for use in bushed areas and over boggy muskeg. The idea was to improve military mobility in the North, which had the potential to generate year-round performance spin-offs for civilian industry. "What we learn today, because we have to if we are to be properly defended, we can apply tomorrow to peaceful pursuits," wrote columnist Cyril Bassett, commenting on the civilian applications of the DRB's northern military research.[51] In a similar search for means to improve the efficiency and effectiveness of the armed services, defence scientists also sought information about the physical characteristics of ice and snow, the uses of refrigeration and freezing, and the suitability of northern vegetation and animal life as sources of food and sustainable nutrition.

As the problems associated with living and fighting in northern Canada and the Arctic went hand-in-hand for senior military and defence officials, scientists and engineers across the country adapted and designed experimental research projects to address the DRB's advertised needs. Kingscote shifted gears, expanding his entomological research program, originally conceived strictly for agricultural purposes, in order to obtain military research funding. In April 1948, he received the first instalment of one of two DRB grants, which he held until 1957.[52] The first public mention of either of the two grants appeared in the 1948 annual report of the Ontario Veterinary College, in which Kingscote alluded to his department at the University of Guelph as having had "received an extramural research grant of $5000.00 for special projects which are being investigated."[53] Four years later, without providing specifics about the grant, the annual report for 1952–3 named the DRB explicitly. The following year, Kingscote openly reported that the DRB had "continued to support investigations related to insect repellents and attractants. Data elicited so far have been assembled in technical reports published [internally] by the Board."[54] He did not elaborate, leaving the details looming as if to suggest that his military-sponsored research was secretive.

Opened government records now indicate that Kingscote was attempting to develop a digestible pill for insect repellency. At the height of his research in the early 1950s, the Canadian military conducted regular training exercises in northern Canada, often simulating a large-scale invasion of Soviet forces in both winter and summer conditions. Because mosquitoes, blackflies, and other biting insects interfered with military training and affected soldiers' morale during warm-weather exercises, DRB authorities contracted Kingscote to investigate orally administered repellents, hoping to eventually equip and prepare soldiers for intense insect conditions.[55] Few details about Kingscote's research have survived, but the available records indicate that he received funding under DRB Grant 75 for three years. The DRB closed the grant in March 1952, presumably owing to a lack of results. Kingscote's military research did not end, however. He continued to study insect-control methods for the armed services throughout the 1950s.

While Kingscote studied insect repellency in southern Ontario, biting insects continued to torment Canadian soldiers training in the North. In July and August 1950, a small group of armed services personnel, together with representatives of the DRB and the British army, carried out Exercise Shoo Fly to investigate the feasibility of small-scale military operations under typical summer conditions in subarctic Canada. The tactical setting envisaged the dispatch of a specially trained reconnaissance section of the Mobile Striking Force to investigate reports of air landings by small enemy groups in the Churchill and Duck Lake areas. On the exercise, scientists observed the effects of terrain, weather, and biting insects on soldiers and carried out trials of specially designed mosquito clothing and repellents. "There is no doubt that the average individual has a mental picture of the Arctic greatly exaggerating its difficulties and hazards," stated the final exercise report. Summer indoctrination courses at Fort Churchill, the report continued, serve to "build up a nucleus of officers and NCOs in all units able to disseminate balanced and factual information on the Arctic and its problems and thus counteract the exaggerated views so widely held."[56] During the exercise, mosquitoes, blackflies, and deer flies proved numerous but caused relatively few issues for personnel, largely because of the consistently dry and windy conditions encountered. The situation was vastly different during subsequent exercises, during which wet and boggy conditions and milder winds produced hordes of biting insects.

To be sure, military personnel who trained in northern Manitoba learned first-hand about the impact of terrain and weather conditions on ground operations. In mid-1952, the armed services conducted Exercise Deer Fly, a continuation of the summertime training program at Fort Churchill. Three consecutive exercises took place in the Churchill area between June and September with the aim of training small MSF elements to deter enemy lodgements over diversified terrain in northern Canada.[57] Soldiers on the exercise experienced difficulties performing regular duties, as the constant buzzing of mosquitoes and other insects restricted their operational capabilities, affected their alertness, and reduced their overall morale. Spray-on repellents provided reasonable protection, but for a limited time only, and mosquito nets impaired their vision.[58] Also, repellents washed off quickly and easily when workers perspired and thus offered scientists and military personnel limited protection in the fly-infested bush areas near the base.

Interestingly, entomologists at Fort Churchill determined that certain materials and clothing colours attracted biting insects.[59] Experimental research indicated that flies and mosquitoes had a natural attraction to black and dark-coloured clothing; white and light-coloured clothing attracted considerably fewer insects. The attraction to lighter shades of green and yellow was low, whereas a moving object dressed in a somber hue apparently generated lively attention. Weather influenced the number of biting insects during the exercise, of course. Comfortable temperatures combined with low winds produced typical tundra weather and "hordes of insects."[60] Conversely, considerable rain and high winds meant few insects during subsequent exercises Deer Fly II and Deer Fly III.

The short summer season in northern Canada affected the character of the insect research program. DRB scientists proposed a biological study of northern mosquitoes to develop a technique for laboratory rearing, which in theory would enable year-round research on insect control.[61] Entomologist William Beckel conducted seven years of "warfare against the northern mosquito" for the DRB's northern research laboratory before earning his doctoral degree in insect physiology from Cornell University.[62] After completing his work at Fort Churchill, Beckel transferred to the Department of Agriculture, where he continued his entomological research in Chatham, Ontario. His proposed study involving the laboratory rearing of northern mosquitoes was never conducted, but Beckel and Kingscote provided research assistance throughout much of the 1950s.

Entomological research on insect control and the effectiveness of repellents and mitigation techniques was carried out during the winter months as well. Scientists under Kingscote's direction conducted several experiments with various repellents, essential oils, vitamins, hormones, and miscellaneous substances alleged to give protection against biting insects. Field trials of twenty-four different repellents were conducted at Churchill and in Burwash, Ontario, in the late 1940s and early 1950s, and scientists tested more than 100 other substances in experiments involving mice and human research subjects. "All hormones and several other agents have been tested externally on staff members who volunteered for treatment and exposure," stated a DRB environmental research report in April 1950.[63] The full details of the experimental research program were unrecorded, but the researchers tested various substances on the body surfaces of mice and introduced repellents directly into the stomachs of some experimental animals. For "humane reasons," the scientists chloroformed several mice that showed signs of pain and suffering after receiving high doses of certain experimental compounds.

As for the human research subjects, records indicate that DRNL staff and volunteer soldiers participated in experiments designed to test various suppositions about insect tolerance and immunity to bites on the human body. More than 2,200 army personnel participated in training and user trials at Churchill between 1952 and 1954; another 200 or more took part in specialized engineering tests.[64] In one experiment, two research subjects "consumed from sixteen to twenty cups of strong coffee daily for a period of one month."[65] The experiment ultimately gave negative results, and the researchers dismissed high caffeine intake as a plausible means to immunize against insect bites. Another series of studies in 1952–3 explored the possibility that Indigenous soldiers might be better suited to northern military operations than men of European ethnicity. "This race-based testing suggests that the Army was at pains to find men who were naturally suited to military service in such an unforgiving climate," writes historian Andrew Iarocci.[66]

Toxicity and Environmental Degradation

The DDT spray program in northern Canada parallels the history of Gagetown, the Canadian military base in southwestern New Brunswick that served as a controversial experimental testing ground for all-weather research involving chemical defoliants. Canadian military officials selected 420 squares miles of land northwest of Saint John for the Gagetown base site in 1952. The local population and the dense forests of pines, redwoods, spruces, and leafy hardwoods impeded the armed services from initiating operations in the designated area. Economic necessity and a sense of patriotic duty compelled residents to allow the construction of the base, and they watched as military personnel swept in and felled trees on carefully managed woodlots that had yielded a steady harvest for centuries. "Like the people, the woods were in the way," writes historian Joy Parr. "In time, the woods did become useful, as testing grounds for chemical defoliants."[67]

The clearing of trees and brush started in 1953. Clearing and grubbing was done mechanically until 1955, when the military began using herbicides. Over the next twenty-eight years, DND officials hired private contractors to clear the training ground of any trees and brush taller than three feet. Contractors released nine different herbicide mixtures and more than 6,500 barrels of chemicals over approximately 181,000 acres, using aerial spraying from helicopters and planes as well as ground spraying from vehicles and individuals fitted with personal tanks or backpacks. Various methods of chemical clearing continued through 1984, including testing of the controversial herbicide and defoliant Agent Orange.

Military officials saw Gagetown as a viable and versatile geographic landscape suited to the army's purpose. "It's got mountains like Korea, lowlands like the north German plains, mud like Italy," writes Parr, citing the words of the camp's first commanding officer, Major-General J.M. Rockingham: "You could train men here to fight anywhere."[68] Much like the Churchill area in terms of terrain and topography, Gagetown represented a natural laboratory ideal for military training. Workers on site began spraying trees with defoliant in 1956. Compared to mechanical methods, herbicides were a reliable and cost-effective method for removing trees and freeing land for military purposes. Each year, the defoliants killed some trees and caused desiccation in others. Trees regularly grew back, which spurred senior officials in the defence department to increase spraying efforts and authorize the release of more chemicals.

The Gagetown spray program was not exclusively Canadian. When US President John Kennedy approved herbicides for use in Vietnam in 1962, the American military wanted to determine which chemical compounds were most effective as defoliants.[69] On the invitation of military officials in Ottawa, US personnel arrived at the Gagetown training ground, where they commenced an experimental program in June 1966. During testing that month and the following June, military personnel sprayed Agent Orange and other unregistered herbicides to evaluate their effectiveness.[70] The Canadian government has since acknowledged that the tests occurred and extended a one-time, tax-free *ex gratia* payment of $20,000 to affected veterans in 2010.[71]

The damaging ecological effects of herbicides were unclear at that time, and many lauded the chemicals sprayed at Gagetown. Chemical herbicides symbolized modernity in the early post-war period, as David Kinkela points out.[72] The chemical industry epitomized progress, thanks in large measure to the success of DDT and other pesticides. During the Second World War, DDT provided temporary relief by nearly eradicating malaria, typhus, and other rampant wartime diseases. There had been efforts in the past to investigate the ecologies of disease prior to and during the war, but confidence in DDT and other chemical pesticides had reduced the importance and relevance of environmental impact studies.[73] However, the effects of herbicides on humans and the environment would generate widespread public concern in the 1970s, after antiwar protesters condemned US actions in Vietnam and promoted public awareness about the spread and use of hazardous chemicals.

Public Response and the Decline of BW Research

Entomological research involving radioactive materials provoked a modest but noteworthy public reaction in Canada. One of the first news reports about the insect-control research conducted at Fort Churchill appeared in August 1950, some three years into the program. The columnist was sceptical, to say the least: "Of course, the scientists plan to catch their tagged insects after a spell of liberty to find what they have been up to. If we know our mosquitoes, that is a naïve hope. When it comes to enlisting these misbegotten creatures as atomic weapons, every red-blooded citizen should be concerned. We view these extra-curricular activities with doubt, not to say alarm."[74]

Similar criticism appeared four years later. Following reports out of Saskatoon that radioactive mosquitoes had travelled far from their hatching place in the Saskatchewan River, the *Globe and Mail* printed a scathing editorial in June 1954 criticizing the research of scientist John Spinks. "To loyal prairie dwellers this may seem a bit of scientific exactitude not entirely called for," the editorial declared. "We may be pessimistic, but we have a disloyal feeling that science for all its accuracy may be going at this thing in reverse. What this country needs is not people who can spot mosquitoes eight miles away, but mosquitoes which can't do the same to people."[75]

Biological warfare research declined in Canada after 1957, owing predominately to the changing threat against North America. As the danger of a nuclear attack evolved with the introduction of intercontinental ballistic missiles and the launch of Sputnik in October, military officials in Washington based their security assessment on the premise that a Soviet nuclear attack against North America would be suicidal if it did not destroy the retaliatory capability of the US military.[76] Recognizing that even a severe biological warfare attack would not destroy the retaliatory capability of their country, US military planners shifted resources away from BW research, and Ottawa followed suit. Pentagon intelligence also suggested that Moscow wanted to avoid provoking Washington for fear of sparking a nuclear confrontation, and this further reduced the need for an active BW research program in the US. As the possibility of a BW attack against North America decreased, so too did Canadian research into insect vectors.

This did not prevent the Canadian armed services from continuing research into the biology and control of biting insects in northern Canada. In June 1958 the DRB's lead entomologist, Ian Lindsay, announced that government scientists under his direction were in the second year of a five-year research program initiated to find the answer to the "itchy question" of what makes the human body attract mosquitoes. Biting flies were far more than a simple nuisance to the armed services, Lindsay explained in a press interview.[77] He cited aircraft maintenance performed outdoors as an example, claiming that routine mechanical service on a military plane took twice as long as normal when swarms of mosquitoes, blackflies, and other biting insects had to be dealt with. Existing repellents provided four hours of protection, which was far below the effectiveness sought by entomologists in the DRB.

As the decade turned, defence scientists also expressed increasing concern about the threat that disease-carrying insects would arrive to Canada via commercial and military aircraft. In an April 1961 interview with Canadian Press, Lindsay, now head of the DRB's Environmental Protection Section, revealed details about the development of an "aerosol bomb" for passenger aircraft.[78] Canadian scientists had initiated a research and development program for aircraft insecticides after learning that disease-bearing insects had been brought to Canada by ship. Concerned about the risks to Canadian agriculture, government entomologists studied such travelling insects as the potato bug and the cockroach. Meanwhile, scientists in the DRB collaborated with the RCAF to develop an insecticide spray for military aircraft. Reports indicated that insect eggs could survive the cold and altitude of intercontinental flight, and this led to concerns that diseased insects could hatch upon arrival to Canada. The threat of genetic manipulation was enough to convince DRB authorities that genetically modified mosquitoes could survive exposure to severe elements, supplanting preconceived knowledge that associated insect vectors with tropical or temperate climates.

Cartographic research concentrated on the North also brought to the fore distinct topographical similarities linking Canada's northern terrain and landscape with the northern geography of the Soviet Union. The same month that Lindsay announced the aerosol bomb, Norman W. Radforth of McMaster University produced what the DRB referred to as Canada's first muskeg map. As a biologist interested in the composition and distribution of organic terrain, Radforth spent twelve years compiling and scanning thousands of aerial photographs of northern Canada to chart muskeg for the defence department. His map showed high-, medium-, and low-frequency occurrence of muskeg, defined as a layer of organic material or peat between twenty-five and thirty inches deep. Radforth found a high incidence of muskeg in central Newfoundland and in a vast strip of land extending northwest across northern Ontario, Manitoba, and Saskatchewan and into the Northwest Territories. "Because muskeg regions of Canada are analogous to similar regions in other parts of the world, such as the Soviet Union, the British and U.S. military have shown interest in Dr. Radforth's work," stated a press release announcing the map's completion.[79] Military officials in Ottawa and Washington embraced Radforth's findings and approached the North as a vital component of continental defence, but also as a hostile environment to overcome.

Insect-control efforts in northern Canada promised civilian benefits as well. Chemical insecticide research conducted by scientists working on behalf of the armed services contributed to the production of entomological knowledge, which had both military and civilian applications. The use of DDT as an insecticide expanded in many northern communities until anti-war protests in the late 1960s brought attention, indeed ridicule, to the development and use of chemical warfare agents. On 11 July 1968, the Minister of Indian Affairs and Northern Development, Jean Chrétien, announced that DDT use would be discontinued in Canada's national parks.[80] Chrétien stated that although DDT was regarded as an effective insecticide, it also produced undesirable side effects for Canada's ecosystems. Scientific reports about the hazardous ecological impact of DDT residues on animal populations – particularly fish – led National Parks Canada to replace DDT with other, less persistent insecticides. Recent findings had shown DDT residue build-up in animal tissue samples, which became more concentrated along food chains. Authorities with National Parks responded by implementing non-persistent insecticides to reduce mosquito and blackfly populations near campgrounds and visitor service centres. Park officials saw these measures as temporary, hoping to find insect control methods that involved no pesticides at all.

Uneasy about public complaints, the defence department restricted publicity about insecticide spraying prior to the DDT ban. "Units being provided with airspray in support of the Fly Control Programme will not make a general press release on the subject," determined a committee of officials who met at Air Defence Command Headquarters in Saint-Hubert, Quebec, on 23 April 1964. "Past experience has proved that publicity on the operation has brought complaints from populace in adjacent areas, suggesting that the programme is harmful to plants and wild life."[81] The committee advised all participating service units to provide information about the spray program only in response to direct queries; moreover, it would not inform communities near spray sites in advance of any experimental trials. This policy continued for several years, despite letters of concern from citizens and non-government scientists. In April 1965, for instance, biologist Carl Schenk asked DND for information about the use of DDT near Camp Borden. Concerned about the health of spawning populations of rainbow trout in nearby streams, Schenk asked for a detailed map indicting exactly where spraying had been done. [82] Technical Officer A.E. Winmill responded to Schenk, providing the requested map, perhaps owing to Schenk's official position with the Ontario Water Resources Commission.[83] Schenk used the map to conduct analyses of DDT residues in fish and aquatic organisms in streams near Camp Borden, in this way taking action in lieu of a formal federal investigation.

Other correspondence about the air spray program suggests that federal authorities responded differently to health and safety concerns associated with the widespread use of DDT and other insecticides. In August 1966, for instance, medical health officer Raymond Miller of Moose Jaw wrote a letter to the squadron leader of the local Canadian military base. Miller criticized the use of DDT and other chlorinated hydrocarbon compounds for mosquito control, calling the practice of aerial spraying "not desirable in areas occupied by domestic animals, wildlife, or in urban areas; characteristic of most members of this group is their residual quality and the tendency to accumulate in fatty tissues of mammals and birds."[84] Miller recommended various other organic phosphorus compounds considered safer and less toxic for mammals, birds, and any affected populations. After considering Miller's letter, military officials in the federal government decided to continue using DDT for insect control. Correspondence about the decision suggests that the military favoured DDT, because of its resistant properties, over the compounds recommended by Miller.

The eventual DDT ban was slow to reach Churchill. According to local news reports, Cape Merry and Fort Prince of Wales, two federal parks within Churchill's jurisdiction, had their ponds and sloughs seeded with DDT pellets in the weeks following Chrétien's announcement in July 1968.[85] Street fogging, an annual summer practice dating back to the mid-1950s, continued to involve spraying DDT and fuel oil mixtures throughout the community. In May 1969, DRB scientists estimated the mosquito population in the Churchill area at around 5 million per acre during peak season. "If a man bared his fore-arm, from wrist to elbow," noted one southern Ontario newspaper, "he could get up to 280 mosquito bites within a minute."[86] Such anecdotes about the northern mosquito menace reinforced southerners' ideas of a distant land ripe for scientific research. Indeed, insect control in the Churchill area was a concerted response to the natural ecological conditions of northern Canada. Warfare against biting insects in the North began, but did not finish, with the Canadian military.

Conclusion

One legacy of Cold War–era military activity is environmental degradation. The entomological research conducted on behalf of the Canadian armed services is an important reminder of the hidden ecological damage wrought throughout select areas of southern and northern Canada in the name of security and national defence. Scientists and government officials approved the widespread use of synthetic chemicals, military land became the subject of experimentation, and ecological ruination resulted. Unlike in Gagetown, however, we know very little about the immediate or long-term impact of the DDT spray program on the military personnel, civilians, and natural environments exposed to the chemicals. Without specific information about the substance, quantity, and exact geographical distribution of the chemical insecticides sprayed during military testing, a complete assessment of the social or environmental damage cannot be performed. Unfortunately, the information needed to complete a full and accurate damage assessment went unrecorded. The scientists involved in the research followed the standards of the early post-war period – standards set in Canada by senior government officials who authorized and administered state-sponsored research activities conducted in northern Canada and the Arctic.

The federal DDT spray program is a distinct and enduring symbol of government administration in Canada during the early Cold War. Public reaction to insect-control methods, although modest in the political atmosphere of secrecy, questioned and ridiculed government scientists for experimenting with radioactive isotopes. That public distain towards experimental insect-control methods did not influence the military's efforts to investigate and control biting insects in the North suggests how much power and authority government-sponsored scientists enjoyed. Federal representatives did not consult northern residents before releasing chemical insecticides widely into both air and water, and the federal departments involved in the DDT spray program did not investigate the full extent of the experimental work for either social- or environmental-assessment purposes. Collectively, the entomological research experiments investigated in this chapter left an indelible mark on peoples, lands, and ecologies of select locations in northern Canada. The full extent of any social and environmental consequences remain largely unknown.

While the health and environmental impacts of Agent Orange, DDT, and other chemicals should not be understated, current studies suggest that toxicity and contamination are perhaps best understood as social issues.[87] As John Sandlos and Arn Keeling explain, the hazardous effects of toxic chemicals extend well beyond any discernable physical impact on the individual body. Toxicity and contamination also alienate people from land and cultural ways of living, "a form of dispossession and loss of health tied to the inability to safely and confidently use local land and water resources," write Sandlos and Keeling.[88] Their research traces the detrimental impact of land degradation and toxicity endured by the Dene of Giant Mine in Yellowknife, where arsenic contamination continues to threaten water and other local resources. For the Yellowknives Dene, healing the land and providing compensation for lost resources is essential if there is ever to be reconciliation for the environmental injustices to their traditional territories. In this context, the appropriation and abuse of land for the purposes of military training and experimental testing represents a colonial legacy of the Cold War that requires social and environmental remediation. The insect-control efforts and air-spraying program carried out under the auspices of the federal government for the Canadian armed services are certainly no exception.

Ultimately, the entomological research activities undertaken in the North during the early Cold War demonstrate the positionality of senior military leaders and government scientists vis-à-vis environmental research. The DRB's Environmental Protection Program aimed, first and foremost, to devise scientific means to protect soldiers – the human body – from nature. The figurative and literal exploitation of Canada's northern geography was a distinct feature of post-war defence planning in Ottawa and Washington, and senior decision-makers in the Canadian government allowed chemical experimentation on northern lands when developing military science projects for the North. The DDT spray program was not unique to northern Canada or the Arctic; however, the heightened military importance of Canada's northern regions and the difficulties of training service personnel in northern latitudes trumped any concerns about the natural environment. The research was driven by a desire for operational mobility; that desire is what sustained a long and coordinated scientific effort to understand and overcome the natural elements in the North during a rapidly evolving period of the Cold War.

4 The Changing Science of Arctic Warfare

Canada's approach to Arctic-related military research underwent significant changes in the 1950s as federal officials responded to the intensifying Cold War. As a signatory to both the UN Charter and the North Atlantic Treaty, Canada had committed itself to supporting multinational efforts to safeguard Western security and the international order.[1] When the Canadian military rearmed for Korea, the defence department received an influx of funds and a new agenda. "The Korean campaign [has] added active aggression to the so-called cold war," Omond Solandt told the Manitoba Chamber of Mines during an address in Winnipeg in late October 1950. "We have now entered upon what some experts have called the 25% war," he added, "[and] we must win it if we are to prevent the 100% war that we all dread."[2] For experts and decision-makers in the DRB, whose primary mandate was to provide scientific and technical assistance to the Canadian armed services, this meant a heightened focus on research that could produce quick results for the military.[3]

Rearmament for the Korean conflict had immediate and lasting consequences for Canada's peacetime military research program. Communist aggression threatened the international order, and Canadian officials responded by adjusting Ottawa's plans for security and national defence. Long-term research projects undertaken to support the armed services were either put on hold or scrapped in favour of short-term projects that consumed fewer resources and promised immediate returns for the military. This increased the pressure on defence scientists to work and deliver results with tighter time frames.[4] During the earliest years of the Cold War, the DRB supported a wide range of short- and long-term research projects that showed potential for generating current or future returns for the armed services.[5] Korea changed this approach to military research. As the armed services rearmed and entered the conflict in support of the UN coalition forces, the DRB expanded its scientific research capacity in Ottawa, including its staff numbers, prioritizing operational research. Just as in the Second World War, interdisciplinary research would be for studying and improving the efficiency and effectiveness of fighting forces.

A difficult challenge facing UN forces in Korea was frostbite resulting from cold exposure; indeed, the failure of the multinational forces to prepare adequately for military operations in cold regions was a defining feature of the Korean battlefield.[6] Canada and the US had conducted cold-weather training and simulated operations before the conflict broke out; now, soldiers' experiences in Korea drew even more attention to the many problems associated with military operations in cold regions. Consequently, the US Air Force increased its capacity for doing cold-weather science at the Arctic Aeromedical Laboratory after 1953. Located on the premises of Ladd Air Force Base in Fairbanks, Alaska, the laboratory became an important and controversial hub of military-sponsored research. Research specialists from disciplines such as biology, civil and materials engineering, geophysics, physiology, and psychology travelled to Alaska and studied cold-weather problems for the US forces.

In Canada, military training in cold conditions continued at Fort Churchill, and scientists at the DRB's Defence Research Northern Laboratory increasingly participated as field observers, studying first-hand the effects of cold on the physical and mental capacities of soldiers. As evidenced by the NFB's 1954 production *Vigil in the North*, which includes footage of laboratory experiments performed within DRNL and during winter-warfare training exercises conducted outdoors near Fort Churchill, scientific research and technological development was geared towards helping soldiers overcome the twin enemies of the North: "fear and fatigue." The film describes these twin enemies as man's initial response to the Arctic environment.[7] In turn, DRNL scientists studied the human body under military duress, seeking ways to inure soldiers to the rigours of the Canadian North; it could then disseminate that knowledge for cold-weather military operations the world over.

Defence and security in northern Canada remained a high priority. The evolving geostrategic threat of the 1950s affected the role and structure of the DRB's research laboratories, leading to the eventual demise of DRNL at Fort Churchill. Yet soldiers and scientists continued to travel north in large numbers throughout the decade, training and conducting research in northern Manitoba and the subarctic region around Hudson Bay. As the Cold War turned hot in Korea, the push to acclimatize soldiers from southern Canada to the rigours of the North did not abate. The character and extent of military research in Canada underwent significant changes during this period of the Cold War, but the pursuit of operational efficiency and effectiveness in high-latitude cold remained a feature of Canadian military preparations.

Winter Warfare and Operational Research

By the mid-1950s, DRNL was one of more than ten permanent research facilities operated across Canada under the umbrella of the DRB. Senior defence officials chose to open a scientific research facility at Fort Churchill for several reasons. The geography of northern Manitoba provided year-round access to a subarctic climate with challenging terrain and topographical features. The Hudson Bay rail line carried supplies and personnel from Winnipeg to Churchill, and the Port of Churchill provided deep-sea access to the waters of Hudson Bay and the Arctic Ocean during summer months. The RCAF operated the wartime airfield at Fort Churchill, serving communities in northern Canada and providing scientists transportation to locations accessible solely by air. Additionally, the garrison allowed for the close integration of DRNL with the armed services and military test teams that operated in the area. Military training and testing was the Canadian Army's principal activity at Fort Churchill. It assumed a "housekeeping" role for the base because it was the largest and most logistically organized of the armed services.

Shortly after the Second World War, Fort Churchill became the launching point for a large-scale military exercise called Musk Ox. Designed as a three-month trek across northern Canada, the proposed route would see a military convoy (Moving Force) travel north from Manitoba in a horseshoe pattern, reaching Cambridge Bay in present-day Nunavut, crossing the Mackenzie River in the Northwest Territories, and eventually travelling south through northern British Columbia and Alberta towards Edmonton.[8] The exercise marked the culmination of the winter warfare training conducted by the Canadian Army during the war. However, in contrast to the wartime exercises Eskimo, Polar Bear, and Lemming, which received little publicity, Musk Ox received extensive coverage in national and international media.[9] As a test of mechanized movement and air resupply over a proposed 3,200 miles across northern Canada, the cold-weather exercise caught the attention of the Canadian public and international military observers. Politicians in Ottawa lauded the civilian benefits of the exercise, including the research contributions of participating soldiers, who collected snow and ice data and recorded the flora and fauna they encountered along the way.[10]

Exercise Musk Ox rekindled Canada's military interest in northern operations and reaffirmed in the minds of senior officials the challenges and opportunities of Arctic research. In effect, the exercise showed that the Canadian armed services – principally the army and the RCAF – had the technology and personnel to operate and resupply a small ground force in the Canadian Arctic.[11] Maintaining a force capable of carrying out effective and efficient military operations in the winter, however, was a challenge that demanded brains as much as brawn. As Solandt reflected in 1973: "This experience supported the view that an active research programme in the Arctic, which involved all the Canadian Armed Services, would both provide the Services with the very demanding situations that are so hard to provide in peace time training and would at the same time help to provide an effective Canadian presence in the Arctic and thus avoid any questioning of our sovereignty there."[12] Exercise Musk Ox, in other words, showed that scientific research and technological development were vital to cold-weather military preparations and postwar modernization, thus underscoring a key motivating factor behind the creation and early growth of the DRB's Arctic research program at Fort Churchill.

Because of the army's active role in northern Manitoba, DRNL's scientific staff focused on tackling the operational impediments encountered by soldiers on the ground. The army held annual winter exercises in the eastern and western Arctic that, in addition to having tactical value, served as opportunities to evaluate men and equipment. Scientists at Fort Churchill undertook training to live and move with soldiers in the field; this allowed direct participation in military exercises, real-time research assistance, and observational studies of operational performance. Researchers lived and ate in the same tents as soldiers, carried out scientific tests and measurements, and assisted with the movement of men and supplies. Mechanized equipment was rare during army exercises, and all personnel pulled their share of the weight, including accompanying scientists and engineers.

Beginning in the early 1950s, DRNL's Operations Research Section (ORS) carried out regular evaluations as part of the army's northern exercises, in both winter and summer conditions. Operational research originated during the Second World War, when the British government and its military sought every available means to defend the United Kingdom against German bombs in the Battle of Britain.[13] Radar was a relatively new technology, and the British government called upon a group of leading civilian scientists to determine, mathematically, how best to distribute antennae, organize signals, and deploy interception systems to their maximum advantage. The creation of Britain's highly effective air defence system marked the start of a permanent union between scientist and soldier that saw operational researchers evaluate myriad military issues quantitatively, thus generating knowledge to assist with prediction, decision-making, deployment, and a host of related factors concerning all branches of the military and areas of operation.

By war's end, operational research had emerged as a new scientific method designed to provide "commanders and executive departments with a quantitative basis for making decisions regarding the operations under their control."[14] The Canadian Army Operational Research Establishment (CAORE), a division of the DRB, was established in 1948, a child of the wartime success of the discipline and rapid post-war advances in military technology and weapons systems. Operational research thus played a central role in maximizing the efficiency and effectiveness of Canada's fighting forces during the early Cold War.[15]

Given the many practical problems at hand, military personnel lacked the time, resources, and expertise to conduct organized research during Arctic training exercises. In this respect, scientists aided the armed services by observing military personnel and recording the valuable information obtained in the field. DRNL test teams regularly employed questionnaires that asked soldiers about military methods and morale, in this way obtaining useful feedback or suggestions for improvements to tents, portable heaters and stoves, food and provisions, weapons and ammunition, and personal kit, as well as transport vehicles and other military equipment. Basic field research enabled scientists and engineers to uncover problems overlooked in the laboratory and thereby determine which practical military issues required further investigation.

During winter exercises in 1951 and 1952, several soldiers reported difficulties performing routine maintenance on portable cooking stoves issued for northern service.[16] Operational research scientists at Fort Churchill conducted a preliminary survey and determined that the stoves were not faulty, contrary to the opinion of the soldiers, who had simply received the wrong tools for the job. Upon further investigation, workers at DRNL determined that the soldiers lacked proper equipment for the efficient and effective repair of other items as well. The existing small-scale repair kits were too heavy and were also unsuited to military duties in the North. DRNL's technical staff responded by designing and developing an Arctic repair kit that included all the items the military deemed necessary for routine field maintenance. The kit weighed less than two pounds and contained only nine items, including stove replacement parts, repair tools, and parts for fixing broken tents and snowshoes.

Operational researchers at DRNL also conducted elaborate field trials of clothing and equipment, using soldiers as test subjects when developing Arctic fatigues and supplies. In cold rooms, laboratory scientists conducted physiological stress experiments on civilian and military research subjects, including on manual dexterity; these experiments at times had unintended consequences and led to injuries (see previous chapters).[17] Outdoors, DRNL's operational research staff oversaw and evaluated the performance of clothing and equipment, the effectiveness of indoctrination and training methods, and the deficiencies of standard navigational instruments and techniques.[18]

Infantrymen often trekked on snowshoes as far as twelve miles in temperatures between –30°F and –45°F, enduring thirty-five-mile-an-hour winds while carrying fifty-six pounds of weapons and supplies. Soldiers also hauled 250 pounds of equipment across the tundra on toboggans, learning how to move, work, cook, sleep, fight, and survive outdoors in the winter. "Conditions such as these proved the leaders and those they led," wrote Colonel Strome Galloway, associating military leadership in northern Canada with performances of hypermasculinity.[19] Participating scientists moved and lived with soldiers during indoctrination exercises and training, studying such problems as morale, physical and mental fatigue, night watching and sentry duty, and manual dexterity and physical performance. Operational research of this sort also extended to observations of leadership and discipline among senior officers.

Military research at DRNL and Fort Churchill attracted a range of scientists, engineers, and technicians to northern Manitoba. Biologist Cecil Law, a wartime infantry commander, drew in fellow biology students Ward Stevens, a former RCAF navigator, and Winston Mair, who himself had served as an infantry commander during the war. Psychologist Jim Easterbrook, another former navigator, joined DRNL's operational research staff to work on ground navigation and training, as did wartime infantry officer Mace Coffey and scientist Donald Ross. Easterbrook and Ross completed a parachute training course at the Canadian Joint Air Training Centre in Rivers, Manitoba, and jumped several times with airborne battalions during field exercises conducted near Fort Churchill.[20] Wartime disability pensions restricted Law and Coffey to assault helicopter and glider training as well as airborne resupply. Scientists who performed operational research at Fort Churchill kept an eye on the overall capabilities of the armed services in northern Canada, forwarding suggestions for new military equipment and research studies to the appropriate personnel, government scientists, and defence officials.

Of particular concern for DRNL's operational research staff was the efficiency of northern military operations conducted by the Mobile Striking Force, the air-transportable brigade discussed in chapter 1. Operational research scientists conducted a range of experiments to test the combat effectiveness of soldiers facing the peculiar stresses of Canada's northern environment – research deemed valuable for training and improving the MSF.[21] With the assistance of John Mayne and Jim

Johnson from the DRB's operational research team in Edmonton, the DRNL group participated in every northern military exercise the MSF conducted between 1951 and 1955.[22]

Airborne units involved in Exercise Bulldog II served as test subjects for psychological observations. Attached to the Army's Western Command Joint Services Operational Research Unit in Edmonton, Easterbrook and Ross recorded what they referred to as "psychological observations" of paratroopers during winter flight training in the Churchill area.[23] The two scientists became the first civilians to qualify at the Canadian Joint Air Training Centre when they completed their parachute jump course in Rivers, training that enabled in-flight observation of the soldier's pre-jump eating habits.[24] In a series of observational studies, soldiers received a variety of foods in flight. Instructed to select preferred meals, the participating paratroopers completed individual questionnaires prior to arriving at the jump site. Easterbrook and Ross wanted to determine bodily responses to eating while airborne, which they thought would provide useful information for identifying foods suitable for pre-jump meals. Both scientists jumped on exercises with paratroopers and studied the operational problems of parachutists before, during, and after drops.

Exercise Bulldog was a series of joint Army–Air Force exercises held annually in the mid-1950s as winter training for the MSF. During the third exercise, in February and March 1955, MSF units repelled a mock enemy force that had landed near Yellowknife to secure an airfield and infiltrate a key strategic location in subarctic Canada.[25] The defending force launched two airborne assaults against the enemy, using tactical, transport, reconnaissance, administrative, and medical support provided by operational control of RCAF Tactical Air Command, Edmonton. Airlifted from Quebec to Yellowknife via Fort Churchill, the paratroopers rendezvoused with a local defence force upon landing at the drop zone. As historian P. Whitney Lackenbauer explains, No. 7 Company, Canadian Rangers, guided the assembled force to a patrol base near the exercise airfield.[26] Citizens of Yellowknife cooperated in the exercise as well, providing administrative support and assistance as required. Both airborne assaults were considered successful. The exercise demonstrated the effectiveness of the MSF and assisting units while operating under realistic winter conditions in a western part of subarctic Canada.

Arctic Indoctrination

Operational research scientists at Fort Churchill established a strong rapport with the military units and staffs stationed on base, advancing indoctrination and winter-warfare training for military personnel from

Figure 4.1. Princess Patricia's Canadian Light Infantry soldier on Exercise Sweetbriar, 1950. Library and Archives Canada / Department of National Defence fonds / e010750870.

Canada, Britain, and the US and even revising the US Army's Basic Arctic Manual.[27] "Fort Churchill was an excellent location for a team of defence scientists, since the Fort was at the time a large military base where scientists and members of all three Armed Services, as well as US Forces, lived and worked side by side," wrote one-time DRNL superintendent Archie Pennie in 1994. "It was this close association which enabled the [Defence Research] Board to make useful and valuable contributions to Arctic defence science."[28] Effective operational research depended on close civil/military integration, of course. Scientists and engineers required a thorough knowledge of the technical design and performance characteristics of the military material and equipment under study, including tanks, guns, radar, aircraft, warships, torpedoes, mines, and sonar.[29] In the specific context of operational research at DRNL, the work was multidisciplinary by necessity, with physicists, chemists, biologists, and others called upon to investigate the human factors of northern military service. Weapons, vehicles, sensors, and other equipment received attention alongside visual and auditory performance, emotional reactions to cold and isolation, and physical duress and fatigue.

To overcome the question of rank, which was extremely important in military life but factored little among the professional scientific community, Solandt instituted an organizational policy that barred DRB scientists from wearing uniforms or military ranks in peacetime. When the civilian members of the DRB's operational research staff joined the Canadian Infantry Brigade (CIB) in Korea, they wore army uniforms bearing no rank. Economist and operational research scientist Robert Sutherland, who had risen to the rank of captain in the Tank Corps during the Second World War and who still served as a lieutenant-colonel in the reserve force, wore a uniform with no rank in Korea, where military personnel sometimes mistook him as a member of the Salvation Army.[30]

Solandt used every means available to attract civilian scientists to government service and overcome the problems imposed by military secrecy. He made the DRB's research facilities available to graduate students and non-government scientists and encouraged new recruits and senior defence scientists to publish as much of their research as allowed under the existing security restrictions. In 1952, Charles Pope became the DRB's first public relations officer.[31] He vetted research produced and sponsored by the DRB for publication in the open literature, and although DND imposed tight restrictions as a matter of security and national defence, DRB authorities seldom refused requests that results be published or shared publicly.[32] Before Pope's appointment, scientists funded by the DRB had practised their own discretion in determining whether to publish or share research openly. Solandt also initiated an annual DRB symposium, inviting well-known scientists from Britain and the US to Canada and giving younger employees the opportunity to meet and learn from qualified researchers who held senior positions in the military science communities of all three countries.

The severe winter conditions in northern Manitoba meant that scientific research at the DRB's northern laboratory focused on cold-related problems affecting the human body and military equipment. Permanent staff and visiting scientists and research technicians at Fort Churchill performed preliminary studies within DRNL, using the laboratory facilities to establish and refine basic principles before carrying out fieldwork under actual operational conditions. Because field trials focused on the military problems associated with the performance of ground soldiers and equipment, army officials established a special test team for experimental investigations.

In 1955–6, the 1st Royal Canadian Horse Artillery – consisting of one sergeant, eight soldiers, and one bombardier – served as Fort Churchill's test team and assisted DRNL scientists in the field.[33] After a ten-day Arctic indoctrination course, the unit worked on a wide range of operational problems with civilian scientists and research technicians. They completed forced marches with heavy loads, camped outdoors on open terrain, and performed various military tasks under both winter and summer conditions. Bolstered by members of the laboratory staff and local military personnel, the test team also conducted a series of small-scale summertime exercises near the tree line and on the barrens north of Churchill, appraising warm-weather military problems associated with movement, equipment, logistics, and tactics. That same winter, the Royal Canadian Corps of Signals tested several new types of equipment under cold and wet conditions, including a mobile transmitting and receiving station.[34] Designed for parachute drop from transport aircraft, the self-contained radio unit weighed 1,200 pounds and fit atop a sled for movement over snowy terrain. Signals Corps personnel also conducted cold-weather performance trials of radio-teletype circuits, line equipment, and a new army field telephone.

As with laboratory work performed within DRNL, military scientists at Fort Churchill who conducted research in the outdoors focused their efforts on improving the operational capabilities of the MSF. Training regularly doubled for field research, and service personnel often represented the primary research subject. An instructional cadre of officers and non-commissioned officers (NCOs) from the three infantry battalions of the MSF held a series of manoeuvres in the Churchill area during the winter of early 1956, for instance. Personnel representing the Royal Canadian Regiment, the Princess Patricia's Canadian Light Infantry, and the Royal 22e Régiment took part, travelling on foot with equipment and toboggans, cooking and sleeping in portable nylon tents, and constructing snow houses and other temporary shelters.[35] A test team from the Royal Canadian School of Infantry, Camp Borden, conducted their own exercises that same winter, carrying out live-fire trials of infantry weapons under subarctic conditions. The tested weapons included a newly designed rifle, a lightweight machine-gun, a mortar, and an anti-tank weapon developed by DRB engineers called the Heller.[36]

Field research allowed the military to test men and equipment under realistic conditions in a northern environment and enabled DRB scientists and engineers to identify any existing problems or deficiencies in the operational capabilities of personnel and their equipment. During Arctic warfare training at Fort Churchill in January and February 1957, participating units of the MSF attended a series of two-week winter indoctrination courses, designed to teach paratroopers how to live and fight in the harsh and cold climate of northern Canada and the Arctic.[37] Each group of forty to sixty officers received instructional training at the garrison and spent several days outdoors, learning from qualified personnel and gaining first-hand experience in the field. Lectures by seasoned soldiers and operational research scientists covered such topics as cold-weather clothing, personal kit and equipment, rations and food preparation, stoves and cooking, tent pitching and snow shelter construction, navigation by dead reckoning, stowing and lashing of toboggans, snowshoeing, and first aid.

After four days of instruction, candidates departed the garrison and spent eight days travelling, eating, sleeping, and working on the open, subarctic barrens. They underwent a detailed and structured course schedule, complete with load carrying, shelter and meal preparation lessons, day and night marches, day and night navigation trials, construction of defences and fortifications, rifle and light machine-gun firings, and basic survival drills. At the completion of the field training, personnel returned to Fort Churchill and spent the final two days turning in supplies and equipment, discussing lessons learned, and debriefing before returning to regular duties.

By providing soldiers with the tools, knowledge, and capabilities to overcome the demanding challenges of northern service, both physically and psychologically, DRB scientists and engineers played an important role for the Canadian military during the mid- to late 1950s at a time when preparations for Arctic warfare were still a priority for senior officials in Ottawa. In fact, paratroopers in the MSF trained to live and fight anywhere in Canada year-round, and operational research scientists regularly participated in military exercises conducted away from Fort Churchill as well. In January 1957, the 2nd Battalion of the Royal Canadian Regiment undertook a ten-day Arctic training exercise at Butler Lake, Ontario, some twenty-five miles northwest of Kirkland Lake.[38] Four rifle companies, with support elements, travelled to the exercise site via train from London, Ontario, where an advance airborne party had dropped equipment and supplies. Pre-exercise training included tent-raising, stove-lighting, cooking, snowshoe marching, and dead reckoning, but most of the officers and NCOs had little or no experience with winter camping and living outdoors in the cold.

Scientific information, technical knowledge, and Arctic know-how were as important to soldiers as operational capabilities in the field. The Butler Lake exercise taught the value of high-quality tents and sleeping bags, the necessity of special energy-packed food rations, and the dangers of overexertion. Marching without sweating was a difficult but necessary skill to learn; so was cross-country navigation without a magnetic compass; and so was performing the myriad tasks of Arctic warfare:

constructing defensive fortifications in the snow, patrolling with limited knowledge of terrain and landmarks, attacking enemy-held positions, and handling metal weapons in winter fatigues. "There really wasn't much time to feel cold," reflected Major J.W.P. Bryan in a brief post-exercise article. "As a result of this winter training, these soldiers have lost their fear of the cold and have confidence in their equipment and themselves."[39] Exercises of this sort introduced personnel to winter warfare, providing the soldiers and paratroopers who travelled north to Fort Churchill with the experience and knowledge required to understand and appreciate the daunting challenges of northern military service.

Except for minor quibbles over the use of soldiers in certain experimental studies, relations between military personnel and scientists at Fort Churchill were overwhelmingly positive. DRNL regularly requested test subjects for laboratory and field studies about physiological responses to cold, wind, snow, rain, biting insects, and other challenging elements and conditions of northern Manitoba, and soldiers consistently volunteered to do their part. Whether soldiers truly "volunteered" for experimental studies is unclear, but few stories of unease survived, and there is little evidence that military personnel and scientists struggled to get along at Fort Churchill. "Their co-operation and enthusiasm have resulted in a most successful field programme and the practical fulfillment of long months of laboratory research," wrote DRNL's superintendent Pennie in January 1956, referring to his military colleagues. "Their assistance has been invaluable, and clearly demonstrates that civilian scientist and soldier, when working together, can perform as a happy and productive team."[40]

This is not to suggest that issues never arose between civilian staff and military personnel. Surviving correspondence among senior military officials indicates that officers complained about regular requests for volunteer research subjects. With personnel and resources constrained, regular duties trumped scientific investigations, which some of the garrison officers considered important but excessive. As a tri-service, multinational, and largely civilian-operated post, Fort Churchill generated command and administrative control issues for the Canadian Army. "Tact and diplomacy, understanding and sympathy were as important to the maintenance of good order and military discipline as were Canadian Army Orders and the Queen's Regulations," recalled Colonel Strome Galloway, Fort Churchill's last commander. "Naturally, the quest for multi-directional *rapport* had to be entered into by all, and this was done to the credit of by far the vast majority."[41]

As far as DRNL scientists were concerned, careful and accurate investigations were essential to solving many of the operational problems encountered by military personnel in northern Canada and the Arctic, but laboratory studies at Fort Churchill were inadequate on their own, and fieldwork thus became increasingly significant during the 1950s. Small-scale exercises were inexpensive and produced useful results in a timely manner, unlike larger exercises, which required extensive planning, resources, and personnel. Scientists at DRNL also created mobile field laboratories, using basic sleds and vehicle trailers to house and carry recording equipment and other scientific instruments for field trials.[42] Powered by a basic gasoline generator, each portable laboratory functioned as a working unit in summer and winter conditions, enabling real-time studies of operational performance on the subarctic barrens. Some researchers used thermocouples to record body temperatures during physiological experiments on cold-weather acclimatization, for instance. Others used body measurements recorded in the field to investigate the performance of specially designed clothing, footwear, sleeping bags, and other kit under realistic northern conditions. Mobile field laboratories also enabled the expansion of mechanical engineering research, including performance tests of new and improved batteries and lubricants for military vehicles and automotive equipment operating in extreme low temperatures.

In addition to DRNL's internal work program, the laboratory provided space, resources, and supporting facilities to visiting scientists and research teams. The Fifth Commonwealth Defence Conference on Clothing and General Stores was held at Fort Churchill in January 1956, for instance. Military officials and scientists representing Australia, New Zealand, India, Pakistan, South Africa, and the United Kingdom attended the meeting, where they all discussed the latest military clothing and equipment designed for Arctic use by the army, navy, and air force.[43] They also toured Fort Churchill and watched Canadian soldiers perform winter warfare demonstrations nearby on the snow and icy barrens. The live exhibitions included snowshoe marches and rifle and mortar firings. Soldiers also constructed a portable nylon tent, a snow house, and a temporary lean-to shelter made of felled tree logs. Army personnel gladly displayed their new and improved uniforms as well, which included a novel snow mask and durable, lightweight magnesium snowshoes; both had been designed and developed by DRB scientists and engineers. Members of the Vehicle Experimental Proving Establishment test team participated in the exhibition, demonstrating the transport capabilities of the Saracen, an armoured personnel carrier, as well as the Wapiti, a large over-snow vehicle designed specifically for northern Canada's terrain.[44]

The DRB's Arctic research program focused on the environmental, geographical, climatological, physical, and psychological impediments to soldier performance in both summer and winter conditions; however,

136　Frontier Science

Figure 4.2. Soldiers undergoing tests for physiological responses to cold temperatures and windchill at Fort Churchill, ca. 1953–4. *Vigil in the North*, National Film Board of Canada, 1954. Public Domain.

indications of the need for research in regions outside northern Canada grew stronger towards the end of the 1950s. Between 23 September and 4 October 1958, a joint Canada–US party of scientists and military officials travelled north from Ottawa and Washington on a visit to Fort Churchill. Organized at the request of Admiral Rawson Bennett, chief of naval research for the US Navy, the group toured northern military facilities in Canada and the US, surveying military installations for a subsequent report about continental defence in the Arctic. "We have seriously inadequate information on the cold regions comprising the vital strategic area of the North Polar Basin," stated the report. "Expanded Arctic research is essential so that we may extend our northern military frontier to the Soviet Arctic littoral."[45] The report also stressed the strategic value of territory beyond the North American continent, declaring emphatically that "science will permit our use of Greenland as an Arctic sword and shield – a mighty bastion of deterrent power essential to the NATO concept. Its 850,000 square miles of polar Sahara constitute a giant platform for weapons and weapon-delivery systems, including mobile long-range artillery, and for a far-flung detection, tracking and interception complex." As the missile threat against North America escalated in the late 1950s, the heightened military and strategic value of northern Canada and the Canadian Arctic was not lost on senior officials in Ottawa. While the nature and scope of the strategic threat in the North had reduced the need for large standing ground forces, Ottawa's growing involvement in NATO and any outward projection of military power did not replace or diminish the need for personnel trained and equipped to operate in high latitudes.

Changing Priorities at DRNL

The scope and extent of the DRB's Arctic research program underwent significant change in the mid- to late 1950s. The DRB hosted more than 250 scientists from fifteen NATO nations in June 1955, when the Fifth General Assembly of the Advisory Group on Aeronautical Research and Development (AGARD) convened at the Château Laurier Hotel in Ottawa.[46] Ralph Campney, the Minister of National Defence, hosted the delegates along with the Canadian AGARD council member J.J. Green. Shortly thereafter, sweeping changes were made to the DRB's Arctic research activities.

The DRB's Arctic Section ceased to exist in November 1955 when its duties and responsibilities were transferred to the DRB's directorates of physical and engineering research.[47] Within the Directorate of Physical Research, authorities created a Geophysical Section to continue many of the duties of the disbanded Arctic Section. Trevor Harwood headed the new section; Moira Dunbar and J.P. Croal joined his staff in a supporting role. Their research focused on fields such as geology, glaciology, ice physics and forecasting, oceanography, hydrography, meteorology, and navigation. The remaining Arctic activities were transferred to the mechanical and civil engineering sections of the Directorate of Engineering Research. The Mechanical Section assumed work on engines, vehicles, materials, fuels, and lubricants, while the Civil Engineering Section focused on snow, ice, soil, and permafrost, as well as survey and air photograph interpretation.

Despite this reorganization, senior officials in Ottawa still valued the northern research facility at Fort Churchill. Adam Hartley Zimmerman succeeded Omond Solandt as DRB chair in March 1956 and shortly thereafter arranged an airborne tour of northern Canada for military and defence representatives from Ottawa, London, and Washington. The trip included flights over Canada's northernmost islands, tours of

Mid-Canada and DEW Line radar sites, and visits to Fort Churchill and the RCAF station at Cold Lake, Alberta.[48] Zimmerman also travelled overseas to witness British atomic weapons trials at the Maralinga Range in southern Australia, and he worked diligently to maintain the close ties among Canadian and Commonwealth science organizations first established under Solandt's leadership. He spent eleven years as chair and saw to fruition many research and development projects initiated under Solandt, including variable depth sonar for naval security in the Arctic Ocean and the integration of Canada's rocket and satellite programs within NATO's overall defence structure.[49]

DRNL remained active during the Solandt–Zimmerman transition. February 1956 was a particularly eventful month for the research facility, as the Canadian staff hosted government officials and scientists during the Fort Churchill sessions of the Commonwealth Advisory Committee on Defence Science.[50] Representatives from DRNL and other government agencies tied to the DRB delivered research papers on a wide range of topics addressing various Arctic military problems, and the delegates in attendance toured DRNL and learned about the ongoing operational research projects undertaken at Fort Churchill. DRNL also hosted members of the Joint Intelligence Bureau (JIB) that month. Headed by JIB's director Ivor Bowen, the group held sessions in the DRNL conference room before touring the military garrison and spending one day on the open tundra learning about northern defence research from a strategic point of view.

The mid-1950s marked a high point for Arctic research carried out on the ground in northern Manitoba. During 1955–6, the Canadian Army assigned a fulltime test team to DRNL.[51] The commitment marked a milestone in research progress for the laboratory, allegedly because test team personnel served as human research subjects for field and laboratory trials during the summer and winter months. Government officials claimed that the test team exemplified military cooperation and interest in the DRB's Arctic research program, despite some minor existing tensions between DRB scientists and military personnel.[52] Nevertheless, the full-time presence of the army's test team allowed for a marked increase in field research and laboratory studies at DRNL, which authorities in the DRB considered highly valuable for Canadian defence efforts in the North.

As the scope of military research activities changed at DRNL, senior military officials slowly phased out existing projects at Fort Churchill. Canadian defence scientists remained involved in Arctic research, however. Operational research generated added interest after the Korean conflict, and scientists stationed at Churchill occasionally travelled farther north, undertaking research in the Canadian Arctic and Greenland. D.I. Ross, an operational researcher with DRNL, attended the US military exercise Arctic Night at Thule in March 1956, which involved a battalion combat team defending against a mock airborne assault. Ross represented Canada on the exercise, briefing scientists and officials from DRNL, the Canadian Army, and the RCAF on his return to Fort Churchill. Work at DRNL continued for the next decade, but the transiency of the facility increased as permanent laboratory research declined in the early 1960s.

Despite the changes to the DRB's Arctic research program, Fort Churchill remained an active locale for military training and operational research. In fact, hundreds of soldiers from Canada, Britain, and the US completed winter warfare indoctrination courses on location between 1946 and 1964. Senior officer students from the National Defence College and the Canadian Army Staff College also visited the garrison to observe unit training and experience day or overnight Arctic soldiering.[53] Most students undertook training in their early or mid-twenties, as the physical exertion required to complete an indoctrination course made "it undesirable to train any but those who are very fit."[54] Otherwise, training methods did not discriminate. All ranks used the same equipment, underwent the same procedures, and experienced a structured and consistent course.

Indeed, as an asset for the Canadian government, Fort Churchill attracted friendly forces to northern Canada; in this way, it represented one of Ottawa's tangible military commitments to NATO. In the winter of 1963, a company group of 3rd Battalion, Parachute Regiment, flew from the United Kingdom for training at Fort Churchill and spent several weeks conducting tactical exercises outdoors.[55] The year after, the oldest and most senior line infantry regiment in the British Army, 1st Battalion, Royal Scots, trained a company group on the frozen barrens outside the garrison's boundaries, miles from its steam-heated barrack blocks. The training wing at Fort Churchill was relatively small, but its staff were sufficient to conduct the Canadian-led all-arms officer and NCO courses on-site. A major performed the duties of chief instructor, with junior officer and NCO assistant instructors. Before a company deployed for training, selected officers and NCOs arrived early and undertook an advance three-week course to qualify as winter warfare unit instructors.

Low-level training was a hallmark of winter warfare courses at Fort Churchill. Knowing the sheer difficulty that armies, corps, divisions, brigades, and battalions would encounter operationally in northern Canada and the Arctic, the army focused its training efforts on individual soldiers and small group formations.[56] Proper equipment was essential for adequate training, and the army issued two types of tents for subarctic and Arctic use, one with a five-person capacity and a second capable of sleeping ten soldiers. Operational use was a point of debate among senior officers, but the ten-person tent was the standard training unit. With one slot allotted for the instructor, winter warfare courses included nine trainees to a tent. Fort Churchill's training wing had a forty-five-person capacity, accommodating multiple groups at once while freeing the chief instructor for planning and supervision.[57] A run of four courses occurred annually at peak, funnelling several hundred personnel north in preparation for the Arctic battlefield of tomorrow.

A properly constructed snow cave proved the warmest shelter for soldier trainees on the open barrens near the Fort Churchill military base. British paratroopers, five days out of their home base in the United Kingdom, slept in snow caves at −44°F and experienced no ill effects, according to military records.[58] During one training course, participating personnel stumbled upon a large and consistent snowdrift, some 20 feet high and 100 feet long, which they hollowed into living quarters for a whole platoon.[59] The snow cave slept thirty soldiers, included kitchen space and latrines, and had a staircase leading to a sentry post on the upper level. Soldiers on the course also hollowed out a connecting tunnel and garage space to house an over-snow vehicle and other equipment, ultimately constructing a tactical set-up considered superior by Canadian Army standards.

Because the number of activities associated with the DRB's Arctic research program at Fort Churchill increased during the winter months, scientists and service personnel encountered several opportunities to participate in cold-weather survival and operations trials. Soldiers at Fort Churchill were expected to gain confidence and improved abilities in open cold and snowy conditions, but they also trained in wooded areas. Trainees constructed lean-to shelters with downed tree logs and slept on beds made of spruce boughs. They used tents, learned to live and move in forested areas and bushed terrain, and trained to overcome practical issues with snowshoes and load hauling in soft snow. Soldiers found that snowshoes were a nuisance on the open barrens and hard-packed snow, but the army equipped all personnel with full kit and supplies, knowing that proper winter warfare training meant preparing them for all types of subarctic conditions and terrain.[60]

Military research and training at Fort Churchill extended to trials of equipment and weapons. In February 1957, DRB public relations officer Charles Pope emphasized in writing the importance of trilateral research cooperation among Canadian, British, and US personnel stationed at the garrison. "The assistance extended to test teams from other countries – particularly to those from U.S. Army Corps with active research programmes involving

Figure 4.3. MGM-18 Lacrosse missile in cold-weather trials near Fort Churchill, ca. 1958–9. Library and Archives Canada / Department of National Defence fonds / e010782883.

the north – has been extensive during the past eight years," he wrote in the RCAF magazine *Roundel*.[61] During the winter of 1955, for instance, scientists and technicians at DRNL assisted US Army personnel during guided-missile firings conducted as low-temperature performance tests. Similar work occurred annually in the late 1950s at Fort Churchill as the DRB's research in northern Manitoba shifted towards the physical sciences, operational research, and equipment testing. Cold-weather trials of the MGM-18 Lacrosse, a short-range tactical ballistic missile, were conducted on-site during the winter of 1958–9. Developed for close support of ground troops and deployed during the early 1960s, the Lacrosse was a US Army weapon and was never intended nor put into service for the Canadian Army. From this perspective, the Lacrosse trials reflect Fort Churchill's typical utility. Geared towards advancing military preparedness, defence research in northern Manitoba focused on improving the high-latitude, cold-weather capabilities of men and machines alike.

Decline and Closure of DRNL

On 31 March 1964, nearly eighteen years after taking control of the wartime encampment and airfield, the Canadian Army surrendered responsibility for Fort Churchill and began withdrawing personnel and supplies from the garrison.[62] RCAF Unit Fort Churchill closed at the same time, as DND relinquished its responsibilities at the garrison. In the interim years since the army assumed control in October 1946, the base had grown from a small collection of huts built to support northern exercises into a mini-metropolis equipped with housing, recreational facilities, a hospital, a school, chapels, a scientific research laboratory, a shipping and receiving store, and other amenities typical of larger urban centres in southern Canada. DND's on-site investment was geared towards comfort as much as military utility. "Fort Churchill was Canada's most unusual military base from almost every aspect," reflected Flight Lieutenant H.R.R. Noble in the October 1964 edition of the RCAF magazine *Roundel*. "Where else, for instance, could you have attended the marriage of a USAF (SAC) pilot to an American southern belle, and see an honour guard composed of members of the USAF, US Army, US Marine Corps, RCN, Canadian Army and RCAF?"[63]

Between 1946 and 1964, changes in strategic thinking, military capabilities and commitments, and government policy pointed gradually towards the day when the army would abandon its "barrenlands citadel," as Colonel Strome Galloway, Fort Churchill's commander at the time of withdrawal, called the northern military base in a farewell article written for *Canadian Army Journal*.[64] "It is a sad thing," reflected Galloway, "taking leave of the strange white land where windchill and hungry polar bears were two of the occupational hazards faced by those who found themselves members of the Fort Churchill Garrison – a community which fluctuated between the 3000 to the 3500 mark, men, women, and children. It is farewell to the land of Nanuk, a land where only the hardy and wise can survive. It is goodbye to the 'Shining Land,' as the Eskimos have named this area, where the sun reflects on snowy wastes in winter and on endless pools of icy blue water in summer."[65]

After eighteen years of subarctic training courses, the army pulled out of Fort Churchill. By then it had surmised that three weeks of indoctrination provided soldiers with sufficient basic experience and knowledge to be able to live and move in conditions of extreme cold. "At the end of the third week, the student leaves for more temperate parts of Canada, with a confidence borne of the fact that he has, for three weeks, faced and conquered one of the bitterest winter climates he is every likely to encounter anywhere," Colonel Galloway wrote.[66] Three weeks was not always enough, of course. Despite training and indoctrination, some soldiers continued to experience the so-called twin enemies of fear and fatigue associated with performing military duties in northern Canada during the frigid winter months. Military service in northern Manitoba and other subarctic regions of the country did not appeal to everyone. Nevertheless, senior military officials were pleased with the lessons learned at Fort Churchill and adapted army doctrine accordingly.

During a ceremony marking the army's withdrawal from Fort Churchill, Colonel Galloway unveiled a bronze plaque that had been issued to commemorate the eighteen-year winter warfare program. "Fort Churchill, site of a winter training establishment of The Canadian Army, 10 October 1946–31 March 1964," read the inscription, which had been bolted onto the face of a large decorative boulder.[67] Drummers of Fort Churchill's Army Cadet Corps flanked Galloway during the ceremony, which drew a large crowd of military personnel and local civilians. Less than a year later, on 29 June 1965, Defence Research Northern Laboratory closed its doors.[68] Although the DRB would continue to fund Arctic research, work at DRNL ended because senior DND officials had decided that the armed services no longer required a permanent research facility in northern Canada.

Given the DRB's changing role and structure in the late 1950s and early 1960s, the gradual phase-out of DRNL was simply a by-product of the circumstances at hand. General cuts to Canada's defence budget played a significant role, to be sure. In 1962, Prime Minister John Diefenbaker's government adopted austerity measures to address Canada's worsening economic situation. The measures reduced federal spending, increased tariffs on imports, and enabled the federal government to arrange large financial loans from foreign banks.[69] Money for security and national defence, which had accounted for the largest portion of the federal budget, began to decline. In 1964, as Andrew Iarocci observes, "the conventional defence of the North Country was no longer the strategic priority that it had been in 1946," and there was little political will to continue funding Fort Churchill.[70] As the budget for defence declined, the DRB could no longer afford the costs associated with maintaining an active military research program in the North. In retrospect, the circumstances of DRNL's closure say more about budgetary constraints and changing Cold War security concerns in Ottawa than they do about the internal priorities of leading officials in the defence department. The DRB transitioned away from laboratory research in the North out of necessity, closing DRNL and diverting the DRB's limited financial resources elsewhere to support the immediate and evolving research needs of the armed services.

Conclusion

The DRB was only one government agency affected by changes in Canada's national defence effort. The number of scientists working on military research projects in other agencies also increased to meet the changing demands of the Canadian armed services. When scientists from other agencies began to devote all or part of their time to providing direct assistance to the military, be it as project officers on development programs or as scientific and technical advisers on research problems, the DRB assumed greater responsibility as scientific adviser for Canada's entire national defence effort.[71] The unique advisory function of DRB officials and scientists came at a cost, however. In Britain and the US, government research agencies and independent contracts covered nearly all the military research and development needs of their respective armed forces. Military forces in both countries received scientific advice and assistance as a by-product of research. In contrast, Canada's limited military research program did not enable scientists and engineers employed or contracted by the DRB to cover all the research needs of the Canadian armed services. For this reason, the DRB's initial mandate had a dual purpose: the organization focused on research where Canada could make a unique or special contribution to North Atlantic security, while simultaneously ensuring that the Canadian armed services had ready access to the latest military research and resources out of Britain and the US.

When national priorities for military research changed in response to the events in Korea, the DRB's initial mandate was no longer practical. The DRB had little choice but to divert resources away from fieldwork and laboratory research, and this left the DRNL's personnel and facilities in a precarious position. The economic and political circumstances of the time were too much to overcome, and DRB officials decided to support short-term research and development projects that would serve the immediate needs of the armed services. Northern defence remained a high priority for military and defence officials in Ottawa, but technological advances in the 1950s had altered the strategic threat to Canada, and the DRB's military research program evolved accordingly. At the same time, Canada's commitment to NATO and global order created new difficulties for the DRB. As Canada's peacetime military science organization focused increasingly on the immediate operational research needs of the armed services, decision-makers in Ottawa gradually phased out laboratory research. In the process, Arctic research at DRNL declined in significance, and military science at Fort Churchill entered a period of relative stagnancy.

The military and financial circumstances brought about real changes in the DRB's mandate; even so, scientific research involving soldiers and military duties in northern Canada did not stop post-Korea. Military officials and scientists in the defence department remained strongly interested and involved in Arctic research, albeit under quickly changing operational expectations and requirements. As the strategic situation evolved in northern Canada and the Canadian Arctic owing to the escalating threat of bombers and missiles, military research remained important on the ground, but analysts and officials in Ottawa increasingly turned their attention towards the changing operational requirements facing the armed services. Adequate defence in the North meant having soldiers, sailors, aviators, and equipment trained and designed to operate year-round and on all terrain. Forecasting future military needs became increasingly difficult in a world largely defined by drastically evolving technologies, weapons systems, and modes of delivery. For Arctic research scientists funded under the defence budget, this meant a refined but continuing focus on applying science to understand and overcome the various natural, human, and technological challenges impeding operational efficiency and effectiveness in Canada's northern regions.

5 Operation Hazen and the International Geophysical Year

In the summer and autumn of 1952, well-known Arctic traveller and zoologist Thomas Manning, accompanied by University of Ottawa student Andrew Macpherson, sailed a canvas canoe off the coast of Banks Island in the Beaufort Sea.[1] Forced to abandon their plans to circle the coast because of an unusually early freeze, the pair improvised a sled from salvaged barrel staves and walked across the island on an unplanned 200-mile trek. They crossed the island on foot and rendezvoused with an Inuk, who returned them to the mainland by boat. Unfavourable weather held the two researchers in Tuktoyaktuk for several weeks before they boarded a transport aircraft and returned to Ottawa.[2] Their trip marked the first in a long series of geographical and geological surveying attempts in the Canadian Arctic Archipelago – research motivated and made possible because of the intensifying military interest in the Far North in the 1950s and 1960s.

Hired as a consultant for the DRB, Manning received military support to conduct oceanographic research in the Arctic and collect specimens for the National Museum of Canada. He returned to Banks Island the following summer with E.O. Höhn, a physiologist from the University of Alberta, and Captain I.M. Sparrow of the British Army Corps of Royal Engineers.[3] The group flew aboard a RCAF Dakota aircraft from Edmonton to Norman Wells, where they awaited suitable weather before attempting to land on a sandbar near Banks Island.[4] Upon their arrival, Höhn surveyed the island's wildlife by dog team, encountering a colony of some 30,000 snow geese, while Manning and Sparrow surveyed the coast and recorded tidal observations for the DRB.[5] After crossing the island on foot, the pair paddled along the coast and south through the Prince of Wales Strait to Holman Island, the site of a Hudson's Bay Company trading post, where they received an airlift back to Ottawa.

Manning and Sparrow produced confidential reports containing topographical information about Banks Island. Given the island's strategic location north of mainland Canada, British and Canadians officials wanted information about the suitability of coastal inlets and harbours in the Arctic for military purposes.[6] The DRB and the British War Office were particularly interested in knowing what challenges the island's coastal and inland terrain posed for the movement of military personnel and equipment. Could airborne units and paratroopers land or drop near coastal areas in this Arctic locale, and which areas on the island were accessible to the RCAF and the RAF, given their existing capabilities? Ground surveying complemented aerial reconnaissance and military mapping missions in the North, improving the accuracy of topographical information generated by geostrategic analysts in Ottawa and London.

Military-sponsored surveying in the Canadian Arctic culminated in Operation Hazen, a Canadian-led research expedition organized and carried out by DRB scientists during and after the International Geophysical Year (IGY). The IGY was a celebration of global science and international cooperation in such fields as meteorology, oceanography, and upper atmospheric studies.[7] The idea for the IGY originated with British mathematician Sydney Chapman, along with American physicist Lloyd Berkner and space scientist James Van Allen, who together suggested reprising the International Polar Years (IPYs) of 1882–3 and 1932–3. The IPY was the first worldwide coordinated scientific program and would play a crucial role in the founding and early organization of the science of geophysics. Chapman, Berkner, and Van Allen were all advocates for the advancement of science internationally, and they posited that the increased solar activity anticipated between early July 1957 and late December 1958 would be a perfect opportunity for a coordinated worldwide research effort. The scientific program would include participants from more than sixty countries whose fields of research covered a range of environmental sciences. While the scope of the initiative was global, it would focus mainly on outer space, the Arctic, and the Antarctic. A considerable amount of research would be conducted in the Far North.[8] The Soviet Union launched Sputnik, the world's first artificial orbital satellite, under the auspices of the IGY on 4 October 1957.

Sputnik greatly heightened Western concerns about the Soviet missile threat; however, the competitive tensions of the period were not limited to the superpowers.[9] The IGY brought widespread attention to scientific competition among smaller powers as well, as evidenced by Canada's involvement. In fact, the Canadian IGY research contribution was proportionally larger than those of the US and the Soviet Union – so claimed the DRB's Arctic researcher Trevor Harwood.[10] Canada's IGY activities encompassed various studies across fifteen scientific disciplines as well as 190 research stations that monitored geophysical activity and weather patterns from Cape Race in Newfoundland to Sandspit in British Columbia and from Harrow in Ontario to Alter in the Northwest Territories.[11] Scientists and engineers conducted research at twenty-six locations in northern Canada alone, generating knowledge in many fields, including atmospheric physics, glacial ice, and ionospheric magnetism.

Historians of science point to the IGY as an exemplar of high modernism and enviro-military science during the Cold War. As a new target for both geographic knowledge and military defence strategy, northern Canada and the Arctic became increasingly entangled with developments in the geophysical sciences during the late 1950s. Ronald Doel's research has shown that the US military sought the assistance of polar scientists, whose expertise and international networks could contribute significantly to high-level discussions about the relationship between science and foreign policy.[12] Richard Powell has analysed the Canadian experience with the IGY in a similar context. Examining the emergence of new technologies such as intercontinental ballistic missiles and nuclear-powered submarines in relation to the Polar Continental Shelf Project, Powell contends that the federal government attempted to "mobilize a pan-Canadian nationalism in response to perceived American and Soviet incursions upon territorial sovereignty during the IGY."[13] His work sheds light on the complexities of security politics in Canada during the IGY, especially between idealized notions of scientific globalism and strategic considerations for continental defence and territorial sovereignty in the North.

The global impetus for the IGY was non-military, but it was the DRB and other military research organizations that provided the personnel and funds required to plan, construct, and operate many of the IGY's scientific and technical projects.[14] This was certainly the case in northern Canada and the Canadian Arctic, where the IGY reaffirmed the role and utility of military research. The DRB focused Canada's efforts on glaciers, the aurora borealis, and ionospheric research. Founding chair Omond Solandt had prioritized Arctic research when developing the DRB's original mandate, on the basis that Canada's geography and northern expertise were fundamental to post-war North Atlantic security. Historian Jonathan Turner suggests that Solandt attempted to assert "scientific authority" over Arctic research, in this way mirroring and reinforcing Ottawa's efforts to establish Canada's political and territorial sovereignty over the Arctic, where international boundaries were being disputed.[15] When A. Hartley Zimmerman succeeded Solandt as DRB chair in 1956, he carried forward Solandt's vision as well as his management structure

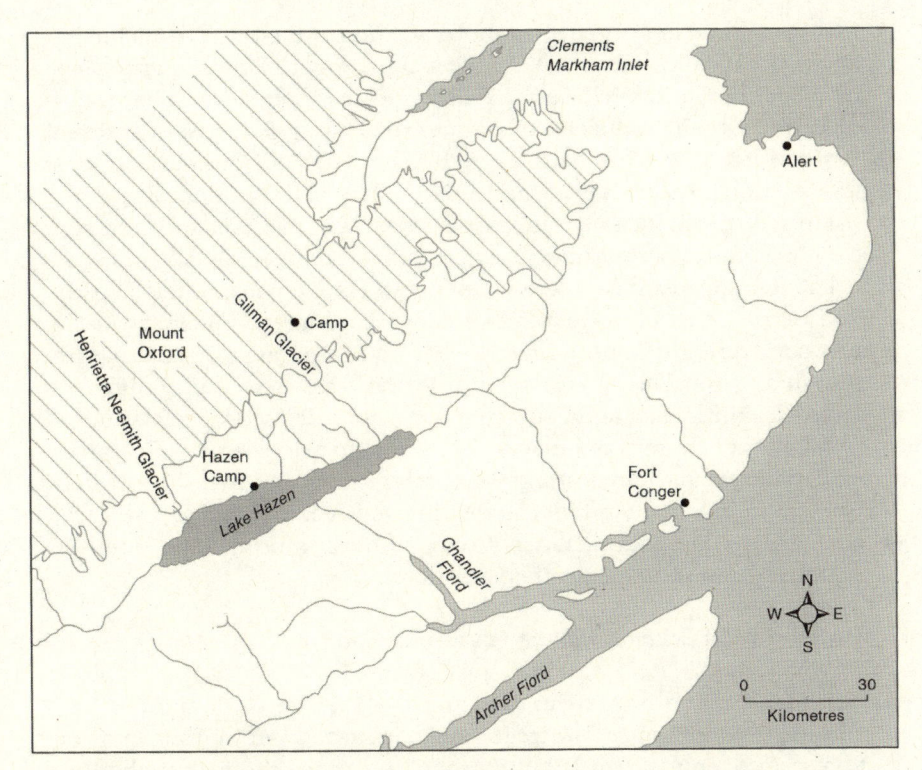

Figure 5.1. Map of Lake Hazen, Ellesmere Island, 1960. Canadian War Museum, George Metcalf Archival Collection, box 58A 1 295, file 295.6, J. Lotz and R.B. Sagar, "Meteorological Work in Northern Ellesmere Island," *Weather* 15, no. 12 (December 1960), 397. Map reproduced from original.

for Canada's defence research program. Zimmerman viewed the IGY as an opportunity to demonstrate the value of, indeed the need for, defence research for Canadian preparedness in the Cold War, both militarily and politically.

The DRB's expansive IGY effort involved scientists and engineers at three existing facilities as well as field studies conducted in northern Canada and the Arctic. Scientists and technicians performed electronics research in Ottawa, ballistics research in Valcartier, and atmospheric research related to the behaviour of the aurora borealis and the ionosphere in Churchill. DRNL saw its program slowly decline in the mid- to late 1950s as simulated climate research increased at the DRB's Toronto-based facility, Defence Research Medical Laboratories; even so, the

DRB's foothold in northern Manitoba positioned the DRNL team to play an active role in Canada's IGY. A rocket-launching pad and research facility called the Churchill Research Range, located some thirty minutes east of Fort Churchill at the end of a gravel road along the coast of Hudson Bay, passing a Mid-Canada Line radar site, enabled metrological tests with sounding rockets during and after the IGY.[16] DRNL scientists coordinated activities at the range and, as discussed later in this chapter, worked closely on northern rocketry research with scientists in the US military.

DRNL also helped coordinate Operation Hazen, a major field expedition carried out on northern Ellesmere Island in the Canadian Arctic between April 1957 and August 1959.[17] Four separate scientific parties travelled to the Lake Hazen area of Ellesmere, where they jointly collected scientific data and generated geographical knowledge about the Arctic islands for decision-makers in the defence department. Researchers carried out glaciological and meteorological studies at the Lake Hazen base as a supplement to wider studies in Arctic geology and archaeology, generating geophysical data and topographical information of military significance to strategic planners in Ottawa.

Environmental Science and the Post-War Arctic

Post-war research projects initiated on the US side of the border had considerable influence on Arctic research carried out in Canada or by Canadian scientists. In 1949, the Arctic Research Laboratory – the first federal US laboratory based in the Arctic, established in 1947 as a branch of the US Office of Naval Research – launched a long-running research program called the Arctic Ice and Permafrost Project.[18] That project covered a range of topics – for example, it developed maps of permafrost distribution in northern Alaska and conducted economic analyses of permafrost's potential for various construction and infrastructure development projects in northern regions. Military officials in Washington expressed keen interest in applying this knowledge to future large-scale developments in the Arctic, as environmental historian Andrew Stuhl explains. This led engineers and scientists to coin the term "terrain intelligence" as a descriptor for the wide-ranging ecological, geological, and geophysical knowledge they helped generate.[19]

Government scientists in the US shared terrain intelligence with their cross-border colleagues and counterparts in Canada, where great interest was shown by the NRC, the Department of Transport, the DRB, and other branches of the federal government. Permafrost research had been a priority for Canadian defence scientists since the early days of the DRNL. In mid-1947, James Croal had arrived in Churchill and commenced permafrost drilling research, first alone and later with the assistance of US Army engineer Bill Crumlin.[20] Federal scientists in Ottawa committed extensive resources to permafrost research four years later with the Canadian Arctic Permafrost Expedition of 1951, a joint venture among researchers from Canada's NRC, Purdue University, and the US Army Corps of Engineers. By combining aerial photography with soil sampling, the survey aimed to develop and perfect new methods of Arctic engineering. Expedition researchers leveraged new modes of post-war transportation to study the Mackenzie River Valley with an eye toward understanding "the relationship between terrain features and infrastructure so as to foster more efficient construction."[21]

The terrain intelligence generated by permafrost research conducted in the 1950s suggested that movement through nature was a critical component for understanding the environment. Researchers made intentional decisions on driving, canoeing, and flying in an attempt to visualize the Arctic environment in its entirety, combining different perspectives and vantage points, and comparing and contrasting thousands of photographs.[22] Meanwhile, senior decision-makers and military officials co-opted terrain intelligence to claim authority over particular Arctic environments, relying on newly acquired scientific knowledge to guide such mega-engineering projects as the Distant Early Warning Line, the string of radar stations constructed between 1953 and 1958 across Arctic North America from the northern tip of Alaska to Greenland.[23]

Government officials and scientists in Ottawa and Washington considered meteorological data and knowledge about Arctic weather patterns to be just as important as topographical information. The weather bureaux of each government funded and operated five permanent weather stations in the Canadian Arctic: Resolute, Eureka, Mould Bay, Isachsen, and Alert. Constructed between 1947 and 1950, the Joint Arctic Weather Stations (JAWS) transmitted regular meteorological recordings to receiving stations in southern Canada and around the world, enabling daily and weekly forecasts. Much like Fort Churchill, each station also served as a communication and transportation hub, providing scientists and survey teams with access to remote locales in the Arctic. Historians Daniel Heidt and P. Whitney Lackenbauer contend that spatial analysis and *place* are both key to understanding the JAWS program.[24] That government and military officials living far away decided to invest personnel and financial resources in Arctic weather stations highlights the overlapping importance of science, sovereignty, and geostrategic interests during the early Cold War. Establishing territorial and cognitive authority over the Arctic required forward-looking data that would be useful for military forecasting – put another way, it would entail the use of intelligence and open information to prepare and strategize for national defence. Real-time meteorological data recorded in the Arctic could indicate, for instance, plausible flight windows and pathways for enemy bombers or the direction and spread of radioactive fallout from a nuclear test conducted on the Soviet side of the North Pole.

Senior officials in Ottawa recognized that military activity and installations in northern Canada and the Arctic would affect the region's Indigenous peoples. During the negotiations that culminated in the Canada–US bilateral agreement to construct the DEW Line, the Department of Northern Affairs and National Resources raised questions about how the construction of the radar system would affect Inuit communities.[25] Historian Tina Loo's account of the history of forced relocation in post-war Canada indicates that the federal government facilitated the movement of Inuit men and women within and beyond the Arctic for training and work.[26] Construction of the DEW Line provided jobs for many; but at the same time, the installation of radar stations in the Far North had drastic social and environmental consequences for Inuit communities living near the radar sites. Even small, short-term radar bases constructed in the North would affect the land and lives of several generations of people in the surrounding areas, as anthropologist Stacey Fritz has noted in her extensive research. Each affected community "has its own particular history with the DEW Line and each individual has their own opinions on militarization in the Arctic," writes Fritz.[27] Confronted with the sweeping changes introduced to the Arctic by military activities, Indigenous residents responded with resilience, finding ways to adapt to the social and environmental changes that were affecting their traditional ways of knowing and living off the land.

High-latitude military installations were a dominant feature of North Atlantic security after 1945. One of the more tangible markers of the US military presence in the Cold War Arctic was the Thule airbase in northern Greenland.[28] Construction of the base began in 1951, the assumption being that the USAF would require a forward operating base for long-range bombers and fighter interceptors. Theoretically, aircraft stationed at Thule would be able to strike at the Soviet enemy or intercept Soviet bombers before they reached North America. But US military planners had broader, more elaborate plans for the region. As Janet Martin-Nielsen's research on Project Iceworm has brought to light, the US military attempted to construct an elaborate nuclear arsenal beneath the surface of the massive North Greenland Ice Sheet.[29] That project ultimately failed, owing to structural weaknesses in the ice, but the extensive ice cores drilled at Camp Century made important contributions to climate research, nonetheless.

Although Denmark maintained formal sovereignty over Greenland, US military interests dominated regional activities in the early Cold War. The environmental physical sciences ensured and projected the military, geopolitical, and colonial interests of both the US and Denmark. Geophysical research was "a multidimensional colonial endeavor" that brought local Inuit communities into sustained contact with the outside world and introduced transformative changes to the region.[30] As historian Adrian Howkins writes of Thule: "One of the most remote communities in Greenland suddenly became one of the most cosmopolitan."[31] Radar installations in northern Canada also brought outsiders to the North in large numbers, affecting Indigenous communities and the natural environment in myriad ways. However, military preparations in the post-war period went beyond air defence and made their way into the complex research field of Arctic geophysics.

Military Surveying in the High Arctic

Long before the IGY and Operation Hazen, several scientists had performed surveying work for the Canadian government in the Far North. Military planners in the US were particularly keen to learn about Arctic topography, and Canadian scientists took every opportunity available to join their US counterparts on research expeditions. During consecutive summers in 1951 and 1952, a small research vessel operated by DRB scientists carried out detailed in-shore explorations in the western Arctic from the Alaska–Canada border to the mouth of the Amundsen Gulf, and north along the east coast of Banks Island to Cape Prince Alfred on the island's northwestern tip.[32] At the same time, the US Navy operated a large icebreaker in the Beaufort Sea. The objective of the twin operations was to obtain oceanographic, hydrographic, and biological information about the Arctic waters to the north of the Canadian mainland.

By the end of 1952, the researchers involved realized that small vessels were inadequate for detailed scientific investigations of the Arctic ice pack. In consideration for future US expeditions in Canadian waters, the DRB withdrew Canada's independent vessel, and the two countries agreed to continue their research jointly.[33] The following summer, US and Canadian scientists aboard the 6,000-ton icebreaker USS *Burton Island* investigated the waters of the Amundsen Gulf. They travelled to the north end of Prince of Wales Strait, a waterway in the Arctic Archipelago between the uplands of western Victoria Island and the east coast of Banks Island.

The joint scientific party returned to the Canadian Arctic in the summer of 1954 and conducted oceanographic and hydrographic research in McClure Strait, the western entrance to the Northwest Passage.[34] The USS *Burton Island* and the US Coast Guard Cutter *Northwind* carried military personnel and scientists during the research expedition. Both vessels departed San Diego in early July and sailed north to Canada's Pacific coast naval base at Esquimalt, British Columbia, where they picked up the Canadian party on route to the Arctic Ocean. In addition to the US and Canadian naval officers who took part, several scientists from the DRB's Pacific Naval Laboratory at Esquimalt, and from the Hydrographic Survey of the Department of Mines and Technical Surveys, cooperated with a group of American scientists directed by Waldo Lyon of the US Naval Electronics Laboratory, San Diego. William Cameron of the DRB's Esquimalt research facility was in charge of the overall scientific investigations, which included charting the ocean floor, currents, and the temperatures and movement of sea ice. "The whole thing is really part of the North American continent's increasing interest in the Arctic, both from a defence point of view and from resources," Cameron said in an interview.[35] Data compiled during the expedition was valuable from a strategic point of view, in part because military officials in Canada and the US required detailed knowledge about the Arctic if they were to deploy vessels accurately and effectively in the event of war.

Cameron was a natural fit for the DRB's research program in the Arctic Ocean. He graduated in zoology from the University of British Columbia in 1938 and worked as a biologist at the Pacific Fisheries Centre in Nanaimo, where he first became interested in oceanography. He joined the RCAF as a meteorologist at the outbreak of the Second World War and later performed anti-submarine warfare research for the RCN. After the war, he trained at the Scripps Institute of Oceanography in San Diego, California, developing contacts on both sides of the border and gaining professional experience that added to his wartime knowledge.[36] Cameron understood the military and strategic value of oceanographic and hydrographic research in the Arctic. The DRB sponsored aerial, ground, and sea investigations north of the Canadian mainland. Ice reconnaissance had a dual purpose. While senior officials in Ottawa, London, and Washington wanted to know about the feasibility and challenges of deploying military units on Arctic islands, they also sought to understand the movement of sea ice and the opening of Arctic waterways. Canadian and US vessels required up-to-date maps in order to manoeuvre quickly and safely in the Arctic Ocean, and detailed information about water passages was valuable for assessing the threat of Soviet penetration in the vast areas north of Canada's mainland.

Navigational research in the Canadian Arctic expanded in the summer of 1954 when the RCN commissioned HMCS *Labrador*, a newly built

6,500-ton Arctic patrol vessel. The *Labrador* had been designed for Arctic waters and served as Canada's largest floating research laboratory. It carried scientists into the Arctic Ocean and enabled several branches of the federal government to conduct oceanographic, hydrographic, and meteorological research. Captain Owen Robertson, the *Labrador's* commanding officer, consulted with DRB scientists regarding special research facilities aboard the ship. In the end, those scientists had access to additional equipment space and to specially designed winches and other gear for facilitating and carrying out scientific investigations. British and US scientists sailed aboard the *Labrador* as well, including the polar geographer Terence Armstrong, an authority on Arctic ice conditions.[37] Senior officials in the DRB commissioned Armstrong, who was affiliated with the Scott Polar Research Institute at Cambridge, for an eight-month study of ice in the Canadian Arctic Archipelago. He reviewed Ottawa's existing methods of sea ice reporting and developed a new system for civilian and military use.

Arctic Ice and the Cold War

The first phase of Operation Hazen lasted from April to August 1957, when a research party of six conducted glaciological, meteorological, seismic, and survey studies on the Gilman Glacier while two geologists worked out of Hazen Camp on Lake Hazen. A party of four meteorologists arrived in August, living and working at the base camp through the end of the winter season in April 1958. The third party arrived that same month, continuing the original studies on the Gilman Glacier, investigating the ice cap north of Lake Hazen, and coordinating an expansive research program concerning the lake basin and other parts of northern Ellesmere Island. The latter program included geological, botanical, zoological, archaeological, and other fieldwork performed by twelve members stationed at base camp. One final party initiated a limited program of glaciological, meteorological, and botanical research on the Gilman Glacier during the summer of 1959, carrying out research on site until transport arrived.

Military support was critical to the planning and execution of Operation Hazen.[38] All three services participated in pre-expedition consultations and offered their support during every stage of the operation, smoothly and efficiently carrying out their assigned duties and assisting the scientific program. Under the direction of geographer Trevor Harwood, the DRB's Geophysics Section served as the co-originating agency for Operation Hazen; Geoffrey Hattersley-Smith, also a DRB geophysicist, led the expedition. Hattersley-Smith specialized in geology at New

College in the University of Oxford. He graduated in 1948 and continued studying in the field, later earning both a master's degree and a doctorate in geology, also from Oxford. In the spring of 1948, he joined the Falkland Islands Dependencies Survey in Antarctica and spent two years directing a scientific base camp on King George Island. He came to Canada in 1951 and joined the DRB's Arctic Section, conducting glaciological research on the Alaska–Yukon border and surveying the Beaufort Sea for the Canadian government. "I had a strong hankering to go to the Arctic," he later admitted.[39] Colin Bertram, Director of the Scott Polar Research Institute, and Graham Rowley, head of the DRB's Arctic Section, facilitated Hattersley-Smith's new job in Canada and gave him the opportunity he sought. Hattersley-Smith first travelled to Ellesmere Island in the spring of 1953 when the DRB organized and funded research expeditions in consecutive years, sponsoring a two-year study of drifting ice islands in the Arctic Ocean.[40]

An ice island is a floating mass of densely packed multiyear sea ice that drifts slowly according to the wind and currents of the Arctic Ocean.[41] Appearing as flat-shaped boxes that can rise to forty feet above sea level, most ice islands are large and solid enough to enable semi-permanent occupation.[42] Relatively flat and of freshwater origin, they drift with the winds and tides of the Arctic Ocean until trapped among the islands of the Canadian Arctic Archipelago. Ice islands usually form by splitting apart from a large mass of shelf or shore ice attached to the northern coast of Ellesmere Island; they then drift in a clockwise circle at a slow pace of 7 kilometres per day. From their starting position, they pass through the Beaufort Sea and drift past the northern coasts of Alaska and Siberia, moving northward along the margins of the Arctic Archipelago towards the Pole.[43]

The study of ice islands grew more significant during the early Cold War. Both the US and the Soviet Union established temporary bases on polar pack ice north of Siberia and Alaska in 1950.[44] The Americans established an ice base in three subsequent years from 1950 to 1953 as part of Operation Ski Jump, which aimed to test and improve techniques for landing heavy aircraft on Arctic ice. For their part, the Soviets established a station called North Pole-2 in April 1950. The establishment of both facilities opened a long period of the Cold War during which several polar stations drifted in the Arctic Ocean.

As historian Adam Lajeunesse demonstrates, US and Soviet activities on ice floating north of Canada concerned officials in Ottawa. For the first time, foreign powers had essentially occupied a sector of the Arctic Ocean that Canadian policy-makers considered autonomous to Canada. International law stated otherwise, however. Most states viewed the Arctic Ocean as an international body of water, which meant that neither the Americans nor the Soviets could be viewed as "occupying" ocean waters within the Canadian sector.[45] Officials and military planners in Ottawa understandably worried about Canada's inability to claim or exercise autonomy over Arctic waters, especially considering the increasing importance of securing and defending the northern reaches of the country.

It was the Soviet presence north of the Canadian mainland that most concerned Ottawa. Soviet ice stations posed a dual threat to Canada's national sovereignty and security, and furthermore, by the onset of the Cold War, Soviet scientists had gained extensive experience in Arctic research. Between 1935 and 1941, the Soviet Union conducted no fewer than four drift expeditions on ships or ice floes as well as a series of aircraft landings on sea ice north of Siberia.[46] In one isolated attempt at polar research, in 1937–8 Soviet scientists established a semi-permanent station on an ice floe. Dubbed North Pole-1, the station came close to the northernmost fringes of the Canadian sector before drifting to the northeast coast of Greenland.

When Soviet research expeditions in the Arctic Ocean resumed after the war, Ottawa took notice. In fact, the Soviet Union occupied four separate ice islands or floes between 1950 and 1955.[47] Scientists operated three of the four stations concurrently and conducted research studies on Arctic meteorology and oceanography as well as on the structure, distribution, and movement of pack ice. While the occupied islands generally drifted outside of Canada's Arctic sector, authorities in DND maintained a close watch on the ice-related activities of the Soviet Union. In May 1954, when a Soviet aircraft flew over a US station on an ice island known as T-3 or Fletcher's Island, officials in the defence department voiced concern over Canada's sovereignty because the flyover was conducted while T-3 was in the Canadian sector. Ultimately, officials in the Department of External Affairs were unable to confirm that a violation had taken place.[48] The issue was a matter of interpretation, and the legal status of the ice island was unclear under international maritime law.

Scientists in the DRB raised some of the loudest alarm bells in Ottawa. Harwood called for an immediate diplomatic response to the Soviet "intruders" and for a stronger Canadian military presence in the Arctic Ocean. He also took issue with the Americans, who were allegedly planning to monitor Soviet activities with scheduled flyovers; he even recommended that the RCAF patrol and occupy ice islands such as the abandoned American T-3.[49] DRB chair Omond Solandt also voiced his concerns about foreign activities on Arctic ice, writing a detailed report that emphasized the scientific research activities the Soviets were conducting on the Canadian side of the Pole. The "presence of the polar station manned by the USSR in this period of tension should invite profound concern in their activities," wrote Solandt, alerting his government colleagues in Ottawa to possible Soviet intentions to dominate scientifically and even militarily the Arctic ice north of Canada.[50] The ice island in question eventually drifted away from Canada and into Danish waters north of Greenland. This incident increased concerns about the possible military uses of Arctic sea ice.

In the wake of the Soviet ice scare, senior military and defence officials in Ottawa considered a host of ideas about how ice islands might be militarized. Some strategic analysts believed that floating stations might enable the collection of magnetic data to support guided missiles; others raised warnings about more imminent concerns. Perhaps most significant, the DRB cautioned that research conducted on ice islands might yield seafloor data to support Soviet submarines and that drifting sea ice could facilitate forward operating bases for Soviet aircraft. The RCAF dispelled many of these concerns through aerial surveillance, confirming that Soviet activities on ice islands north of Canada were largely scientific. A series of RCAF flights established that Soviet ice research focused on meteorological, oceanographic, and zoological studies. Nevertheless, ice-related activity in the Arctic Ocean raised the ire of officials and military strategists in the defence department and prompted a full study of ice islands.

With Canada unable to assert legal jurisdiction over ice islands floating into Canadian Arctic waters, the armed services turned to surveillance for security. The DRB suggested that Canada occupy the abandoned T-3 and issue a public declaration stating Ottawa's intent to appropriate and claim sovereignty over any present or future territories available within the Canadian sector.[51] The federal government never issued the proposed statement. Officials wanted to assert control over foreign activities in the Canadian Arctic, but Ottawa was unwilling to upset international relations to achieve this goal. After all, the US operated far more ships in the region than did Canada, and Washington was likely to reject any clear statement from Ottawa regarding ownership of ice islands. Thus, the Canadian position remained cautiously pragmatic, and the defence department maintained close watch of Arctic waters and islands north of Canada through aerial surveillance and ground surveying.

It was US concerns about Arctic security that prompted the DRB's initial field activities in northern Ellesmere Island in 1953 and 1954. Increasingly aware of the strategic importance of the Arctic Ocean, government and military officials in Washington paid close attention to the possible role and importance of ice islands and proposed a joint research venture to their counterparts in Ottawa. Taking advantage of the fact that the US could easily mount such an operation with its search-and-rescue aircraft out of Thule Air Base in Greenland, senior Canadian officials agreed to the proposed bilateral research party, and the two countries initiated a formal investigation of the origins of ice islands floating in the Arctic Ocean above mainland Canada.[52] The agreement required a measure of military and strategic foresight, because the initial discovery of ice islands was intriguing from a scientific perspective but less so from the perspective of security and national defence.

When Hattersley-Smith first travelled to Ellesmere in the spring of 1953, he flew north with a small party and studied the island's north coast and ice shelves.[53] That May, he and Robert Blackadar of the Geological Survey of Canada recovered records left in the Arctic nearly five decades earlier by US explorer Admiral Robert E. Peary.[54] The two scientists found the records in a cairn erected at the peak of Cape Columbia, the northernmost point on Ellesmere Island. Having been airlifted to Alert by the USAF, they were investigating whether ice islands floating in the Arctic Ocean stemmed from the ice shelf on Ellesmere's northern coast. Canadian and US scientists theorized that thirty-five ice islands then recently observed by aircrews flying over the Arctic had been calved by the ice shelf at the top of Canada. "From the summit of Cape Columbia peak we recovered Peary's 1906 records and a piece of the silk ensign which three years later he flew at the North Pole," Hattersley-Smith reported in a message to DRB headquarters sent from the Alert weather station. Peary had departed Cape Columbia in 1909 and trekked over Arctic ice floes, later claiming that he had reached the North Pole.[55] Hattersley-Smith and Blackadar copied the records along with a cache of documents left in 1920 by Danish explorer Godfred Hansen at Cape Aldrich, the northernmost point of North America. The two scientists copied both sets of records for replacement in the cairns and brought the originals to Ottawa, along with the piece of Peary's flag, on their return that August.[56]

Specialists, transport, and supplies provided by the armed services enabled government and civilian scientists to travel north and work in distant parts of the Arctic during the early Cold War. Military forces from both sides of the North American border supported the joint Canada–US expeditions to the Beaufort and Chukchi Seas between 1948 and 1954, enabling oceanographic research in waters previously inaccessible to civil scientists.[57] Between 1954 and 1957, the cruises of the HMCS *Labrador* gave researchers access to the northern reaches of the continent, furthering scientific investigations considered imperative to the military and resource development interests of the Canadian state. The association of scientists and the armed services was mutually beneficial, as the logistics of establishing a scientific research station entirely by air in the Arctic provided useful training for all parties involved. Participating personnel also tested military food, clothing, and equipment in the field, identifying practical issues and making suggestions for improvement.

Operation Hazen

Operation Hazen illustrates the role and value of military sponsorship in supporting and furthering the Arctic sciences in 1950s Canada. Before and during the IGY, 408 Reconnaissance Squadron of Air Transport Command carried out aerial mapping of the Lake Hazen area and northern Ellesmere Island.[58] Expedition personnel used aerial photographs generated by the RCAF to assist field investigations, conduct data analysis, and prepare final reports. Long flights in difficult operational conditions forced pilots and aircrews to learn and adapt on the job, especially considering that the RCAF had no previous experience with landing military aircraft on northern Ellesmere Island. Credit for the first landing on a glacier in the interior of northern Ellesmere Island went to Wing Commander J.G. Showler of 408 Reconnaissance Squadron, who landed a ski-wheel DC-3 Dakota aircraft on the Gilman Glacier, at an altitude of 4,000 feet, in early March 1957.[59] Accompanied by Harwood, Showler also landed on the ice of Lake Hazen, where the light snow cover, relatively calm conditions, and absence of snowdrifts suggested that the sheltered area was ideal for future landings.

DRNL at Fort Churchill served as the main staging point for Operation Hazen, receiving and organizing food provisions, clothing, and general stores during the planning stages of the expedition. Clothing provided by the Canadian Army included olive-drab parkas and windproof pants, Arctic mitts, mukluks, balaclava helmets, and snow goggles. Scientists also used military-issue tents, rucksacks, light toboggans, and thermos bottles, as well as other items and food rations from army supplies stored on base. Sergeant D. Engels, Royal Canadian Engineers, was a valued member of the Hazen expedition team. His practical experience and mechanical knowledge were an asset, especially among a group of scientists who lacked the skills to operate and maintain the ground equipment required to live and work on northern Ellesmere Island.[60]

Government and civilian scientists in Canada knew very little about northern Ellesmere Island before Operation Hazen. Researchers on the British Arctic Expedition of 1875–6 reached the island's north coast but did not travel inland. Adolphus Greely, the leader of the ill-fated Lady Franklin Bay Expedition of 1881–4, had come across Lake Hazen in the spring of 1882.[61] Other expeditions and research teams had visited the area in the early twentieth century, but much of the scientific knowledge about the area derived from observations recorded during the Greely expedition. In late April 1957, Engels and Jim Lotz were on board the first C-119 Flying Boxcar aircraft of Air Transport Command to land on Lake Hazen. A second aircraft landed an hour later, carrying the remaining scientists and equipment. The two Flying Boxcars carried around thirty-five tons of stores, fuel, and equipment to Lake Hazen, along with six scientists and two dog teams.[62] The two aircrews established a base camp and a small settlement on the lake, about one mile from its northern shore and half a mile from Johns Island, where Air Transport Command also flew in a RCAF Shoran crew, who set up a supporting station. The expedition later relocated to the lake's north shore and established Hazen Camp, equipped with several pyramid tents, two prefabricated huts, two vehicles, ration boxes, fuel drums, scientific instruments, and various supply crates.

During the first phase of Operation Hazen, a ski-wheel Dakota aircraft made three landings at 3,500 feet, some twenty-five miles northeast of base camp, where scientists conducted glaciological, glacial-meteorological, geophysical, geological, and survey studies during the summer of 1957.[63] The RCAF also made several parachute drops, delivering additional supplies to the six-member research team. The operational plan called for navy assistance as well, but since HMCS *Labrador* was resupplying DEW Line stations that summer and was unavailable, Canadian officials arranged for the US Military Sea Transportation Service to support Operation Hazen. Lieutenant-Commander James Croal, who served as logistics coordinator and liaison officer for the operation, joined USCGC *Eastwind* at Thule Air Base in Greenland. Under Captain R.F. Rea, the icebreaker steamed north from Thule in mid-August and anchored at Chandler Fiord, some eighteen miles from Lake Hazen and only nine miles from the expedition base camp. The *Eastwind* was the first supply ship to penetrate that far north, although she carried limited cargo. Expecting difficult conditions and restricted sailing, planners cut cargo requirements to a minimum, and around 15 tons of fuel, oil, food, scientific equipment, and other stores remained at Thule for airlift the following spring. Nevertheless, at anchor, two helicopter pilots made several flights, delivering eight men, twenty-six dogs, food, supplies, and equipment for Hazen's winter party.

The second phase of Operation Hazen saw four scientists live and work at base camp through the winter months, obtaining the first continuous meteorological record for an inland station in the Canadian Arctic Archipelago. Occasional RCAF flights airdropped provisions, but the research team had no outside human contact until the end of March. More than six months after the initial summer party departed aboard the *Eastwind*, a USAF C-130 Hercules aircraft landed on the frozen lake and delivered supplies directly. One month later, an eleven-person research team landed and commenced the third phase of Operation Hazen. Major T.L. Hoy, a Canadian Army psychologist, examined the four scientists who had wintered at base camp as part of a study of men in isolation. He determined that they were in "excellent health and fine spirts," according to a report printed in *Canadian Army Journal*.[64]

Phase three of the operation included a series of seismic studies, geological investigations, and ground surveys carried out near the Gilman Glacier and elsewhere on northern Ellesmere Island. During the summer of 1958, a research party made sled journeys over the ice cap of the interior to expand the expedition's geophysical work and continue meteorological observations. Others stayed and worked at base camp, performing archaeological, botanical, geological, and zoological studies throughout the summer. Evacuation assistance arrived in mid-August, when the USS *Atka*, another icebreaker, supplied courtesy of the US Military Sea Transportation Service, carried expedition personnel from Ellesmere to Thule Air Base.[65] Five scientists remained at Hazen to complete their work and secure the base camp for the winter, leaving shortly thereafter via airlift on board an RCAF Canso aircraft. Their departure marked the end of the Canadian-led IGY fieldwork at Lake Hazen, although small research parties carried out additional work in the area during subsequent summers in 1959 and 1960.

Military surveying in the Arctic islands continued into the next decade. In early May 1962, a team of eight soldiers and seven civilian pilots and maintenance personnel flew north to Resolute on Cornwallis Island for a third straight season of field surveying in the Arctic islands. A second team of six soldiers departed Ottawa one month later, travelling to islands in the District of Keewatin, northwest of Fort Churchill. Sponsored and led by the Canadian Army's Survey Establishment, both teams carried out advanced topographical surveys using a newly designed electronic tellurometer.[66] First used to measure distances in the field by army survey teams in 1959, the portable battery-powered microwave instrument significantly increased the amount of area surveyed and enabled military cartographers to produce highly accurate maps of Arctic and subarctic Canada.

Captain H.C. Honeyman, a British Army exchange officer from the Royal Engineers, commanded the survey team at Resolute, while Captain D.M. Matheson of Ottawa commanded the survey team working out of Fort Churchill.[67] The first team set out to survey approximately

Figure 5.2. Operation Hazen field party stand in front of a temporary shelter, Ellesmere Island, ca. 1957–8. Library and Archives Canada / James Patrick Croal fonds / e010771720.

50,000 square miles over a four-month period, covering the Cornwallis, Bathurst, Amund Ringnes, and Lougheed islands, in addition to the Grinnell Peninsula, the western portion of Devon Island, and some of the southern parts of Ellef Ringnes Island. The second team surveyed an area of approximately 20,000 square miles, ranging some 550 miles northwest of Fort Churchill. In addition to the northern summer fieldwork commanded by Honeyman and Matheson, the Army Survey Establishment deployed nine parties in 1962, who used electronic tellurometers to undertake large-scale mapping of several major cities and to revise outdated maps of military camps across the country.[68]

The belief in Inuit cold tolerance persisted well beyond the DRB's acclimatization research projects. On return from the Lake Hazen expedition on northern Ellesmere Island, the DRB's Geoffrey Hattersley-Smith and anthropologist Moreau Maxwell of the Canadian National Museum theorized that an extreme cold wave had driven Inuit out of northern Ellesmere Island 700 years earlier. "Even Eskimos Get Cold," proclaimed one press report, citing the research conducted during the Lake Hazen expedition.[69] Hattersley-Smith and Maxwell reportedly identified thirty-three abandoned Inuit sites near Lake Hazen, including traces of former dwellings and artefacts indicating an earlier presence. As Malcolm Brown and other medical scientists in Canada struggled to understand human responses to cold, perceptions of Inuit as both frost-bound and content in the North loomed large in the southern imagination.

Hazen's Military and Strategic Value

Operation Hazen and the DRB's research and surveying activities on northern Ellesmere Island were possible because both Canadian and US services provided logistical support. Airlift provided by RCAF Air Transport Command was particularly fundamental to the operation; the DRB's scientific research on location would not have been conducted otherwise, not the least because commercial flying companies lacked the technical capability the RCAF could provide. From a knowledge acquisition perspective, it was fortunate that a small group of Arctic specialists in the DRB, with the encouragement and foresight of senior management, were able to exploit the RCAF's airlift potential, and that the defence department absorbed the associated costs and commitments. It is "fair to say that no other Canadian organization would have filled the gap," Hattersley-Smith wrote in 1973.[70]

But did the circumstances justify the means? Was Operation Hazen valuable to Canadian security and national defence in the Arctic? From the perspective of the scientists and military personnel who sponsored or participated in the field activities, the operation demonstrated Canadian sovereignty on the country's northernmost "frontier," and the field stations it established and the research parties it deployed proved to be of great value to the DRB. Although some government officials questioned the expense and military commitment required to complete the operation, the results of the field observations and research carried out on northern Ellesmere Island found their way into unclassified reports and several scientific journals.[71] Hattersley-Smith would later write that in a period marked by significant strides in Arctic research, "the Defence Research Board operations played their part in keeping Canadian scientists within the main stream of advance; in introducing to field work younger scientists who went on to make their mark; and in the exchange of scientific information with other countries on an equal footing through the media of the literature and conferences."[72]

The DRB's fieldwork on northern Ellesmere Island was Canada's primary Arctic research contribution to the IGY's international scientific activities. Indeed, it was the IGY that drove and enabled Canada's first interdisciplinary government-supported field research program in the

High Arctic. The scientists who performed research as part of Operation Hazen benefited greatly from the military support received and from the opportunity to conduct and discuss fieldwork with scientists from other disciplines. Noting the importance of the interdisciplinary fieldwork and the new ideas it generated, in 1958 one US visitor lightly referred to the operation base camp as "Hazen University."[73] The success of Canada's contribution to the IGY generated momentum for additional Arctic research, and Ottawa in 1959 began providing similar logistics and financial support, but on a larger scale, under the aegis of the Polar Continental Shelf Project.[74]

The findings of Operation Hazen had practical value for the Canadian military as well. New research in meteorology, oceanography, glaciology, geology, and biology provided a scientific base for future Arctic studies and also generated geographical knowledge useful to military planners and strategic analysts in Ottawa. Regarding security and national defence, researchers observed and collected data about snow, ice, and terrain conditions affecting land mobility and aircraft operations in the Arctic. They also recorded weather and climate conditions, including oceanographic and sea ice activity useful for understanding the impact of natural conditions on surface and submarine navigation.

Several scientists from beyond Canada took part in the DRB research activities on northern Ellesmere Island, which proved quite advantageous from the Canadian point of view. Government, military, and academic researchers from Britain, the US, and Denmark participated in a range of independent and joint research initiatives to investigate the suitability and geostrategic vulnerability/value of the island and nearby waters for mounting and supporting military operations in the High Arctic. A.P. Crary of the 1954 field party later became Chief Scientist for US Antarctic field programs and provided, according to Canadian records, a valuable working-level contact inside American polar operations.[75] The DRB also developed and maintained a close association with the US Army Snow, Ice and Permafrost Research Establishment, the forerunner of the Cold Regions Research and Engineering Laboratory, which cooperated well with Ottawa in terms of information exchanges and occasional joint research projects.

The DRB also invited scientists and researchers from other government departments and academic institutions to carry out geophysical, geological, and biological studies as part of its operations. Geological survey officers intermittently participated in various field operations between 1953 and 1963, for instance, collecting data in order to assess oil prospects in the Arctic islands. Also, the Entomology Research Institute of the Department of Agriculture conducted research around the Hazen

Camp between 1961 and 1968, making that area one of Canada's most active biological research locations. Although distinct from the unique and deliberate military motivations of Operation Hazen, these secondary and tertiary research contributions generated additional knowledge about Arctic ecosystems and added to Ottawa's information pool about northern Ellesmere Island and the Canadian Arctic Archipelago.

International scientific cooperation in the field was especially advantageous for the DRB's small cohort of Arctic researchers, who participated in Operation Hazen and returned to Ottawa with the experience and connections to continue their work. Reflected G. Hattersley-Smith in his 1973 account of DRB activities on Ellesmere Island:

> Perhaps the most important result of nearly twenty years activity in northern Ellesmere Island has been to maintain within the Defence Research Board a nucleus of Arctic specialists in the fields of geophysics, military geography and logistics, whose knowledge can be tapped on problems of military commitment in the Arctic. By having an active research group the Defence Research Board has ensured profitable exchange of data and consultation with counterparts in the United States and other countries, particularly at a time when the Arctic has assumed a greatly increased importance in political terms. This expertise within the Board assisted rapid adjustment to the recent surge in Arctic activity.[76]

In the early 1960s, following the construction of the DEW Line, at a time when extra attention was being paid to the escalating threat of ballistic missiles, DND gradually withdrew from its ground- and sea-based commitments in the Arctic. DRB operations in the Arctic islands suffered as a result, especially owing to the loss of long-range airlift capability and the eventual disbandment of 408 Reconnaissance Squadron, RCAF, whose crews had carried out special photographic flights on request and were alert to movements of ice islands and changes in Arctic ice conditions. Military geography work on and near northern Ellesmere Island continued, albeit on a smaller scale, as DRB scientists stationed at Tanquary Fiord deployed field parties to conduct exploratory oceanographic studies of nearby waters that were navigable by submarine and possibly surface vessels as well.

Radio Propagation Research and the DRB

Although the advent of intermediate- and intercontinental-range ballistic missiles in the mid-1950s increased the importance of the continental defence partnership between Canada and the US, officials in Ottawa remained committed to North Atlantic security projects undertaken with scientists and military forces from the United Kingdom. British scientists were particularly keen to collaborate with Canadian colleagues as part of a large-scale contribution to the IGY's research activities. In February 1956, DRB officials hosted the Commonwealth Advisory Committee on Defence Science.[77] Scientists from several Commonwealth countries gathered in Ottawa and Toronto and at Fort Churchill to discuss collaborative efforts for advancing the military applications of science. Long-distance radio communication was high on the agenda. Geographically situated in an area of maximum aurora activity, Fort Churchill provided a unique location for scientific research into magnetism and the upper atmosphere.[78] The North Magnetic Pole was an occasional source of interference for ground navigation and radio communication equipment operated by military personnel during training.

During the Battle of Britain, when German bombers attacked daily across the English Channel, Allied radar operators in England noticed an unusual disturbance appearing on the screens of their radar sets.[79] It was speculated at first that radio frequency signals sent out by the Germans were deliberately jamming the British radar network; however, further investigation revealed that cosmic rays beaming from the sun and other solar bodies were causing this phenomenon. It was US physicist and radio engineer Karl Jansky who first detected radio waves emanating from an astronomical object, in August 1931, when he observed radiation from the Milky Way while working at Bell Telephone Laboratories in New York City.[80] The field of radio astronomy expanded after the war as scientists began studying celestial objects at radio frequencies.

Under the stress of magnetism generated by the aurora borealis, both conventional and military radio sets often malfunctioned in northern Canada and the Arctic. Canadian scientists became acutely aware of the disturbance issue during the Second World War when personnel operating in the North encountered a host of technical difficulties. "In addition to the problems presented by the cold and snow of the barrens," wrote Omond Solandt in May 1946, "the magnetic disruptions and the aurora introduce difficulties in navigation, communication and the use of radar that are of primary importance to the Services."[81] Recognizing the importance of short- and long-distance radio communications for soldiers, sailors, and aviators operating in high latitudes, the DRB supported an expansive research program intended to develop special equipment for overcoming the disruptive phenomenon and improving radio transmission and reception in handsets and larger instruments. Aiming to resolve the problem of magnetic interference and equip the armed services with the techniques and instruments for effective and reliable communications, scientists in the DRB studied radio wave propagation and signals transmission to improve the performance of radar and other equipment in northern Canada.[82]

The DRB invested heavily in communications research, seeking ways to improve transmission technologies. Scientists at the Radio Physics Laboratory, one of two research units that comprised the Defence Research Telecommunications Establishment (DRTE) in Ottawa, developed a technique for transmitting messages over long distances by reflecting radio signals off tiny meteors located approximately sixty miles above the earth's surface. Despite the atmospheric issues known to affect regular telecommunications, a team of DRTE scientists led by Peter Forsyth developed a technique called JANET that reportedly enabled the successful transmission of radio signals up to one thousand miles.[83] The *Saint John Telegraph-Journal* reported in December 1952:

> Since 1945, Canada has made her first full-fledged effort to know the Arctic. She has mapped it, manoeuvred military units in it, established air and weather bases and radio stations, [and] wrestled steadily with the mysterious barriers the northern lights or aurora borealis throw up against radio communications. The war on [the] northern lights has finally reached the point where scientists are predicting that within a few years they will be able to send radio signals right through rather than skirting the obstacle.[84]

Scientists, engineers, and technicians at DRTE created one of the largest radio propagation programs in the world, as Edward Jones-Imhotep discusses in his work on the history of ionospheric research in northern Canada.[85] Staff of the Radio Physics Laboratory coordinated a wide network of ionospheric radio and radar research at stations stretching across Canada and into the Arctic. Under Forsyth's direction, radio physicists in the DRB carried out investigations using an antenna and receiver system specially designed to study the behaviour of radio signals.[86] They monitored and charted the behaviour of radio waves in relation to the ionosphere, the outermost layer of the earth's atmosphere, testing and retesting experimental equipment to work out the kinks and improve radio communications in the Far North.

In another area of communications research, scientists from the DRB and Cambridge University worked closely with associates and technical officers of the British Ministry of Supply and the Royal Air Force on methods to improve aircraft signals. Research scientists from Cambridge had developed a system called Single Side Band that enabled pilots operating high-speed aircraft to maintain reliable voice contact with a central control point over distances ranging beyond 1,000 miles.[87] Researchers with the DRB assisted trials involving a system of flights travelling across the Atlantic between London and Ottawa, and then across Canada from Ottawa to Vancouver. Operators at the DRB's Shirley Bay site provided transmitting and receiving services during the trials, which ultimately proved successful.

DRTE scientists and engineers used the term "sweep frequency top-side sounding technique" to describe the employment of a satellite to investigate the structure of the ionosphere's upper levels.[88] Earlier results obtained with ground-based sounding equipment yielded information about the lower levels of the ionosphere only, largely because a dense reflecting layer of the ionosphere hid its uppermost portion from scientific investigation. Ground-based radio waves penetrated through the dense ionosphere but became lost in outer space, returning virtually no information to scientists on the ground. Apart from the Soviet Union, only Canada was able to undertake high-latitude radio propagation research of this sort. No other country experiences aurora conditions on a similar scale, and Churchill is located at an area of maximum aurora activity.

Rockets, Satellites, and the Cold War

The DRB's Defence Research Northern Laboratory at Fort Churchill underwent a series of drastic changes in 1956. As the military and strategic requirement for an Arctic research laboratory declined, authorities and senior scientists in the defence department recast DRNL as a multipurpose research facility for atmospheric research.[89] Located in an area of maximum aurora activity that experienced magnetic and ionospheric disturbances, Fort Churchill was an ideal research venue for the IGY's subarctic program. Besides being serviceable by rail, air, and sea, it could provide accommodations, medical facilities, and advanced scientific research equipment. Recognizing these benefits and features, DRB authorities offered the use of DRNL facilities to the IGY's International Committee, which accepted the offer and selected Fort Churchill as one location in a worldwide network of sounding rocket research ranges.[90] Rocket trials took place at the Churchill Research Range, a joint Canada–US establishment operated by the Office of Aerospace Research of the USAF. The US Army Corps of Engineers constructed the actual rocket range, selecting an open site within driving distance of the military garrison, while the US Army Ordnance Corps and the US Army Signal Corps shared the responsibility for any subsequent research and rocket launchings. The latter two agencies took up shop at DRNL, operating most of their communications, telemetry, timing, and Doppler velocity and position equipment from the Fort Churchill base.

After the 1956 transition, DRNL functioned primarily as a ground-support facility for atmospheric research.[91] Scientists and administrative personnel at the laboratory assisted in the firing of rockets and provided technical support to visiting researchers, both military and civilian. Most of the sounding rockets launched at the range returned information about the aurora borealis, and technical assistants at DRNL operated the ground instrumentation required to observe and record both real-time and delayed data associated with each test.

Besides carrying out a long list of specialized research projects at several locations in northern Canada and the Arctic, employees of the DRB became heavily involved in monitoring and tracking the scientific activities of the Soviet Union in the late 1950s. When the Soviets launched Sputnik, a group of scientists at the Radio Physics Laboratory began developing satellite-tracking equipment.[92] In fact, Canada was one of the first countries to report accurately the orbital track of an artificial satellite – a significant achievement for senior officials in Ottawa at the start of the Cold War space race.

During the mid- to late 1950s, the DRB hosted a series of rocket launches at the newly constructed research range. Initiated through a bilateral cooperation program with the US, the launches tested Canadian equipment on one Nike-Cajun rocket launched by a team of US scientists and combined rocket/balloon (rockoon) instrument packages.[93] Operation Frost Jet – the cold-weather trials of a supersonic anti-aircraft guided missile developed by US Army engineers – took place at Churchill early in 1956.[94] The Royal Canadian Artillery Guided Missiles Trials Troop carried out the tests with US Army technicians, placing the heavy equipment, preparing the machinery and launch site, and firing the actual missile. Participating scientists conducted the trials as a technical performance measure for the missile and ground radar equipment. Research teams also used the site to launch sounding rockets and conduct earth-based observations in various scientific disciplines.

A sounding rocket was a device launched to an elevation above the lower edge of the ionosphere, which is the boundary between the earth's lower atmosphere and the vacuum of space. Scientists launched rockets to 100 kilometres or more above sea level, using special instruments to "sound" the ionosphere from above and collect atmospheric data that were otherwise unattainable. John Chapman and other DRB electronics specialists had attempted to use electromagnetic radiation signals to study ionospheric properties from the ground, but the ionosphere's dense inner core prevented accurate readings and often reflected signals back to earth that did not yield new information. At the time, short- and long-distance radio communications failed regularly in northern Canada because of interference from the North Magnetic Pole. Chapman and his DRB colleagues hoped to create a map of the ionosphere called an ionogram that could be used to predict ionospheric conditions, thus generating useful information for adapting and improving radio emission and reception.[95]

Scientists at the range conducted atmospheric research, launching several sounding rockets to study the behaviour of the aurora borealis and to record meteorological data specific to northern Manitoba and subarctic Canada. Commanded in the late 1950s and early 1960s by Colonel Jerry Flicek, USAF, and staffed by military personnel and civilians living and working at Fort Churchill, the range was a venue for novel and futuristic research: "Here," wrote Colonel Galloway, "on the icy barrens where people live so close to the very, very primitive in many ways, it was exciting to know that the Space Age was also within their gates and that among their associates were men who looked far beyond the frozen horizon which limited the view of most."[96] In the beginning, the US Army was responsible for range operations, but a fire destroyed the main facility in early 1961. As construction workers set out to restore it, the US government adopted a new policy that placed all upper atmospheric research in the hands of the USAF.[97] The new and enlarged launch site reopened on 1 November 1962, thanks in large part to US and Canadian soldiers who assisted the rebuilding efforts.

Canada's participation in the IGY was possible largely because of the resources made available for science through the DRB. Indeed, a substantial number of qualified Canadian scientists had experience directly applicable to some of the IGY's core research areas. Scientists with the DRTE participated in several specialized activities in addition to the rocket and satellite program at Fort Churchill, helping advance international space research efforts. Canada earned an appointment to the initial UN Committee on the Peaceful Uses of Outer Space (COPUOS) in 1958 and has served on it ever since.[98]

After the US launched Explorer 1 that January, Canadian scientists proposed the development of a novel satellite with advanced transmitter/receiver technology: the S-27, or what would later be renamed the Alouette. Zimmerman and the DRB's senior leadership supported this project, which would enable continuous sounding of the ionosphere from above. The satellite was designed to orbit the earth every ninety minutes and transmit signals with ionospheric data, operating like an airborne radio station. DRTE described the project as "a dramatic example of international collaboration in space science and an illustration of the close association between DRB scientists and their NASA colleagues."[99] The US National Aeronautics and Space Administration

(NASA) accepted a Canadian proposal in early 1959, and DRB authorities ironed out an agreement to put DRTE's satellite into orbit using a Thor-Delta rocket, launched from Vandenberg Air Force Base near Lompoc, on the Californian cost, in late 1961.[100] Delays hampered the initial launch, but Alouette 1 departed the earth's atmosphere successfully on 29 September 1962. The launch was a significant achievement in Canadian space history. The technical capabilities of Canadian science were matched at that time only by the superpowers and surpassed those of all other Western nations except the US.[101]

After the IGY, Canada continued to pursue scientific work related to radio technology, the aura borealis, and satellite development. Scientists at the Churchill Research Range conducted the first flight of a Black Brant sounding rocket in 1959. Thereafter, rockets became increasingly important to the continued exploration of the upper atmosphere – research that fit wider Cold War aims of using technology to explore and understand the universe. Lisa Ruth Rand's thought-provoking work on the history of "space junk re-entries" demonstrates that outer space represented an enviro-technical challenge to international order in the Cold War, for example. The nuclear re-entry of the Cosmos 954 satellite, described in chapter 1, "collapsed geographical boundaries and brought far-flung states, communities, and environments on either side of the Iron Curtain into dangerous proximity."[102] Rand likens outer space to other extreme environments such as oceans and the polar regions, calling attention to the global (and extra-global) Cold War struggle for technological domination over nature – a struggle that, in the end, failed. From this perspective, rocketry research had both civilian and military applications. Ottawa funded rocket and satellite programs in part to obtain soundings from passes over northern Canada and learn about the aurora borealis, thereby gaining information deemed useful for developing special equipment to overcome existing communications issues in high latitudes. Research into atmospheric magnetism and radio telecommunications was particularly significant for North American air defence in the Arctic, and the changing character of defence research in northern Canada and particularly at Fort Churchill raised questions about the future of DRNL.

As the debate over nuclear weapons acquisition intensified in Canada during the early 1960s, media outlets in Manitoba speculated about the possibility of nuclear testing at the Churchill Research Range. A US rocket team under the command of Colonel Flicek operated at Fort Churchill, regularly testing weather probes in the northern skies over Hudson Bay. "If high altitude nuclear testing becomes part of their task," commented Winnipeg reporter Fred Cleverley in a front-page article in February 1963, "information gathered at Fort Churchill can contribute to the man-in-space

Figure 5.3. Governor General Georges Vanier waving to a crowd outside RCAF Fort Churchill, 1961. Library and Archives Canada / National Film Board fonds / e011177469.

program and to world-wide control of nuclear weapons."[103] Range activities enticed speculation about nuclear weapons testing for good reason. Canadian and US scientists who fired sounding rockets during and after the IGY determined that the high-altitude air over Churchill was warmer than air at similar heights over White Sands, New Mexico, where the US Army tested rockets and missiles extensively during the early Cold War. Warm air was ideal for achieving the rocket heights thought necessary to safely test nuclear materials. In theory, rockets launched from Churchill would deliver a small-yield atomic bomb for detonation in the lower atmosphere. As terrifying as this idea sounds in hindsight, the looming spectre of nuclear war spurred Western scientists to design and test various fallout scenarios; controlled detonations over Churchill were no exception. Fortunately, the proposed testing never took place.

In this context, scientific efforts to explore and understand the military and strategic value of Ellesmere Island and the skies above Churchill reinforce the argument that Canadian scientists and policy-makers approached northern Canada and the Arctic as a frontier during the Cold War. US missile and aircraft manufacturer Martin Marietta celebrated Antarctica's first nuclear power plant in 1962, advertising the success of a small reactor that supplied power for the 1,000 scientists who worked at the US Antarctic Station McMurdo.[104] Built on the bare volcanic rock of Hut Point Peninsula on Ross Island, McMurdo opened in December 1955 and to this day is the largest Antarctic station. McMurdo typified Cold War–era approaches to building military infrastructures in relatively remote regions of the world with harsh or extreme environments, where support ships, submarines, aircraft, and other military hardware could be powered by a small nuclear plant. As Julia Herzberg, Christian Kehrt, and Franziska Torma explain in their work on the history of extreme climatic environments, the "hubris of employing technology to overcome obstacles of ice and snow on a global scale, and becoming independent of seasonal weather and climate, is characteristic of the dominant attitude toward nature during this era."[105] Their research and the growing body of international literature on the environmental history of the Cold War illustrates that robust military technologies and federal financial support for expensive research expeditions drove attempts to conquer the poles during the mid-twentieth century.

Ottawa's military approach to northern Canada and the Arctic during the first half of the Cold War era has striking parallels to the wider history of techno-military engagements with the polar regions. The Arctic and Antarctica represented "natural laboratories" for military research and development; state-driven goals on both sides of the Iron Curtain motivated new pursuits for scientists and soldiers working under extreme climatic conditions. Acquiring useful military knowledge about ice, snow, and topography was important to strategists and military planners in Washington, Moscow, and Ottawa. Technological attempts to control remote and hostile environments mattered as well, although the military and geostrategic significance of such pursuits often outweighed any scientific value, at least from the state's perspective. Consider the history of state-funded research to control weather in the US: domestic attempts to manipulate and deploy precipitation for agricultural and economic purposes showed potential for military applications abroad.[106] Canadian scientists did not attempt to control and weaponize the atmosphere during the Cold War, but experts in the DRB flirted with biosphere science and attempts to alter the Arctic environment for social, economic, and military purposes.

In fact, the possibility of deliberately changing the climate in northern Canada and the Arctic fascinated DRB scientists in the late 1960s. Operating under the patronage of the US Department of Defense, in 1967, NASA issued contracts to industrial corporations engaged in satellite R&D. NASA sought out companies developing satellite technology for military purposes with the aim of using the latest technology to assist US military operations during the Vietnam War. Under the code name Project ABLE, one program attempted to create satellite reflectors capable of lighting the Vietnamese jungle at night to expose enemy soldiers and positions. Word of the idea spread quickly, and scientists in Canada proposed using the Churchill Research Range to develop similar technologies for use in northern Canada and the Arctic. DRB scientists wanted to develop and use satellite space reflectors to concentrate sunlight on the North and initiate a gradual warming of Canada's northern climate, supposedly for cultural and economic development. The proposed idea received significant attention in the local Churchill newspaper: "If, by means of Space Reflectors the North could be warmed and the hours of daylight increased, Canada could truly open up her North to absorb easily a hundred million more inhabitants, making her a major world power, [and] turning her Arctic Wilderness into a veritable Northern Oasis."[107]

The futuristic promise of satellite-reflection technology never came to fruition, but the failed idea sparked a search by Canadian defence scientists for a solution to the imagined challenge of northern amelioration. In 1968, the DRB's E.R. Hope translated an article published by Soviet scientist P.M. Borisov in *Bulletin of the Atomic Scientists* that detailed an extensive plan to warm the polar latitudes. Borisov proposed the construction of a fifty-mile-wide dam across the Bering Strait, followed by a mass drainage of the Arctic Ocean into the Pacific, to create a 35-degree rise in temperature of the polar latitudes within three years.[108] The magnitude of Borisov's proposal was so great that even the Soviets posited that it would require the combined resources of the Soviet Union, the US, Canada, and perhaps other supporting countries. In response, Canadian scientists suggested a different, more cost-effective approach. Scientists in the DRB proposed using thermonuclear reactors to heat the incoming tide of Hudson Bay and create an artificial "Gulf Stream effect," thereby inducing a controlled temperature increase to warm the Arctic Ocean and melt the polar ice cap.

176   Frontier Science

In an era before climate change in the Arctic came to represent international concerns over global warming, many Canadians welcomed the prospect of climate amelioration in the North. In a speech to university students in Winnipeg in 1968, former prime minister John Diefenbaker claimed that the North would remain unpopulated until scientists found the means to overcome Canada's cold climate. Echoing the DRB's proposal, Diefenbaker cited plastic-domed cities and the nuclear warming of Hudson Bay tides as enviro-technical possibilities for the deliberate amelioration of northern Canada.[109] As a by-product of this idea, DRB scientists claimed that the reactors would produce vast amounts of affordable electricity for northern Canada, where heating costs were typically high and power was in short supply. The North would not be the sole beneficiary, however. In theory, amelioration of Canada's northern climate would result in greater precipitation across the central Canadian prairies and the dry US midwest. Although the idea never came to pass, here too, the prospect reinforces that dominant Cold War–era attitudes towards control of nature factored into the thinking of defence scientists and political leaders in Ottawa. Northern Canada and the Arctic continued to represent a research and extractive frontier for outsiders and their government sponsors.

Conclusion

As tensions grew between the US and the Soviet Union during the early Cold War, the two superpowers established semi-permanent stations on ice islands floating north of Canada and Ottawa increasingly viewed the Arctic as a potential geographic point of vulnerability.[110] Expressing concern over the increasing US and Soviet presence in the Arctic, military and defence officials prepared for imminent dangers across the vast expanse of Canada's Arctic Archipelago. Ottawa viewed the ocean waters north of Canada as part of the Canadian Arctic sector, yet Canada had no legal claim to international waters, and the proximity of foreign expeditions posed a real threat to Canadian security in the North. While aircraft and transportable units offered some protection against land-based incursions, military and defence officials worried that large ice floes could facilitate unwelcome activity.

During the first three decades of the Cold War, as the Arctic grew in strategic significance, security considerations about the region affected Ottawa's approach to military research in northern Canada. Government authorities in Ottawa and Washington believed that adequate defence in the North meant preventing the Soviets from establishing a permanent or semi-permanent military presence, be it on land or on floating sea ice. Joint flights undertaken by the RCAF and the USAF over the Canadian Arctic tracked ice floes that might serve as landing strips and refuelling stations for Soviet bombers, which lacked the range for the round trip from the Soviet mainland to the cities of North America. Operation Hazen extended and expanded DND's efforts to understand the geographical challenges and possibilities of Arctic Canada from a military perspective, combining glaciology, meteorology, and ground surveying to generate topographical and geophysical data for the armed services.

The focus on the Arctic in this period is perhaps most evident in the proliferation of maps oriented over the North Pole.[111] Air-age globalism underscored how close to North America the Soviet Union in fact was; clearly, America was vulnerable from the north. Thus the Arctic became a frontier space of both strategic and scientific importance – a laboratory for intellectual pursuits with implications of local and global significance. But the IGY was neither the culmination nor the sole illustration of high modernism and enviro-military science during the early Cold War. As the nuclear arms race intensified in the mid- to late 1950s, scientists the world over paid increasing attention to the rising occurrence and health hazard of radioactive contamination stemming from atmospheric weapons tests. As northern Canada once again became a focal point for the Cold War sciences, Ottawa developed new concerns.

6 Nuclear Fallout and the Northern Radiation Study

In January 1963, a Whitehorse newspaper reprinted a short article that first appeared in *The Beaver*, a northern affairs quarterly published by the Hudson's Bay Company.[1] "Is A-Test Fallout Poisoning Caribou – And the Eskimos Who Eat Them?," read the headline, altered from the original title written by author and biologist William "Bill" Pruitt of the University of Alaska–Fairbanks.[2] Pruitt described caribou and reindeer as "hot spots" of radioactive contamination in the northern regions of Canada, Alaska, Scandinavia, and the Soviet Union. Atmospheric nuclear explosions, he suggested, had spread radionuclides over the circumpolar region, contaminating lichens and sedges, two primary caribou foods. His concern was that caribou were consuming vegetation contaminated by radioactive fallout and thus passing potentially harmful amounts of radioactivity to the unsuspecting humans who killed and ate northern game as part of their regular diet. "For example, what few data have been gathered on man indicate that all people who eat much caribou or reindeer meat have higher whole-body radiation counts than people who do not eat caribou or reindeer," Pruitt wrote, claiming that the entire northern food chain was prone to radioactive contamination.[3] As evidence, he referenced research studies showing increased radiation levels in Indigenous people from Alaska and Sweden – results that he claimed justified a systematic study of radioactivity in the Canadian North.

Pruitt's article captured the attention of German-born doctor Otto Schaefer, who lived and worked in northern Canada. Schaefer practised medicine at the Charles Camsell Indian Hospital in Edmonton, where he directed the Northern Medical Research Unit.[4] He obtained this position in 1962, after working ten years on rotation at Aklavik, Pangnirtung, Whitehorse, and other locations in the Northwest Territories and Yukon. Schaefer spent several months each year travelling in northern Canada, gathering information about the general health of northern residents and administering vaccinations and medical services where required. He worked with Inuk interpreter Etuangat Aksayook, who managed their supplies and dog teams, navigated their medical rounds, facilitated their meeting and sleeping arrangements with Inuit families, cooked their meals, and taught Schaefer and his wife Didi to converse in Inuktitut.[5] Schaefer became an advocate for Inuit health and expressed concern about the reported risk of radioactive fallout in the North. In a January 1963 letter to P.G. Mar of the Department of National Health and Welfare, he urged medical authorities in Ottawa to investigate the problem.[6] Schaefer proposed an experimental research study involving the collection and analysis of northern vegetation, animal meat, and human bone. A systematic study of samples obtained from northern Canada, he suggested, would indicate the actual public health threat of radioactive fallout.

Schaefer's proposed study hinged on the paternalistic, settler-colonial practices of the post-war state. He confirmed the commitment of his medical staff to "send in future any bones removed at operation or autopsy of Eskimos, with a detailed history in regard to locality they come from and prevailing diet." He also proposed asking doctors at Frobisher Bay (Iqaluit), Inuvik, and Churchill, as well as all institutions receiving Inuit patients in southern Canada, "to try to obtain some parts of skeletons of deceased Eskimos." Schaefer emphasized the value of bone samples obtained from Inuit living near the west coast of Hudson Bay and those connected with the reindeer herd east of the Mackenzie Delta:

> It was my experience in the North that there was no difficulty whatsoever to obtain permission for [postmortems] of Eskimos, as they are most eager to know [about the effects of radiation] themselves. Because of this, and indeed until not long ago, they did crude forms of [postmortems] themselves, and still are doing them with their dogs. By the way dogs and also reindeer carcasses would also be easily obtained from Inuvik.[7]

Operating under the belief that Indigenous peoples in northern Canada and the Arctic required federal support and protection against the escalating danger of nuclear fallout, Schaefer expressed no hesitation about collecting human bone samples for an investigative study.

Officials in the health and welfare department were intrigued by Schaefer's letter. In response, Peter Bird of the department's Radiation Protection Division (RPD) told Schaefer about a nationwide fallout study undertaken by scientists in the federal government. Data compiled from precipitation and soil samples at Resolute, Coral Harbour, Inuvik, Fort Churchill, Whitehorse, and Yellowknife had indicated lower fallout counts than the national average. Nevertheless, owing to the suggestion that northern vegetation concentrated fallout to a degree not found in other regions of the country, Bird confirmed the RPD's interest in a systematic study of Inuit, whose diet presumably consisted of plant-eating caribou. Radiochemical analyses conducted on specimens from Indigenous people in Alaska and northern Finland had indicated fallout concentration levels higher than found in southern populations, he explained. He then outlined a detailed research plan for northern Canada. Concerned that indiscriminate sampling was an ineffective method for assessing radiation hazards in the North, Bird told Schaefer that representatives of the Canadian Wildlife Service and Northern Health Services had surveyed human and animal population distributions as part of the nationwide fallout study. The next step was systematic sampling of humans.

Nuclear Fallout and the Global Response

Otto Schaefer's letter arrived in Ottawa at an opportune time. The scientific reports about increased fallout levels in Alaska and Finland spurred Bird and his RPD colleagues to extend the nationwide fallout study and devise a special program for northern Canada. Established in January 1950 to study and safeguard the health of radiation workers, the Health Radiation Laboratory of Canada's Department of National Health and Welfare – later renamed the RPD – expanded its efforts after a series of hydrogen bomb tests in the Pacific Ocean in 1954.[8] The first successful test of a thermonuclear device had taken place two years earlier, in November 1952, during Operation Ivy, when US scientists detonated "Mike Shot" on the small Pacific island of Elugelab at Enewetak Atoll in the Marshall Islands.

The hydrogen bomb was the most critical technological development of the Cold War because it drastically increased the destructive capacity of nuclear weaponry. Aided by first-rate intelligence obtained by agents in Britain and the US, the Soviets developed their own nuclear bomb after the Second World War, and Moscow allocated abundant resources to advance nuclear weapons research in its own laboratories and those of its satellite states.[9] Public health officials in Ottawa were aware of the increased scale and intensity of nuclear and thermonuclear testing, and they worried about the spread of fallout and radioactive debris over Canadian territory and large segments of the civilian population. Developments in fallout detection equipment had enabled large-scale studies in Britain and the US that showed increasing radiation exposures on a per capita basis, further heightening concerns in Ottawa and prompting additional research.[10]

Historical geographer Jonathan Luedee documented the RPD's northern fallout monitoring program in a 2021 article in *Journal of the History of Biology*. Using Gabrielle Hecht's concept of *nuclearity*, which describes the techno-political processes and rhetoric tied to nuclear knowledge production, Luedee argued that "the presence of radionuclides in the flesh and bones of caribou did not mean that northern Canada would be treated as a nuclear space."[11] Scientists associated with the RPD struggled to understand the problem of radioactivity in northern Canada using pre-existing frameworks, methods, and technologies. In turn, government officials looked to researchers, wildlife managers, and northern residents to investigate the health risks of radioactive fallout, producing and using language that, in Luedee's estimation, defined northern Canada as a "nonnuclear space."[12]

The logic employed by scientists and officials from southern Canada to address the presence of radionuclides in the North reinforces the view that public health overlapped and intersected with settler-colonialism in the post-1945 period. By the mid-1960s, the RPD's expanded research program included systematic studies of the public health risks associated with effluents from nuclear reactors, natural background radiations, and overexposure to medical X-rays. The growing fear of increased exposure to radiation among the Canadian public motivated an initial nationwide fallout study, which was designed to assess the risks of radionuclides and forecast future trends for the country. Reports about high concentrations of fallout in the North reaffirmed the study's importance and convinced the RPD to investigate the food-chain theory postulated by Pruitt and other scientists familiar with the hazardous effects of atmospheric nuclear testing. In this way, health officials in Ottawa leveraged existing medical networks and policies in Canada to access human specimens for biochemical analysis. The perceived burden of government administration in the North extended beyond individual care to whole-of-population health. Inaction on radioactive fallout came to be a non-option as the scale and intensity of nuclear testing escalated.

As historian Toshihiro Higuchi notes, the myth that radioactive fallout posed a negligible public health hazard collapsed when the US Atomic Energy Commission (AEC), the civilian successor to the Manhattan Project, opened the Nevada Proving Grounds in January 1951.[13] Early trials in Nevada created a public relations nightmare for nuclear officials, not from any mass hysteria, however, but from widespread radiological readings recorded across the US. Sensitive X-ray and Geiger-Müller counters detected even slight changes to radiation levels in the environment, including in the air filters at Kodak's Rochester plant, which began intercepting radioactive particles in the wake of atomic trials conducted at the Nevada site.[14] After Kodak threatened to sue the US government for potential damages, the AEC recruited Merril Eisenbud, a toxic materials expert with extensive experience investigating industrial regulations, to address complaints about nuclear debris and the spread of contaminated air over continental North America.

As director of the Health and Safety Laboratory in New York, Eisenbud created a continent-wide fallout-monitoring network with the support of AEC facilities, universities, the US Weather Bureau, and the Canadian government. By 1952, around 100 ground stations across the US and Canada were serving as collection and analysis centres for airborne dust and radioactive particles. Traces of radioactivity drifted northeast after the first series of nuclear trials in Nevada, appearing in snow that fell over eastern regions of Canada and the US. Part of the plume from one test explosion even collided with a storm system moving eastward and deposited radioactive debris on the ground in areas stretching from Chicago to Quebec City.[15]

Radioactive fallout knew no political borders, as evidenced in the early 1960s when Canadians reacted to public health reports out of the US that connected wheat grown in Minnesota with radioactive contamination. After the passage of the US Atomic Energy Act in 1954 opened to the door to the privatization of nuclear industries, the Rural Cooperative Power Association constructed an atomic power plant at Elk River, some forty miles northwest of Minneapolis.[16] Completed in 1962, the Elk River Station generated nuclear power to support rural development, but early studies of fallout and residual radiation from the plant showed alarming concentrations of radioactivity in nearby sources of drinking water, as well as in milk and food products from North Dakota and Wisconsin. News of the potential health risk quickly jumped the border, and this intensified public calls for a nuclear test ban in the prairie provinces that had elected Prime Minister John Diefenbaker and many of his Progressive Conservative allies in 1957.[17]

When the uproar over the unusual contamination spread north into Canada, Ottawa found itself in the difficult position of having to choose between the security needs of its Western partners in NATO and the disproportionate burden of increasing levels of global fallout. As historian Daniel Heidt explains, Howard Green, the external affairs minister in Diefenbaker's government, struggled with the nuclear file.[18] An ardent anti-communist, Green had long supported bilateral and multilateral military cooperation with the US and other Western allies. But the 1958 report of the UN Scientific Committee on the Effects of Atomic Radiation (UNSCEAR) – established in 1955 – and subsequent concerns about contaminated milk and wheat led him to believe that Canadians were particularly vulnerable to the residual dangers of nuclear trails and atomic warfare. Green acted in October 1959, introducing a UN resolution that called on member-states and supporting agencies to advance international cooperation in fallout research. When he briefed Cabinet on his proposal, he suggested that contamination of Canadian wheat would drive foreign buyers away and cause irreparable damage to the national economy.[19] At a time when the Americans, the British, and the Soviets were in Geneva talking about a test ban, the Canadian resolution passed unanimously on 2 November, illustrating the growing international concern over the public health risks of radioactive fallout.[20]

Circumstances worsened on a different front for Canada in late August 1961, when news reports out of Moscow confirmed that the Soviets had resumed nuclear trials. In a reversal of Cold War diplomacy, the US and Britain, two countries with large nuclear-testing programs in the 1950s, turned the tables on the Soviet Union, using the fallout argument to bring world opinion to bear on their communist rival.[21] The Kremlin might have believed that the Soviet Union had the sovereign right to conduct nuclear trials and release radioactivity upon the globe, but the Western powers strongly disagreed. Fearing the resumption of Soviet tests and any residual response from the US, Canadian officials once again took their concerns to the UN, introducing a new resolution calling on the UNSCEAR to urge the international exchange of fallout data. Canadian representative Paul Tremblay argued that Canada, by virtue of geography and its proximity to the Soviet Arctic, risked receiving heavy concentrations of radioactive fallout.[22] Although some US officials took issue with the resolution because it singled out fallout to the detriment of Washington's publicity campaign against Moscow, it overwhelmingly passed in the General Assembly on 27 October, and with opposition solely from the Soviet Bloc, Cuba, and Mongolia.

The Nuclear Arctic

Concerned about the increased risk of exposure for the many people living downwind of the landlocked Semipalatinsk Test Site in present-day Kazakhstan, the Soviet Union's primary testing venue for nuclear weapons, Nikita Khrushchev diverted as many planned tests as possible to the Novaya Zemlya archipelago in the eastern Soviet Arctic.[23] Upon declaring the archipelago a nuclear test site in July 1954, the Soviet state forcibly deported 536 Indigenous Nenets and all other inhabitants of the island from their homes and villages. The military detained the deported residents about halfway up the archipelago in the village of Lagernoe.[24] Denied the right to hunt or work, the detainees slept in windowless huts, enduring harsh treatment for two years before the state completed their deportation to the Russian mainland.

Around one fifth of all Soviet nuclear trials took place on or directly above Novaya Zemlya, including the October 1961 detonation of a 50-megaton-yield thermonuclear bomb popularly known as Tsar Bomba, the world's most powerful explosion of a nuclear weapon in history.[25] The detonation was conducted close to the abandoned settlement at Lagernoe on the southern tip of the island, and the residual shockwave produced from the immensely powerful blast circled the globe three times, dispersing radioactive fallout worldwide and shattering windows as far away as Finland and Norway. By comparison, the largest US test was the 15-megaton Bravo weapon detonated in March 1954. In the end, the Soviets conducted 224 nuclear detonations on Novaya Zemlya, including 88 atmospheric tests.[26] Nuclear historian Robert Jacobs writes that several studies have since detected extensive radionuclides permeating the region's ecosystem.[27] The seas surrounding the archipelago also served as military dumping sites for nuclear waste, and climate change has melted glaciers containing deposited fallout, further contaminating the immediate region. The militarization of the Soviet Arctic extended beyond nuclear trials, though. The logic of mutual assured destruction also prompted Moscow to position nuclear weapons and nuclear-powered ships and submarines as close to the North Pole as possible, thereby strengthening the retaliatory capacity against a possible strike from the other side. By the 1980s, as Charles Emmerson observes, the Arctic Ocean had become a "Soviet Nuclear Bastion."[28]

The aggressive Soviet atmospheric testing program concerned public observers the world over. When Khrushchev first announced the planned test of a 50-megaton bomb, US and British officials quickly dismissed what would become the world's largest nuclear detonation as a propagandist bluff by the Soviet leader. A worldwide fallout scare ensued nonetheless, leading officials in Canada, Denmark, Iceland, Japan, Norway, and Sweden to introduce a multilateral emergency UN resolution calling on the Soviet Union to refrain from conducting the bomb test for the sake of all countries located in the direct path of the anticipated fallout.[29] The planned thermonuclear trial also received universal condemnation in the Western news media, which framed the bomb as both a danger to international peace and a crime against humanity. "If ever aimed at Toronto and detonated in the lower atmosphere," warned an article in the *Globe and Mail*, "the fireball would shoot flames over an area that would take in St. Catharines, Niagara Falls, Hamilton and Oshawa."[30] According to Higuchi, the "*Washington Post* even ran an obituary notice on its editorial page for the 'unnumbered hundreds of thousands' who would ultimately die because of the Soviet fallout."[31] While the Tsar Bomba was ultimately one of the "cleanest" nuclear weapons ever detonated, the explosion still generated vast quantities of radioactive carbon that lingered in the atmosphere and contaminated the earth well into the future.[32]

Nuclear weapons testing and military build-up in the Arctic created a highly polluted and radioactive environment. Fallout from trials at Novaya Zemlya contaminated the landscape, and large quantities of nuclear waste ended up in the Arctic Ocean. Several nuclear reactor accidents occurred among Soviet ships based in the port city of Archangelsk, and nuclear submarines often broke down at sea.[33] On the opposite side of the North Pole, the logic supporting nuclear build-up in the Arctic was identical: strategic planners in Washington also believed that positioning weapons and delivery systems close to the dividing line was crucial to ensuring first- and second-strike capability in the event of a nuclear confrontation. As a result, the environmental and social consequences of Arctic militarization and associated projects were often similar on the two sides of the Pole.

Senior US officials and military leaders in the Pentagon developed a deep "fascination with the Arctic" during the early Cold War, as historian Ronald Doel has shown in detail.[34] Although much of this interest spanned the Arctic, Alaska, not surprisingly, was a focal point for government and military activity.[35] Military build-up was not the only form of intervention in the region, however. Several high-profile projects and economic development schemes also characterized state and private activity in various locations. One such scheme was an extraordinary attempt to geo-engineer the Arctic landscape using nuclear explosives in the late 1950s. Under the code name Project Chariot, physicists at the AEC led by Edward Teller wanted to use nuclear power to create a deepwater harbour at Cape Thompson, near Point Hope. "While ostensibly peaceful in purpose," writes polar historian Adrian Howkins, "this project might be seen as another attempt to demonstrate the superiority of the capitalist system over communism in the conquest of nature."[36] The plan ultimately failed, owing predominantly to significant opposition from Inupiat communities, activists, and dissident scientists.

Northern Canada in the Atomic Age

Arguably, the convergence of nuclear fission and war presents the single greatest human challenge to have emerged from the scientific and international developments of the twentieth century. In the late 1920s and early 1930s, New Zealand–born British scientist and Nobel Laureate

Ernest Rutherford of the renowned Cavendish Laboratory at Cambridge was the world's leading nuclear physicist. Driven by curiosity, he conducted fundamental research and carried out physics not to produce material outputs but rather to advance human understanding of the universe. "Through his enormous energy and penetrating intellect," writes historian Graham Farmelo, "Rutherford prosecuted what you might call a kind of romantic science."[37] This open approach to science transcended international borders and politics. Rutherford disdained industrial applications – and money, unless those funds supported his scientific research. In 1938, however, the year after Rutherford's death, with war about to break out, physicists in Berlin discovered nuclear fission. The harnessing and possible release of nuclear energy as a weapon instantly eroded any lingering romanticism among the world's nuclear physicists. Rutherford and his contemporaries had overthrown Newton's laws and established the field of quantum mechanics; now, Robert Oppenheimer and the physicists of the Manhattan Project and other Allied nuclear programs began conducting research in service to the state.[38]

After Hiroshima and Nagasaki, "Big Science" consumed post-war physics. The era of fundamental research had passed, replaced by large-scale, government-funded research projects that showed promise for producing military applications. British scientists William Penney, Christopher Hinton, and John Cockcroft – the "Atomic Knights" – transitioned from fundamental research to applied weapons work.[39] In 1946, Cockcroft helped found the Atomic Energy Research Establishment at Harwell in Britain. On the other side of the Atlantic, the opportunistic Edward Teller supplanted Oppenheimer as the US government's leading adviser on nuclear weapons.[40] In the early post-war years, after leading the Manhattan Project and witnessing the destructive power of the atomic bomb, Oppenheimer opposed development of the hydrogen bomb and endured shaming and humiliation at the hands of a state tribunal that questioned his patriotism.[41] Teller, by contrast, advocated further atomic research and urged Washington to fund and develop more powerful nuclear weapons to counter the Soviet threat.

Although Canada never developed its own nuclear weapons, military and defence officials in Ottawa actively supported the US and British programs during the early Cold War. Canadian reactors at Chalk River in the upper Ottawa Valley supplied the US with natural uranium and recycled plutonium for military research purposes.[42] The history of Canada's involvement in British efforts to develop and test atomic bombs is less well known. In cooperation with Penney, by then the British government's chief superintendent of armament research and head of atomic bomb production, the DRB's C.P. McNamara investigated various locations in Canada that might be suitable for testing British nuclear devices. Together, they decided on northern Manitoba, detailing their reasons why in a report titled *The Technical Feasibility of Establishing an Atomic Weapons Proving Ground in the Churchill Area.*[43] The top-secret report identified an area 100 kilometres southeast of Churchill at the mouth of Broad River as a testing site for Britain's first operational nuclear bomb, a 25 kiloton weapon called the "Blue Danube."[44]

McNamara and Penney envisioned using the site as a proving ground for first- and second-generation nuclear weapons, proposing a series of twelve live-fire detonations between 1953 and 1959. Scientists with the British Atomic Weapons Research Establishment estimated that each explosion would contaminate a 500-metre area and render the immediate location unusable for future tests.[45] Canadian officials were not keen on using a different site for each detonation, although historians disagree about whether Ottawa intended to authorize the proving ground.[46] The British ultimately decided against the proposed site because of northern Canada's cold and disagreeable climate, turning their attention instead towards the temperate Montebello Islands off the Pilbara coast of northwestern Australia.

Setting aside the debate over authorization, the proposal to establish a test site for atomic weapons near Churchill indicates the degree of naivety and ignorance that underpinned scientific approaches towards northern Canada during the early Cold War. Consider this disturbing and insensitive report written by McNamara and Penney:

> An explosion on or near the ground at Broad River at a time when the wind lies in the direction between southwest and northwest will not contaminate any area of any importance. The worst that can happen is that at Cape Tatnam [near York Factory on the western shore of Hudson Bay], there will be a very slight contamination due to fall-out, but the risks of this are acceptable, having regard to the very small probability of there being any fall-out contamination at tolerance level combined with the very few inhabitants in the area. (The two or three people in the area could be moved for about one week.)[47]

McNamara and Penney used data from earlier nuclear trials conducted by the US at the Nevada Test Site to draw conclusions about the potential health and environmental hazards, claiming that the proposed blast area would be "lethal for a day or two but a rapid decay of radioactivity will occur." Using callous language to justify their choice of location, the two scientists described the proposed Broad River testing site as a "waste land suitable only for hunting and trapping." The authors also claimed that nuclear trials would leave the nearby town of Churchill and adjacent military base unaffected, arguing that the prevailing winds would carry any residual fallout away from the inhabitants and drinking water of the area. They acknowledged the possibility of contamination travelling as far as 160 kilometres downwind but stated that the trials would neither affect the Beluga whale fishing near the Churchill River nor "contaminate any area of any importance."

Secrecy would be fundamental, the report declared. "Every effort must be made to keep secret the nature of the trial before the event," the authors wrote, emphasizing the need for security and concealment. "Once detonation has occurred, there will be little hope of keeping secret the fact that an atomic explosion has taken place. Some cover name must be invented to explain why men and equipment are being taken into the Base, but if the Base is a Research Station, such as Churchill, it seems possible that no special attention will be called to the preparations."[48] Considering the extensive efforts that Canadian officials took to downplay and dispel Soviet concerns about military activity at Fort Churchill in the early 1950s, it is not surprising to see anxieties about nuclear secrecy reflected in the words of McNamara and Penney.[49] For senior military leaders and decision-makers in Ottawa and London, protecting North Atlantic security meant hiding the nuclear trials from the Kremlin and the public.[50]

If the top-secret report prepared by McNamara and Penney is any indication, Canadian and British officials did not show the same concern about protecting the ecosystems of northern Manitoba and Hudson Bay. Without question, any amount of nuclear testing near Churchill would have done drastic short- and long-term damage to the region. A single detonation would have decimated all flora and fauna near ground zero, with residual fallout from the blast blanketing large parts of northern Manitoba and Hudson Bay. The proposed proving ground was on subarctic permafrost topped with boggy muskeg, small lakes, and swamps.[51] Given the poor drainage of the Broad River watershed and the relatively dry climate and icy conditions, the testing site would have absorbed and trapped radioactive particles for much longer than the comparable trials conducted in Nevada. Any radionuclides not carried away and dispersed by the prevailing winds would settle and sit undisturbed, contaminating the area's shallow and fragile ecosystem for years.[52] Moreover, mosses, lichens, and other northern vegetation concentrate fallout to a degree not found elsewhere in Canada because of the dehydrated soil, an ecological fact that radiation scientists would be unaware of until the late 1950s.

As John Clearwater and David O'Brien observe, the remedial activities of the US military provide some indication of the effort required to clean up radioactivity in a northern location.[53] On 21 January 1968, a

USAF B-52 bomber carrying four Mk-28 thermonuclear weapons crashed seven miles from Thule, Greenland. The bombs detonated on impact and dispersed radioactive debris over a large area. After the crash, the US military excavated more than 6,700 cubic metres of contaminated ice and snow. Service personnel packed the contaminants in shipping containers, and the military transferred the excavated materials to the US for burial as low-level radioactive waste. There is no evidence that Ottawa or London planned to remediate the proposed proving ground near Churchill in a similar way.[54] This same approach characterized US, British, and Canadian research involving biological and chemical warfare agents. Military leaders and scientists from all three countries had collectively disregarded the negative environmental consequences of those agents when planning and conducting experimental trials with chemical simulants.

Radiation Science and Fallout Research in Canada

Scientific reports published in the late 1950s motivated Canadian authorities in the Radiation Protection Division to expand the nationwide fallout study (described at the start of this chapter), with a focus on northern Canada and the Arctic. In 1958, Norwegian scientist F.T. Hvinden reported higher concentrations of strontium-90 (Sr-90) in reindeer bones than in the bones of sheep grazing in the same area of Norway. That same year, botanist Eville Gorham of the University of Toronto collected and dried samples of lichen, moss, and angiosperm while working with the Freshwater Biological Association in Ambleside, United Kingdom.[55] Using samples from Britain's Lake District, Gorham determined that mosses and lichens had absorbed high accumulations of radioactive fallout. Lichens obtained moisture and nutrients directly from the air rather than from soil, a fact that Gorham used to explain the high concentrations of radiation he encountered in his research.

Two years later, in the fall of 1961, the International Atomic Energy Agency (IAEA) organized a meeting of scientists studying radioactive contamination in northern Europe and Scandinavia.[56] Leading scientists took part in the first of a series of annual meetings concerning the environmental and biological effects of nuclear fallout. Biologists and physicists working on the various aspects of radioactive contamination presented research findings and discussed possible ways to contain the spread of radionuclides and reduce the risks to public health. Similar discussions took place across the Atlantic as scientists in the US turned their attention towards Alaska; this eventually led Bill Pruitt and his colleagues to shine a spotlight on the Canadian North.

US radiation science expanded in the early 1960s as scientists explored peaceful uses of nuclear fission. National research initiatives such as Project Plowshare suggested that nuclear energy offered untapped benefits for the medical profession, earth and ice removal, and the power industry. One research initiative conceived under Plowshare was Project Chariot, a 1958 proposal by the AEC to detonate a series of nuclear devices on Alaska's northern slope in order to construct an artificial harbour near Point Hope (discussed earlier in this chapter). A grassroots protest led by sceptical scientists, concerned conservationists, and a community of affected Inupiat grew into a coordinated environmental movement against the planned explosions, and the AEC put the project on hold in 1962.[57]

Research conducted as part of Project Chariot sparked concerns about radioactive fallout in Alaska. During the first year of the project, scientists from the US government's Hanford Laboratory in south-central Washington State found high concentrations of radionuclides in lichens gathered from the proposed testing area at Cape Thompson. Established in 1943 as part of the Manhattan Project, the research facilities at Hanford housed the world's first full-scale plutonium production reactor.[58] Scientists used plutonium manufactured at the site to develop the first atomic weapons, including the bomb detonated over Nagasaki at the end of the Second World War. Hanford researchers also studied samples of caribou meat obtained from the Chariot test site. Samples supplied by Bill Pruitt and his colleague Peter Lent led to a published report documenting the impact of radionuclides on local vegetation, birds, and mammals. The report included data on levels of Sr-90 and cesium-137 (Cs-137) in samples of thirty-five Alaskan caribou. In May 1963, a different group of scientists collected caribou samples in conjunction with Project Chariot and published a study indicating low concentrations of iodine-131 in the thyroids of the tested animals.[59] Spurred by these mixed results, laboratory scientists at Hanford visited several Indigenous communities in Alaska, where they used transportable radiation monitors to obtain whole-body counts from local residents.[60] Concerns about increasing fallout levels justified a systematic study of caribou-eating Alaskans, or so the scientists argued. Their findings about radiation in the northern regions reinforced those of other researchers in other countries in this period.

Opinions about the dangers of radiation varied among scientists in the late 1950s and early 1960s. Some argued that increases in radiation from nuclear explosions and other unnatural sources were a danger to public health and that even traces of radiation from isotopes such as Sr-90 and Cs-137 produced cancer or resulted in genetic changes passable to the next generation. Linus Pauling, a chemist and Nobel Laureate, predicted in the early 1960s that Soviet tests alone would result in "tens of thousands" of cancer deaths and "grossly defective" births.⁶¹ Pauling drafted the "Hiroshima Appeal," a pivotal document issued after the Fifth World Conference against Atomic and Hydrogen Bombs in August 1959.⁶² An influential voice, he urged Britain, the US, and the Soviet Union to conclude the Partial Test Ban Treaty, which entered into force on 10 October 1963.⁶³ Other scientists declared Pauling an alarmist, claiming that radiation was a public health hazard in exceedingly high dosages only. "No, it is largely a false scare," said Lauriston Taylor, chief of radiation physics for the US National Bureau of Standards.

Under the direction of Peter Bird, federal scientists in Canada's Radiation Protection Division belonged to the latter group.⁶⁴ Bird and his associates believed that only long and continued exposure to ionizing radiation posed a public health risk.⁶⁵ One of their main goals was to develop a reliable way to measure radiation dosages received by the general Canadian population and thereby understand an important aspect of public health in a period marred by atmospheric nuclear weapons tests. In this regard, to quell public concerns, federal authorities downplayed reports about increased radioactivity in the North. "Apart from whatever other international distinctions Canadians may have, we are acknowledged as living in one of the 'hottest areas on earth' so far as nuclear fallout is concerned," journalist Peter Worthington wrote in May 1963. "And 1963 is expected to be a record radioactive year for Canada, as the fruits of last year's nuclear tests in Russia descend from the stratosphere and troposphere and are showered over the land in the form of spring rains."⁶⁶ That same month, A.H. Booth, a senior scientific officer in the RPD, attempted to reassure a concerned Canadian public by telling journalists that federal scientists believed fallout posed no health hazard. Tests conducted in Alaska showed no abnormalities, Booth claimed. He added that authorities in Ottawa were aware of the fallout data compiled in Alaska and Sweden.⁶⁷

Such reports led Baker Lake resident G.W. Elliott to ask the federal government for clarity about the radiation threat in his community. Writing on behalf of the Baker Lake Residents Association, Elliott expressed his concerns and those of his neighbours. Booth responded with a letter that described the government's monitoring efforts in the North. He assured Elliott that radioactivity in northern Canada was far below dangerous levels: "Although people who eat a great deal of caribou meat may take up relatively more of the fallout elements, the amounts of these will still be well below the level at which there could be even the slightest risk to health. This Division is keeping a close watch on all aspects of the fallout situation. If any risk to health should arise, this would be immediately made known."⁶⁸

Nevertheless, scientific reports from south of the border inflamed concerns in Ottawa about the hazards of radioactive fallout in northern Canada. In May 1964, a group of scientists connected to the US Atomic Energy Commission released a report indicating increased doses of radiation in Indigenous Alaskans from Anaktuvuk Pass, Kotzebue, Point Barrow, Point Hope, and Fort Yukon.[69] The report cited reindeer and caribou meat containing unusually high concentrations of nuclear fallout as the source of the increased radiation doses in the selected test group. Measurements made during the summer of 1963 showed increases ranging up to 50 per cent in the average "body burden" of radioactive Cs-137. Reindeer and caribou feast on lichens that absorb fallout "like blotting paper," the scientists explained, noting the resumption of nuclear weapons testing in late 1961 as the likely culprit. The group recorded a concentration of 1,240 nanocuries in a man at Anaktuvak, which was exceedingly high but still below the 3,000 nanocuries recommended by the International Committee on Radiological Protection (ICRP) as a maximum permissible body burden for any one individual.[70] Canadian authorities in the RPD served on the ICRP and the UN Scientific Committee on the Effects of Atomic Radiation, generating and receiving knowledge as "world experts" in the field of radiation exposure.[71] By then, the Partial Test Ban Treaty had banned atmospheric nuclear weapons testing. However, explosions conducted before October 1963 generated intense concerns about increasing fallout levels and convinced scientists in the RPD to investigate the health hazards in the North.

The Northern Radiation Study in Theory

The RPD's special northern radiation study initially entailed collecting and analysing meat samples obtained from caribou and moose as well as collecting and analysing urine and bone samples obtained from northern residents. Undertaken cooperatively by various branches of the federal government, including the Department of Northern Affairs and the Department of National Health and Welfare, the program relied upon the assistance of medical authorities across the country. "With regard to [human bone] samples from the Charles Camsell Hospital itself, some samples would be welcome," Peter Bird wrote to Otto Schaefer. "Remember, however, that we are particularly concerned with getting specimens from the younger age groups and I do not know anything about the age distribution of persons on whom autopsies are conducted at the Charles Camsell Hospital."[72]

Opened in August 1946, the Charles Camsell Indian Hospital was the largest of twenty-two new medical institutions established in Canada during and after the Second World War. A marker of post-war reform to national health services, the Edmonton facility housed First Nations and Inuit patients who required treatment in the South. "Indian hospitals not only promised to contain disease," writes medical historian Maureen Lux, "but also assured concerned Canadians that theirs was a humanitarian government that extended the benefits of modern health care and a 'fair deal' to the long neglected."[73] In fact, federal policies continued to marginalize Indigenous people. The Department of Indian Affairs refused to provide transportation services for deceased patients if the costs exceeded the price of burial at the place of death, for instance. This policy was particularly harsh for northern communities because the high price of flights often exceeded what the family members could afford to pay. As Lux explains, the policy also provided Indian Health Services with access to bodies for autopsy without having to secure consent from the families of the deceased.[74] Indigenous communities lost family members because of incomplete record keeping and unmarked graves – a grim reality of the government's paternalistic and authoritative medical practices, which gave scientists in the RPD access to human samples from hospitals in Canada that received patients from the North.

Bird's research proposal involved collecting human bone samples from across northern Canada. Two weeks after he wrote to Schaefer, Bird wrote a detailed memo to his government colleague P.E. Moore, the director of medical services for the Department of National Health and Welfare. The memo outlined the difference between Sr-90 and Cs-137, the two isotopes linked with radiation hazards in the North. "Because strontium-90 is a beta-particle emitter which concentrates in bone it is necessary to obtain bone samples at autopsy," Bird explained. "Also, because of the long retention time in bone tissue, greatest emphasis should be placed on samples from the younger age groups where rapid and continuous bone formation is taking place."[75] Cesium-137, by contrast, concentrated in soft tissue and merely required the proper technical equipment for adequate detection in living subjects. To determine the sample requirements for his proposed program, Bird asked Moore to obtain preliminary data about the number and age distribution of autopsies conducted on Indigenous people at hospitals in Churchill, Fort Smith, Yellowknife, Inuvik, Whitehorse, Edmonton, and Winnipeg. He also asked for correlating data about northern residents sent to Ottawa for medical treatment, information he thought necessary for the study.

In Bird's mind, two factors necessitated a large-scale study of the public health hazards of radioactive fallout in northern Canada. First, he wanted to determine whether living in the North subjected residents to greater exposure levels. Second, if northern residency increased exposure to dangerous radionuclides, he wanted to determine whether there was a direct correlation with the postulated food chain theory. Both factors motivated a systematic study involving the collection and examination of northern vegetation, caribou bone and meat, and human bone obtained during autopsies.[76] Bird proposed a research program focused on samples obtained from Yukon and the districts of Mackenzie and Keewatin. He included northern Alberta, Saskatchewan, and Manitoba because animals and humans, to the best of his knowledge, moved throughout the Northwest Territories west of Hudson Bay, but also because hospitals at Fort Churchill, Winnipeg, and Edmonton could serve as important collection centres. Comparing human samples of caribou-eating and non-caribou-eating residents in the North was a potential indicator of fallout levels, Bird thought. He listed several northern communities as target locations, including Baker Lake, Churchill, Fond du Lac, Fort Smith, Fort Franklin, Bathurst Inlet, and Inuvik.

To assist Bird, Moore consulted medical authorities at different locations in northern Canada. He contacted medical pathologists involved in the collection and examination of children's deciduous teeth at the Whitehorse General Hospital, seeking information about ongoing radiation studies in the North. He also communicated with superintendents responsible for medical services in Yukon and the Mackenzie District, asking for the number and age distribution of autopsies conducted on Indigenous people in recent years. "Numerous autopsies are carried out in Whitehorse Hospital, a fair number of which are on Indian children and we could easily collect vertebral body samples when requested to do so," replied one superintendent from Yukon, who forwarded Moore's request to regional medical authorities.[77] Six weeks later, scientists in Ottawa began receiving the data they wanted. W.L. Falconer, the superintendent for the Foothills region of southern Alberta, provided a list of autopsies performed at the Charles Camsell Hospital between 1958 and 1962, and O.J. Rath, the superintendent for the central region of Canada, provided a list of autopsies performed on Inuit at various northern hospitals between 1960 and 1962. The records provided by Rath listed twenty-five autopsies performed on Inuit ranging in age from two months to seventy-two years.[78]

Race-based perceptions played into Bird's conceived study. "The hospital at Whitehorse may be able to supply bone samples representative of dietary habits more closely related to those elsewhere in Canada," he suggested, associating White people with southern Canada and Indigenous people with the North. "The hospitals in Edmonton and Winnipeg serving the Indians and Eskimos may be in a position to supply supplementary bone samples, or in some cases, fill in gaps in any collection program involving the above-mentioned hospitals," he continued, in reference to the smaller northern communities he identified as key locations.[79] Bird's proposed study also reflected a deeply racialized understanding of northern Indigenous life. Convinced of the subsistence dietary habits of Indigenous peoples in the North, Bird suggested that Indigenous children from northern communities represented a uniform sample for obtaining comparative data about radioactive fallout in northern Canada.

Bird's reasoning was simply an extension of the federal government's long-standing preoccupation with Indigenous populations in the North. During the Second World War, Inuit men living in Quebec's Ungava Bay region worked on the US Air Force base at Fort Chimo, a military facility constructed in 1941 on the west bank of the Koksoak River in the Kuujjuaq area of present-day Nunavik. The prospect of wartime employment had swelled the local population, which numbered around 400 by the mid-1950s.[80] US officials turned Fort Chimo over to the Canadian government at end of the war, and Ottawa closed the base in the late 1940s, converting the facility into a meteorological station for the federal transportation department. In August 1954, Canadian officials relocated fifteen Inuit men from Fort Chimo to work at Fort Churchill.[81] Federal reports about depleted caribou herds and declining prices for furs spurred the relocation. As historian David Meren has shown, Canadian authorities envisioned a complex and racially charged "modernization" scheme involving the transfer of yaks from India to create an agricultural subsistence for Inuit of northern Quebec.[82] Federal authorities also considered transferring Alaskan reindeer to the Ungava Bay region, with the added intention of boosting the local population with Inuit relocated from the Mackenzie District in the Northwest Territories. Removing Inuit from Ungava was intervention by other means – a measure reflecting the authoritative assumptions that underpinned state policy at the time.

Paternalistic settler-colonial medical practices also influenced the northern radiation study. In his original proposal, Bird reasoned that Indigenous people brought south to Ontario for medical purposes represented a suitable sample group that would be ideal in the sense that it would avoid unnecessary travel costs and minimize publicity about the study. "Potential participants in the more extensive program should be made aware of the nature of the program and their possible involvement," he wrote, indicating that he at least partly recognized the ethical concerns at play.[83] He did not elaborate on this statement, however, and it is unclear whether the participants he referenced included non-Indigenous test subjects. Nonetheless, the Canadian medical authorities who championed Bird's proposal targeted Indigenous patients for the study. In a letter to Colonel H.M. Jones, the director of "Indian Affairs" for the Department of Citizenship and Immigration, and B.G. Sivertz, the director of northern administration for Northern Affairs and National Resources, Moore wrote: "Because of the importance of this program not only to the health of Northern Canadians but to the rest of us in the South, it would be most helpful if any Indians and Eskimos from the suspected areas outlined in Dr. Bird's paper, who might be coming to or through Ottawa for one reason or another, could be persuaded to allow estimations to be made with the [whole-body analysis] machine under the supervision of the staff of the Radiation Protection Division."[84]

The machine referenced was a large, whole-body examination device developed by engineers in Ottawa. Personnel in the RPD designed and constructed a heavily shielded enclosure to house scintillation detectors, which had been specially engineered for quantitative analyses of gamma radioactivity in living persons.[85] This equipment enabled scientists to examine the uptake of Cs-137 and I-131 in selected individuals; in theory, the resulting findings, along with urinalysis, would serve as a measure of base radiation dose. Records from September 1965 indicate that five Inuit and two White people from northern Canada transferred south for medical treatment underwent whole-body examinations at the RPD's laboratory on Brookfield Road in Ottawa's Confederation Heights district.[86]

Human radiation experiments were seldom performed before the Second World War, but their prevalence increased rapidly in the post-war period as new funding opportunities and the escalating Cold War arms race incentivized experimental research. Historian Kate Brown documents a long and horrifying list of experiments in her eye-opening book *Plutopia*. In the 1950s, medical scientists working on federal research grants in the US gave radioactive drinks to more than 800 pregnant mothers at Vanderbilt University and the University of Iowa, while doctors administered radioactive iodine to newborns at university hospitals in Nebraska, Tennessee, and Michigan. Similar experiments were conducted at the National Reactor Testing Station in Idaho, where scientists had volunteers drink radioactive milk, swallow plastic capsules laced with fission products, and inhale radioactive gases. Department of Defense researchers, meanwhile, administered whole-body radiation exposures to indigent Black cancer patients at the University of Cincinnati between 1960 and 1971. Doctors at the University Medical Center there forged consent forms and exposed uninformed patients to dangerously high radiation doses that induced vomiting and intense pain.[87]

Although the exact details are unclear, it is possible that more than the seven northern residents described above underwent whole-body examinations during the RPD's research study in Ottawa. Earlier in 1965, the RPD's H.A. Procter mailed a letter encouraging medical authorities in the North to notify him of any Inuit receiving medical treatment in Ottawa.[88] Peter Bird did the same in a letter to H.B. Brett, the superintendent for medical services in the Mackenzie District. Bird's letter indicated that the RPD had asked government departments and agencies responsible for administration in northern Canada to provide information about the movement of Indigenous patients into or through the Ottawa area.[89] His goal was to carry out whole-body measurements on as many people as possible who either lived or travelled in the North. Bird also told Brett that a portable whole-body counter was under development and that medical authorities in Ottawa planned to conduct an expanded northern research program with the new equipment. While correspondence of this sort continued intermittently through the year, the available records provide no indication of the exact number of northern residents subjected to whole-body examinations.

The Northern Radiation Study in Practice

Fallout research first conducted in the US influenced the RPD's northern radiation study. A media storm exploded south of the border in the late 1950s when journalists reported traces of radioactive substances in Minnesota wheat, Sr-90 in Iowa milk, and Zn-65 in oysters collected at the mouth of the Columbia River. Fear of radioactive food, compounded by rising cancer rates and a dawning awareness of the possible long-term health effects of widespread environmental contamination, greatly shook public confidence in the AEC.[90] Three years earlier, in August 1959, the US Department of Defense had published a report about the health and nutritional status of Indigenous Alaskans that prompted biochemist Arthur Schulert of Vanderbilt University to examine the ecological distribution of Sr-90 in Alaska.[91] Schulert studied food grown along the Alaskan coast, fish and caribou, and samples of human bone and urine obtained from Indigenous peoples. He acquired the human samples for his study from the Arctic Health Research Center of the US Public Health Service in Fairbanks, where hospital surgeons treated Indigenous patients and obtained bone samples from children and adults during thoracic resection or autopsy.[92] Schulert published his research in the journal *Science* in April 1962, mere months before Canadian scientists in Ottawa initiated a special investigation of radioactive fallout in the North.

As Washington took steps to quell concerns and reassure Americans that radioactive fallout from nuclear weapons testing posed a negligible threat to public health, scientists in Ottawa grew increasingly curious about the effects of radiation on home soil. RPD authorities carried out several separate but interrelated analyses of the public health risks of radioactive fallout in northern Canada. Bird argued that gathering quantitative data about the concentration of fallout in food was key to understanding the transfer of hazardous radioactive isotopes to humans.

Bird's study required intergovernmental cooperation and support. Scientists in the federal agriculture department and the meteorological branch of the transport department facilitated the collection of the various test samples required for the study, including air, precipitation, soil, wheat, and fresh milk. Researchers analysed samples of milk for I-131 immediately following periods of atmospheric nuclear weapons testing. Milk sampling was conducted at sixteen locations across Canada, precipitation and air sampling at twenty-four, soil at twenty-three, and wheat at nine. Scientists also analysed samples of caribou and reindeer meat, as well as samples of human urine and bone obtained from northern Indigenous residents. Urine sampling was done at selected locations in the North; samples of human bone were drawn arbitrarily from participating hospitals and northern medical centres. Detailed instructions written by RPD personnel stipulated how human bone samples were to be collected, stored, and shipped. The scientists and pathologists who collected animal meat and human bone sent the stored samples to Ottawa, where government chemists conducted radiochemical analyses at federal research laboratories.

The exact number of human bone samples collected and analysed for the northern radiation study is unclear. Beginning in April 1964, *Canadian Medical Association Journal* published a seven-part series about radiation protection in Canada. In one of three articles that Bird wrote for the series, he cited low mortality rates as a reason why relatively few specimens of human bone were obtained from the younger age groups involved in the study.[93] He said the sample size was too small to analyse statistically but reported that the values obtained from bone-forming individuals showed consistently higher levels of radioactivity than indicated in the adult bones tested. Bird published this information in May, thus confirming that RPD scientists received an unspecified number of human bone samples during the first year of the study.[94]

The surviving records contain no information to either corroborate Bird's information or estimate the number of human bones collected and analysed during 1963 and 1964; even so, the disturbing correspondence of Canadian medical authorities demonstrates a considerable effort to obtain bone samples from Indigenous people in the North. "Two weeks ago an Old Crow Indian had a thoracotomy for removal of a hydatid cyst," Otto Schaefer wrote in a letter to medical superintendent O.C. Gray of the Camsell hospital. "This Indian told us that he had lived much

RADIATION PROTECTION DIVISION
DIVISION DE LA PROTECTION CONTRE LES RADIATIONS

Department of National Health and Welfare
Ministère de la Santé nationale et du Bien-être social
OTTAWA, CANADA

RADIOACTIVITY ASSAY PROGRAM

Instructions for Human Bone Collection

1. Specimens should, in general, be taken from the lower end of the vertebral column, and where possible, each specimen should conform to the following guide:

Age	Number of Vertebral Bodies	Approx. Wet Weight (gms.)
0-11 months	15 - 20	30 - 60
1 - 2 years	10 - 15	48 - 80
3 -20 years	6 - 10	60 - 150

   However, if the recommended amount cannot be obtained, amounts which are at least 1/2 of this will be acceptable. If available, specimens up to 250 gms. wet weight in the teen-age group are specially wanted for replicate studies.

2. For the Northern Study, other bone types, e.g. rib, long bone, femur, etc., will be acceptable providing the wet weights conform with the above guide.

3. Remove adhering flesh (gross dissection), place in cotton bag and label with autopsy number, hospital and date of death. Fix for 48 hours or more in formol-saline (metallic salts interfere with analysis). Use labelling technique which will remain legible after fixing.

4. Drain off excess formalin, wrap in newspaper and place in polythene bag. Check to see that identification is legible, if not, replace.

5. Send, Express, Collect, (Air Express from the North) to:

   Room B.105,
   Radiation Protection Division,
   Department of National Health and Welfare,
   Confederation Heights,
   Brookfield Road,
   OTTAWA 8, ONTARIO.

   Use gummed label supplied and fill in address of sender.

6. Complete the Data Sheet on the reverse side of this page with as much detail as possible. Mail the completed DATA SHEET in the envelope supplied. No postage is required.

7. Notify the Radiation Protection Division when supplies are nearly exhausted.

Figure 6.1. Radiation Protection Division instructions for collecting human bone samples, 1964. Library and Archives Canda, National Health and Welfare, RG 29, vol. 2886, file 851-1-27, pt. 1.

of his life on caribou meat. Rib pieces removed at that operation would have been of great interest if we had shipping instructions."[95] Schaefer's correspondence with Gray made its way to Ottawa, prompting authorities in the RPD to address an apparent inefficiency in the study and write detailed instructions for the collection of human bone.

Equally disconcerting is the official published record in Ottawa. On 4 September 1964, federal health minister Judy LaMarsh announced that government scientists had recorded high doses of radioactive Cs-137 in Inuit living in northern Canada.[96] "In view of the fact that Canada has the highest rate of radioactive fall-out in the world, and in view of the dangerously high levels of radioactive caesium 137 recently found in Alaskan Eskimos, has the government considered testing Canadian Eskimos for similar danger?," asked New Democratic Party member William Howe during Question Period.[97] LaMarsh replied: "Mr. Speaker, the radiation protection division of our department has been making a special study of the problem in Canadian Eskimos in co-operation with the northern health services of the department and with the Department of Northern Affairs and National Resources." Preliminary departmental reports published in October 1963 and April 1964, the minister explained, included data on Cs-137 content in samples of caribou meat, reindeer meat, and "urine samples obtained from residents in the north whose diet contains appreciable quantities of caribou meat."[98] The data indicated higher levels of Cs-137 in Inuit who ate caribou meat, LaMarsh said, adding that radiation levels were higher among northern residents than in southern Canada.

In actuality, the reports cited by the minister contain little information about the samples obtained from Indigenous people in the North. While RPD personnel maintained a complete record listing the number, age group, and type of each test sample collected and analysed, the quarterly and annual reports published by the federal health department do not contain specific information about the origins of each sample.[99] Except for a list of cities with contributing hospitals like Edmonton, Calgary, Saskatoon, and Winnipeg, the reports provide virtually no information about the collection procedures employed during the study.

Similarly, LaMarsh divulged only partial information about the RPD's northern radiation study while making her remarks. Preliminary estimates of the total amount of Cs-137 in the bodies of tested Inuit, based on urine analysis, had led government scientists to conclude that the level of radioactivity detected in the sample population was below the amount deemed unsafe, explained the minister. It is unclear whether LaMarsh knew the full details of the study, but she made no mention of the fact that scientists in the RPD had planned to extend the study by obtaining human bone samples from Inuit. "The highest levels noted are slightly below the amount recognized as permissible for continuous lifetime exposure," LaMarsh said, before concluding her remarks. "Levels are expected, I am very pleased to be able to say, to show a gradual decrease provided, of course, that nuclear testing is not resumed."[100]

While neither Bird nor the Canadian government openly published the number of human bone samples obtained during the first year of the northern radiation study, archival records opened under an access-to-information request indicate that government chemists examined no fewer than ten samples of human bone collected after 1964.[101] The first arrived in Ottawa on 19 March 1965 from doctor W.R. Buchan of the Whitehorse General Hospital, who shipped two samples of bone obtained from deceased persons over the age of twenty.[102] Another two arrived on 1 June from the Inuvik General Hospital, where an individual listed as Dr. Rooks sent two samples of bone obtained through separate autopsies of a one-year-old and a fifteen-year-old.[103] The last of the recorded bone samples arrived in Ottawa five months later. On 22 November, radiochemist E.R. Samuels sent a receipt to a Dr. Wilbush of the Inuvik General Hospital, who had shipped an unspecified number of bone samples obtained from deceased children under the age of twenty-three months.[104]

Many factors affected the number of human bone samples obtained during the study, but none more revealing than the perceived limited availability of Indigenous subjects. In September 1965, H.A. Procter wrote a letter reminding medical authorities in the Foothills, Central, and Eastern regions of Canada to obtain and supply human bone samples for analysis in Ottawa. "Regional Superintendents are requested to again remind all doctors who may be carrying out autopsies on northern residents of the necessity to collect bone samples," wrote Procter, urging additional support for research involving the presence of Sr-90 in civilian populations.[105] In a separate letter written specifically for the Eastern region's director general of medical services, Procter asked about the availability of northern patients for use as living test subjects. "Results from recent urine samplings indicate that there is a continued increase in Caesium 137 levels and the Radiation Protection Division is very anxious to have persons who have recently resided in the North visit their laboratory in Ottawa for total body counting," Procter wrote.[106] Desperate to obtain people for whole-body examinations, authorities in the RPD offered to pay the expenses involved in transporting "suitable northern residents" to Ottawa.[107] They hoped that northern patients transferred south to hospitals in Toronto and Montreal would serve as test subjects in Ottawa before returning north.

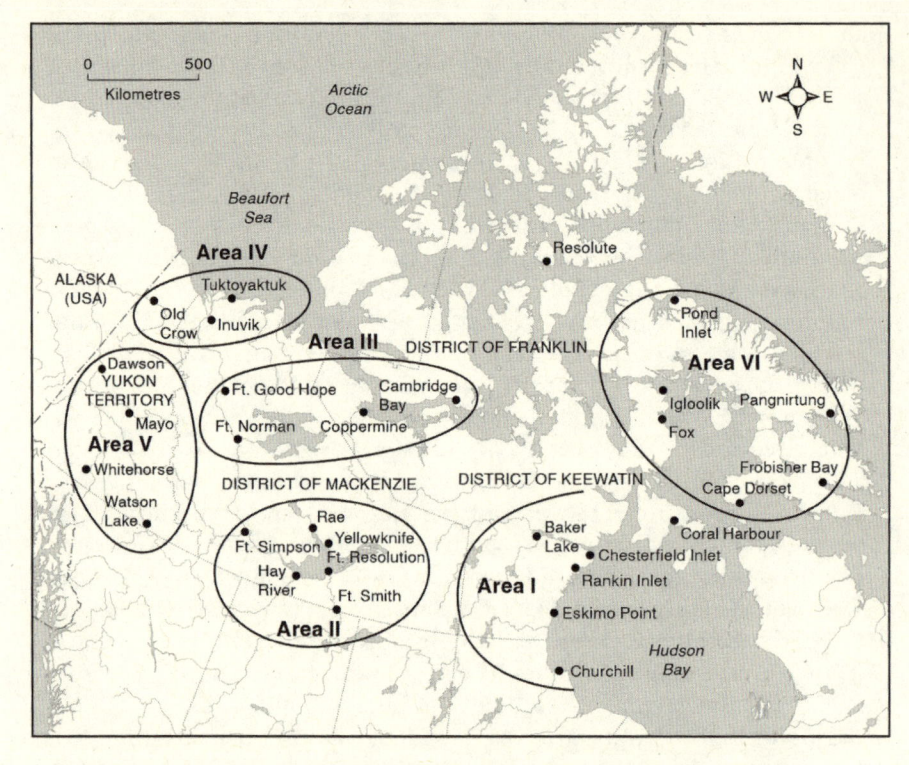

Figure 6.2. Map of radioactive fallout sampling locations in subarctic and Arctic Canada, 1966. Canada, Department of National Health and Welfare, *Data from Radiation Protection Programs* 4, no. 9 (Ottawa: Radiation Protection Division, 1966), 10. Map reproduced from original.

The complex history of reproductive health services in northern Canada provides further indication of the blatantly unjust settler-colonial policies and practices that informed the consent process for obtaining experimental test samples from human subjects. Health care providers in the Arctic sterilized Inuit women without their knowledge or consent in the 1960s and 1970s, as noted by medical anthropologists Patricia Kaufert and John O'Neil.[108] Hysterectomies and reproductive operations were often part of the treatment for depression, tuberculosis, cancer, and other medical conditions. Furthermore, as historians Erika Dyck and Maureen Lux point out, the physicians who performed these surgeries failed to explain the medical purpose or intent to the women they treated. "Seeking and obtaining health services seemed to coincide with unsolicited reproductive health care interventions by a system that continued to operate under eugenic laws in some cases and in the absence of any law in regions outside of British Columbia and Alberta," Dyck and Lux explain, building on the work of Kaufert and O'Neil.[109] Physician bureaucrats in Ottawa made autonomous decisions about health services for Inuit based on arrogant and flawed assumptions about northern Indigeneity. Medical authorities in the federal government were convinced that some Inuit women were incapable of making intelligent choices about their own health and claimed that contraceptive procedures were necessary to save their lives. This protected government physicians from any legal liability and allowed for surgeries on patients who did not consent. Given the authoritative and paternalistic approach of medical practitioners in the North, it is little wonder that Inuit represented a ready test sample suited to the RPD's wider research agenda.

The Natural Background Radiation Study

Health officials in Ottawa were not the only scientists to collect and analyse samples obtained from Inuit in the Canadian North. On 29 May 1964, W.V. Mayneord of the Institute of Cancer Research (ICR) in London, England, wrote to Peter Bird requesting samples of Inuit bone for a natural background radiation study concerning the Arctic. In 1959, Bird had supplied the ICR with bone samples obtained from residents in northern Canada. Mayneord, without specifying the number of Inuit bone samples received, said that preliminary analyses had indicated high alpha activities of polonium-210, a naturally occurring radioactive substance. "It seems possible that, as with fission product fall-out, the polonium is collected by Arctic lichen and reaches the Eskimo by way of caribou meat," he wrote, linking the study with existing theories about the northern food chain. "We quite appreciate that human autopsy material may be impossible to obtain," he continued. "If, however, any should be attainable, I would suggest that the main tissues of interest are: bone, teeth, gonad, liver, kidney, lung and spleen. Otherwise, teeth and placenta from living individuals should also provide a useful means for comparison with 'normal' population."[110]

Bird and his colleagues in the RPD happily agreed to support the ICR's study and provide human samples obtained from Inuit residents in northern Canada and the Arctic. RPD scientists planned and coordinated the collection program with the assistance of the federal departments of health and northern affairs. The collection program for the ICR was distinct from the northern radiation study, but the RPD used its pre-existing structures and networks to collect human specimens from the North and ship the samples across the Atlantic to London. Scientists in the RPD had trouble obtaining test samples from the outset, prompting P.E. Moore to contact medical authorities in northern Canada and reiterate the significance of the collection program initiated for the ICR. Moore wrote in a December 1964 letter distributed to all regional superintendents in the North:

> Regarding the special study of strontium-90 in human bone specimens and polonium-210 in human tissues, it is regrettable that, to date, not a single specimen has been received by the Radiation Protection Division. It is of the <u>utmost importance</u> that we gain some information on the levels of strontium-90 and polonium-210 in the bones and tissues of our northern residents, and our national medical staff who are carrying out autopsies, must be instructed that bone and tissue specimens be collected from any autopsies carried out by them, and private practitioners must also be encouraged to do likewise.[111]

Moore's letter had an immediate impact. Within three months, the first human bone samples obtained from autopsies of Indigenous people in the North arrived at Ottawa. It is unclear whether the RPD shipped the samples overseas, but scientists in Ottawa conducted radiochemical analyses on the human bones received.

Undeterred by the limited number of bone samples obtained for analysis, medical scientists in the federal government turned their attention elsewhere. In the mid-1960s, as historian Tarah Brookfield shows, teeth obtained from Inuit communities in northern Canada and the Arctic showed high concentrations of radiation.[112] Unlike bone, teeth offered a stable structure suitable for determining the uptake of radioisotopes in the body. Medical scientists performed tests of radioactive elements retained in teeth during calcification, hoping to draw a direct correlation between diet and the uptake of radioisotopes in the human body.[113] In January 1963, for instance, a preliminary study conducted by Vancouver's Dental Health Services reported that chemical isolation procedures performed on teeth developed a decade earlier contained radioactive elements, including Sr-90. The scientists also identified radioelements in teeth obtained from grazing cattle, findings that justified a large-scale investigation of public health, they argued: "If analysis of teeth gives a better indication of the uptake of strontium-90 by humans than analysis of bone, then it would be important to establish 'teeth banks' throughout the country for radioactive analysis."[114] In other words, human bone samples were difficult to collect and examine, and the ready availability of teeth appealed to scientists as a stable indicator of radiation hazards under any conceivable concentration of atmospheric fallout.

This same logic informed how RPD scientists perceived and approached human sampling for both the northern radiation study and the ICR's background radiation study. While human bone samples obtained from northern residents remained a priority, scientists in Ottawa increasingly encouraged medical authorities in the North to supply soft tissue and urine samples for radiochemical analysis. The first soft tissue samples arrived in April 1965.[115] Over the next four months, government scientists received twenty-four samples obtained from northern hospitals and medical centres in Edmonton, Whitehorse, Yellowknife, Inuvik, and Chesterfield Inlet. G. Gray of the Camsell hospital shipped samples of lung, spleen, kidney, liver, and rib obtained from the deceased body of a two-month-old baby.[116] Additional samples obtained from the bodies of children, teenagers, and young adults continued to arrive periodically for the remainder of the year. No fewer than fifteen total samples of human tissue and organ arrived from persons twenty-five years or younger, although not all records included age or demographic information.[117]

The open and available records indicate that scientists in the RPD received a greater number of human urine samples and placenta from pregnant women. During the collection program, the RPD issued more than fifty receipts for placenta samples that had arrived from the above-named locations as well as from Coral Harbour, Baker Lake, Rankin Inlet, Eskimo Point (Arviat), and Fort Churchill. Urine samples arrived in even greater quantities and from locations farther north in the Arctic because outpost nurses required no special training to obtain and ship samples from Inuit patients. Nurses at Frobisher Bay (Iqaluit), Igloolik, and other northern centres received urine-collection bottles, subject information forms, and detailed instructions for obtaining and shipping samples.[118] Liquid preservatives added after collection maintained the integrity of the urine samples obtained, enabling scientists in Ottawa to examine samples received on delay from relatively distant locations in northern Canada and the Arctic. It is unclear whether Ottawa forwarded any of the tissue or urine samples to London, as Mayneord had requested for the ICR. Nevertheless, the mere fact that authorities in the Canadian government agreed to supply scientists in London with human test samples obtained from northern residents adds another disturbing layer to the violent and oppressive history of settler-colonial medical practices in Canada during this period of the Cold War.

Radiochemical analyses of this sort fit a wider pattern of scientific research performed in Canada. In early January 1965, medical professor Joseph Sternberg of the University of Montreal wrote a letter to Jean Webb in the maternal health division of the federal health department. Sternberg informed Webb that medical authorities in Ottawa had approved the formation of a research team for studying radioactivity in placenta samples.[119] So-called field investigations on human placentas had taken place in Montreal, and Sternberg expressed an interest in receiving some twenty-five placentas from Inuit women in northern Canada and the Arctic. Webb connected Sternberg with P.E. Moore, who informed Sternberg about the RPD's ongoing studies. Moore agreed to provide his research team with placentas obtained from Inuit, but only after scientists in the RPD had completed their work for the northern radiation study and the ICR. The correspondence ended with Moore's response, and it is unclear whether Sternberg received the placenta samples he had requested. There is ample material to suggest that Sternberg pursued the proposed international radiation study, however. He published medical studies about irradiation and pregnancy in the 1960s and 1970s as part of an active research program on the environmental and health hazards of radioactive fallout.[120]

Conclusion

The RPD's northern radiation study serves as a complex and disturbing symbol of science and medical governance. Physician bureaucrats in Ottawa developed scientific research policies based on the assumption that the bodies of deceased civilians represented an open, readable, and suitable pool for obtaining test samples of human bone. Participating pathologists obtained human bone samples from several ethnic groups, age demographics, and regions of the country, indicating that a wide and sophisticated research network had been created and was being maintained. Unbiased research practices gave way to a targeted collection program involving northern residents when scientists working outside Canada raised concerns about increasing fallout levels in the North. At a time when worries about the hazardous effects of radiation gripped the Canadian public, northern Canada represented a steady source of data for government scientists in Ottawa.

Responding to nuclear anxieties and speculative research, medical authorities in Ottawa embraced a paternalistic role and attempted to safeguard the health and welfare of northern residents with little concern for the individual and familial rights of the people who constituted a ready collection of test samples. When researchers encountered difficulties obtaining samples of human bone for analysis, authorities in Ottawa adapted the collection program and aggressively pursued test samples of human tissue and organs.[121] Samples of urine and placenta obtained from patients in the North enabled medical scientists in Ottawa to perpetuate and grow the northern radiation study in Canada; the same people agreed to collect human test samples for Arctic research conducted overseas. Medical research policies and practices in the North allowed participating physicians, pathologists, scientists, and health bureaucrats to obtain and analyse human samples without having to secure the family's consent and while avoiding any legal repercussions.

The RPD's concerns over radioactive fallout hazards in the North conformed to Cold War anxieties about atmospheric nuclear weapons testing in the 1950s and 1960s. International scientific research posited a direct and traceable link between the concentration of radioactive fallout in northern vegetation and the uptake of radionuclides in the human body. The food-chain theory postulated by Bill Pruitt and other leading scientists convinced medical authorities in Ottawa to expand a pre-existing nationwide fallout study and devote resources to devise and carry out a systematic investigation of radiation hazards in northern Canada. At the same time, growing international concern over radioactive fallout resonated with medical scientists and health officials in the federal government, who saw an opportunity to assert paternal authority while adhering to the responsibilities of governance and the administration of health services in the North. Northern Canada and the people living there represented a tangible asset for medical authorities and government scientists in Ottawa – a group that used its position and power to conduct Cold War–motivated research in the name of national health and safety.

Conclusion: Reflections on Northern Canada, Military Research, and the Cold War

The various interrelated military research projects carried out in northern Canada and the Arctic during the Cold War years documented in this book conformed and contributed to settler-colonial narratives of domination and control over nature, further excluding and marginalizing Indigenous people and communities from contemporary understandings of the North. Western science and technology, when wielded fully, could not only replicate but also, in the minds of newcomers, improve upon Indigenous methods for living, moving, and working in a geophysical space perceived as largely inhospitable in the southern imagination.[1] This belief in the power and prestige of outside knowledge, combined with performances of virile masculinity and toughness in extreme cold, positioned the science-backed soldier from southern Canada as the protector of the nation. The often White male body that braved and conquered the harsh Arctic in past explorations was now ready to serve and defend the sovereign interests of his country in the same territory. Canadian soldiers from the South, having used science to overcome the twin enemies of fear and fatigue in the North, embodied the physical and mental capabilities to achieve effective and efficient military operations in a region of growing strategic importance. This was the conviction of the military and defence officials in Ottawa, who championed Arctic research to address Canada's security interests and military preparedness needs during the earliest decades of the Cold War.

The idea that Canada essentially was/is a northern nation defined by such distinguishing characteristics as cold, snow, winter, tundra, and the aurora borealis – termed the "core myth" by Shelagh Grant – gained popularity and strength during and after the Second World War.[2] The production and distribution of polar maps represented visually what geophysical investigations of the past century had come to explain: the position of the North Magnetic Pole was on Canadian territory, and the aurora borealis blanketed Canada more than any other country in the world.[3] From this perspective, the North represented a natural/national resource suited to the specific needs of scientists and government authorities in Ottawa.

In reality, though, the northern experiences of soldiers and scientists from southern Canada failed to meet the expectations conjured in the minds of outside observers. As a geophysical environment, the North frustrated and resisted the settler-colonial practices that had enabled the widespread dispossession of Indigenous lands and resources in southern Canada.[4] Nevertheless, an unrelenting belief in the military value of science and engineering informed how senior defence officials in Ottawa approached the North as a "frontier" that was ideal for demonstrating Canadian expertise. Despite the myriad challenges encountered by soldiers and scientists new to the North, the DRB remained convinced that Arctic specialization was key to fighting Canada's Cold War.

Senior military and defence officials in Ottawa perceived and approached northern Canada as a hostile climate to overcome, but the North also represented an asset for reinforcing Canadian defence interests. Fort Churchill enabled Arctic research projects deemed valuable to continental defence and military preparedness, serving the Canadian government and the armed services as a gateway to the North and as a signifier of Canada's overall contribution to North Atlantic security in the early Cold War. Between the late 1940s and the mid-1970s, scientists sponsored by Canada's defence department and other government agencies used federal funding and resources to carry out experiments concerning the people and the environment of northern Canada and the Arctic.[5] Senior military and defence officials were keen to protect soldiers from Canada's high-latitude conditions, which they perceived as harsh and unforgiving. Science seemed to offer several solutions to the many military challenges encountered in the North, and the strong belief in Canadian cold-weather expertise manifested itself among defence scientists, who in turn instilled this belief in senior decision-makers at DND.

During the Second World War, Canada's senior military officials recognized the increasing strategic significance of science and set out to create and sustain a strong national defence research program for peacetime. Substantial government interest in the Arctic developed during the war; the two decades that followed it saw permanent and consequential attempts to alter and improve northern landscapes and lives. As tensions between the Soviet Union and the US dominated the geostrategic circumstances of the early Cold War, the two superpowers competed to demonstrate scientific authority and the Arctic became a global stage for displaying dominance over nature.[6] While the two sides often differed in their approaches to Arctic research, their shared pursuit of scientific and political prowess led to a highly militarized Arctic, and that influenced Ottawa's approach towards safeguarding Canadian security and sovereignty in the North.[7]

Developments in Canadian science laid the foundation for Canada's post-war defence research program, and meanwhile, around the entire planet, scientific and technical advances of wartime weaponry raised the spectre of atomic war. Convinced of the need to strengthen Canada's defences in the North, senior officials in Ottawa committed resources and personnel to bolster the state's administrative and territorial presence in a region considered increasingly significant for security and national defence. Northern Canada and the Arctic also represented a scientific frontier, and one extremely valuable for reinforcing Canada's position among the North Atlantic security partners. Decision-makers in Ottawa supported Arctic research, leveraging the North into political and economic capital, in part to access the resources and benefits of US and British military power. DRB records reflect a deep institutional interest in northern geography and climate knowledge. Ottawa approached Arctic warfare as Canada's special or unique contribution to the North Atlantic security partnership, and scientists connected with the DRB thrived in an atmosphere of heightened anxiety that fostered the union of science and military affairs.

Fort Churchill was particularly useful for establishing cognitive and territorial claims over the North. This was especially important during the immediate post-war period, as senior officials in the Canadian government responded to US wartime activity in northern Canada. The possibility of a post-war attack in the region heightened uncertainty in Ottawa over Washington's position. Knowing that the US would defend its interests in the North with or without Canada, officials in Ottawa developed policies to protect and promote Canadian autonomy. Canada had "not gained independence from London in order to relinquish it to Washington," as Hugh Keenleyside remarked in 1949.[8] He shared Prime Minister Mackenzie King's concern that Canada might separate from Britain only to lose power or acquiesce to the US. The history of military research at Fort Churchill and elsewhere in northern Canada and the Arctic suggests that science was useful for addressing such dominant Canadian issues during the Cold War.

Science also addressed real concerns and anxieties over Soviet activity in the Arctic. The demands of continental defence in the post-war North overstretched Canada's limited defence budget, of course. Safeguarding Canadian interests in the North thus depended on gaining and sustaining access to the resources and capabilities of Britain and the US. For officials in Ottawa, Fort Churchill and the Defence Research Northern Laboratory were tangible assets that represented Canada's physical, financial, and scientific contribution to continental defence and Western security. This was important, and not just from the practical standpoint of security; competition mattered as well. Because the Arctic served as a stage for demonstrating scientific prowess, both the US and the Soviet Union invested heavily in the northern sciences to demonstrate their dominance over nature. A close study of Canada's military research activities in the North reveals that the competitive aspects of the Cold War also influenced Canadian policies for security and national defence. The union of scientific and military research in the North signalled Ottawa's firm commitment to the post-war security efforts of its two closest allies. Canadian officials accepted the idea that support for military research was essential for ensuring Western superiority and democratic values in the fight to win the Cold War.

While the origins of military science in Canada predate the Cold War, the strategic threat that emerged after the Second World War required a concerted response involving multiple branches of the federal government, medical authorities, and scientists across the country. Operating under the assumption that cooperation with Britain and the US was essential for security and national defence, senior officials in Ottawa created an independent but integrated military research program to assist the armed services in peacetime. In a system designed to generate and protect knowledge deemed valuable for Western military planning, scientific research was particularly useful as a threat-assessment tool. Scientists employed or contracted by various branches of the federal government generated information about the new and evolving weapons systems of the Cold War and how to defend Canadians from external threats to national security and public health. Military leaders and government scientists in Ottawa also fostered professional relationships with international partners, in this way strengthening Canada's position among the North Atlantic security partners while reinforcing the country's intellectual capacity to counter and overcome the practical concerns created by the tense ideological and geopolitical hostilities of the period.

Although it is difficult to draw clear links between research and policy, scientific knowledge about subarctic and Arctic Canada influenced how senior officials in Ottawa perceived and approached security and defence issues in the North.[9] Scientific knowledge about northern Canada and the Arctic, and particularly about the impact of Canada's harsh climate on men and military operations, provided information perceived as valuable to Canadian interests as well as to NATO's military-preparedness needs in North America. As the Cold War intensified and the pace of nuclear weapons trials increased in the 1950s and early 1960s, radiation scientists expressed increasing concern about the susceptibility of the North and the people living there to fallout and radioactive contamination. Canada's northern geography thus continued to serve as a natural laboratory for military and non-military research, motivated by distinct but interrelated Cold War concerns. Military and defence officials, government scientists, and medical authorities collectively leveraged the North into political capital.

Because Arctic research seemed to hold value for understanding and overcoming the myriad challenges of high-latitude military operations, defence authorities distributed federal funds to attract civilian scientists to military research, and the armed services drew relevant knowledge from copious sources of scientific information. Columnist Blair Fraser wrote in November 1954 after visiting Fort Churchill:

> No one proposes that an entire Canadian army, or even a whole Canadian division, should be made up of men who have all taken courses in Arctic warfare. But it is planned quite seriously to have enough officers, NCOs and seasoned men to stiffen, reassure and instruct any Canadian unit that might ever have to wage war in hard winter conditions anywhere. That is the whole purpose of Fort Churchill, and the basic principle of northern defense policy in Canada. The RCAF is flying up there simply to learn how. And the soldiers are learning to live there, both indoors and out.[10]

This high demand for military knowledge supported and sustained the structures required of a functioning defence research organization in peacetime. Apparently, senior officials in Ottawa approached Canada's unique security and defence issues of the early Cold War under the assumption that science was both a problem and a solution.

Military-funded science carried out in northern Canada and the Arctic during the first half of the Cold War had unintended consequences, however. Under the umbrella of the DRB's Environmental Protection Program, Canadian scientists conducted experiments that had adverse effects on northern ecosystems and human research subjects. At the time, environmental protection had nothing to do with the science of protecting the environment of northern Canada. For scientists in the DRB, environmental protection referred to studies involving the *protection of man* against the physical and psychological hazards of a particular operational environment.[11] Science, that is, was a tool for mitigating and overcoming the many challenges of northern geography: weather, climate, terrain, darkness, and isolation – or so was the belief and intent of the Canadian defence scientists who undertook Arctic research. They saw themselves as helping generate new knowledge and prepare the armed services for high-latitude warfare in peacetime.[12] The entangled histories of science, geography, and settler-colonial governance in postwar Canada thus point to the importance of studying military research in the context of northern Canada and the Arctic, which underscores an important element of Canada's Cold War experience.

What constitutes ethical research practice and equitable human treatment? That question is embedded in another key theme of the preceding chapters, and could also be framed as follows: What were the social, environmental, and political implications of the military-sponsored research activities? The complex and interrelated northern research projects sponsored by the federal government during the first three decades of the Cold War and described in this book illustrate the human and social consequences of selected research activities carried out to prepare the Canadian military and generate knowledge deemed useful to the national interest. As the armed services entered and later exited the Churchill area at their own convenience, northern Manitoba and its Indigenous populations were strongly affected; in addition, Canada's military presence in that part of the country extended Ottawa's reach farther north, from subarctic into Arctic Canada. Malcolm Brown's experimental research in the north of Hudson Bay attempted to yield results useful for acclimatizing White soldiers to the North, an indication of the grip of the colonial project and the intertwined histories of medicine and military research in Cold War Canada. Here, scientific pursuits motivated by the apparent requirements of military preparedness intersected and overlapped with paternalistic state approaches to control over both nature and Indigenous health in the North.

Medical scientists tackled the many problems of northern acclimatization by studying human research subjects; other scientists turned their attention to the environment. During the summer months, wet and boggy conditions in the North produced swarms of biting insects that affected military training. To combat northern Canada's bug problem, Canadian and US scientists carried out various entomological research experiments with the goal of eradicating mosquitoes and blackflies in isolated locations. The eradication of biting insects had a dual purpose: military officials thought that soldiers could train more effectively and efficiently in a bug-free environment, and achieving control over ecological activity in isolated environments theoretically would provide a military advantage to both Canada and the US if the Soviet Union attempted a land invasion in the North. With the support of the Canadian armed services and particularly the RCAF, entomologists sprayed chemical insecticides near various military bases in Canada. The experiments led to further insect-control studies involving the tracking of biting insects tagged with radioactive markers – research that affected persons and ecosystems near military bases scattered across southern and northern Canada.

Cold War–era military research projects carried out in northern Canada extended beyond scientific experiments involving soldiers and their operational environment, however. During the International Geophysical Year of 1957–8, some scientists at Churchill turned their gaze towards the skies, performing research into atmospheric magnetism and the behaviour of the aurora borealis; others boarded RCAF transport aircraft and flew hundreds of miles north, contributing to year-round research conducted on the northernmost island in the Canadian Arctic Archipelago. The scientists who surveyed the northern coast and inland areas of Ellesmere Island as part of the DRB's Operation Hazen generated geophysical knowledge deemed necessary for military preparedness and continental defence in the Arctic. Here, too, the real and potential military demands of the period shaped how officials in Ottawa perceived and approached northern Canada as a frontier, as representing both the challenges and the potential of post-war activity in the North.

But the militarization of Arctic research in Cold War Canada was not confined to the northern reaches of the country. As Ottawa's military and defence priorities evolved in the mid-1960s, the Canadian Army withdrew from Fort Churchill and the DRB's Arctic research program underwent a drastic reduction in physical and financial resources. Nevertheless, military interest in Arctic research remained strong. Scientists and research technicians at the DRB's Toronto-based medical research facilities, which supported and enabled Wilfred Bigelow's hypothermia experiments, turned to laboratory simulations to study the effects of adverse weather and extreme climate conditions on soldiers and military equipment. The continuation (and expansion) of the DRB's Environmental Protection Program in southern Canadian laboratories underscores the extent to which military demands in post-war Canada prolonged experimental "Arctic research." To be sure, the DRB's long-term fascination with soldiers' performance under environmental stress indicates how a dominating feature of Canada's Cold War experience preoccupied several scientists and decision-makers who operated under the auspices of the defence department.

Non-military scientists who carried out Cold War–era research projects were themselves motivated by geographical perceptions of northern Canada and the Arctic. Concerned about increasing nuclear fallout levels and the spread of radioactive contamination in the North, medical authorities in Ottawa planned and carried out a systematic investigation of the health hazards of radiation in northern residents. The creation of an independent Radiation Protection Division within the Department of National Health and Welfare, supported by medical authorities treating patients in and from the North, enabled government scientists to carry out the special northern radiation study. State-driven medical practices and policies permitted the collection and analysis of urine, tissue, and bone samples from living and deceased persons in the North, illustrating the extent to which health concerns raised by nuclear weapons testing during the Cold War encouraged medical scientists in Canada to act from a position of power and authority.

In line with the broader political and government interventions in society that marked the period between 1945 and the late 1960s, hundreds of scientists in Canada embraced research in service to the state. From this perspective, science did not rise above the geostrategic politics or complex sociopolitical dynamics of the Cold War. On the contrary, science and scientific expertise defined and supported state goals in the international battle for ideological supremacy and political orthodoxy. In Canada, government-sponsored scientists functioned as a form of soft power, leveraging their post-war prestige to defend liberal-democratic values on the front lines of the Cold War even while championing their own personal and professional goals. Canadian science was militarized in this period because of the Cold War, and in that climate, a multitude of like-minded scientists wielded their knowledge and expertise to obtain research funding and thereby pursue both individual goals and the perceived national good. The Cold War, much like the two previous global conflicts of the twentieth century, paid dividends for science and scientists in Canada.

After writing this study, I have concluded that access to information is vital for investigating the scientific research practices and policies that informed and shaped Canada's Cold War experience. Military concerns prompted and legitimized state intervention in the North, demonstrating the influence of anxieties induced by the Cold War on federally funded science and medical research activities in Canada during a transformative period in world affairs. Yet had it not been for access-to-information rules, many of the stark and complicated issues discussed in the preceding chapters might still be buried among thousands of archival documents and closed government records. I make no claims that this has been a definitive account of the history of military and non-military research in northern Canada and the Canadian Arctic during the Cold War, but I do contend that the government, military, and institutional documents that inspired and informed this book's core analysis reinforce the critical importance of preserving and unearthing historical records for current and future research.

As historians debate how the Cold War affected the various scientific fields, the power and reach of the military-industrial complex and its national iterations becomes increasingly clear. The lure of federal research funding and the opportunity to perform science in service to the state enticed scientists and engineers on both sides of the Iron Curtain to perform research with real or potential military applications. Military-sponsored scientists in Britain and the US encountered greater public scrutiny as activism against weapons production and warfare research grew stronger in the 1960s, resulting in scandals that ensnared some high-profile scientists and their profession. In Canada, the psychologist John Zubek, who subjected hundreds of military and civilian volunteer research subjects to experiments in sensory deprivation and physical confinement, committed suicide after public allegations linked his DRB-funded research to torture and unethical techniques for police interrogation.[13]

Others conducted military research in relative silence and avoided public scrutiny altogether. In fact, the DRB's organizational structure installed several leading Canadian scientists in unelected yet powerful positions in the defence department.[14] Working alongside top military officials and senior civil servants, Wilfred Bigelow, Malcolm Brown, John Spinks, Alan Burton, and several other scientists and top academic researchers in Canada advised on the use of federal funds allocated for military research. In this manner, the Cold War and Canadian science developed a reciprocal relationship. Cold War concerns and defence priorities shaped how scientists pursued and performed research, and scientists with direct ties to Ottawa shaped how government institutions and the armed services understood the value and utility of novel research. Indeed, the relationship between scientific expertise and liberal-democratic values was a dominant force in Canada's Cold War experience.

Like the health authorities who embraced a professional adventurism to work in the North, DRB-funded scientists, particularly before the rise of southern-based laboratory research in the late 1950s, saw military research in northern Canada and the Arctic as a multilayered opportunity to support the federal government and advance scientific knowledge. *The Arctic Frontier* – an edited collection jointly published by the Canadian Institute of International Affairs (CIIA) and the Arctic Institute of North America in 1966 – is a tangible and historically contingent representation of this very idea.[15] Science was a tool for defining and demonstrating Canadian nordicity during a time of increasing national reflection and change, when the prospects of national development in the North gave renewed and sustained relevance to the untapped potential of the "last frontier."

Concurrently, the history of military research in the post-war North serves as a warning against definitive statements about the impact of the Cold War in Canada. The social and environmental consequences of the government-sponsored scientific research activities documented in this book were a result of multiple complex factors that resist generalization. As historian Adrian Howkins explains, "the history of the Cold War reveals that the nature of the environment alone does not determine political or military histories."[16] That northern Canada underwent a drastic militarization during and after the Second World War was only partly a result of its geography and outside perceptions of its harsh, barren, and scientifically challenging environment.

Historical study of Arctic research in Cold War Canada must, therefore, include both military and non-military topics. Although the "systematic consolidation of nature as a military entity" certainly took place in Canada, as evidenced by the northern scientific activities of the DRB, government-sponsored science also sought solutions to non-military and non-strategic problems.[17] If we are to accept Joy Parr's process of "corporeal embodiment," then all human interactions with sciences and technologies are inherently environmental.[18] This applies to Canada's experience with environmental science in the post-war North as well, where anxieties and concerns generated by the Cold War factored into decisions to support Arctic research. The scientific research activities sponsored and carried out by National Defence, National Health and Welfare, Agriculture, and other federal departments deserve attention on par with wider perceptions of civil/state relations and science in the post-war period.

We must also keep in mind that Arctic research advanced the individual professional careers of participating scientists, engineers, and doctors while meeting the needs of officials in Ottawa. Researchers received financial support, published findings, and took up both academic and government positions. In turn, the federal government learned about geography, climate, population, security, governance, and development in northern Canada and the Arctic. While the complexities of state-sponsored research are difficult to untangle, the integration and cooperation among government officials and civilian scientists provides important lessons for understanding the Cold War's immediate and lasting impact in Canada.

The scientific and technical activities sponsored by the federal government in northern Canada and the Arctic during the first half of the Cold War are particularly useful for explaining the complex interplay between policies developed for science and administration. Highly acclaimed scientists from several academic institutions in Canada, Britain, and the

US received financial and institutional support to travel to the Canadian North and study such problems as cold tolerance, insect control and eradication, biological and chemical warfare, and glaciology and topography. The stated aim of this research was to develop methods for reducing and overcoming impediments to military personnel and equipment operating in high latitudes and extreme cold environments, but the singular and overlapping agendas behind the militarization of Arctic research in Canada were much more complex.

It would be disingenuous to suggest that science alone deserves blame for any unintended negative consequences stemming from the military and non-military research projects undertaken in northern Canada and the Arctic during the Cold War. Scientists, broadly speaking, represented one voice or position among various others. Multiple actors, government and civilian alike, made important decisions in response to the difficult (and at times advantageous) circumstances of the period. It is important to explore and explain, rather than blame or condone, the people and actions described in the preceding chapters. We can derive important lessons from the critical appraisal of Cold War research in northern Canada and the Arctic, nonetheless. Scientists who worked full-time or contractually for the federal government received the position, power, and resources to negotiate, design, implement, and oversee both state- and self-serving research experiments. The sweeping changes introduced to government bureaucracy and the defence department in the 1970s stemmed from a critical review process that identified scientific management structures as flawed.[19] Despite the establishment and growth of a largely successful defence research program, business and public administration models prevailed.

The years between the late 1940s and early 1950s were a time of intense scientific and technical research in Canada; however, military science plateaued after the Korean conflict as the defence budget entered a period of decline. The perceived value of science to defence remained, but as a result of shifting priorities regarding Canada's international role, the DRB began turning from fundamental to applied research. The strategic significance of the North also shifted in considerations of Canadian security and national defence. Scientific research projects in northern Canada and the Arctic remained important to the federal government, although the scope and nature of research in the North became increasingly responsive to the crises of atmospheric fallout and hazardous radioactivity.

Canada's experience with Cold War research in the North did not play out in a vacuum. Historical scholarship about post-war research throughout the Circumpolar North illustrates the open and dynamic contexts of space; constructed identities of place transformed within wider networks of social relations and knowledge exchange.[20] The many identities of the North as both a social and physical place are visible in the records of the various government branches that operated in the region. Both imaginatively and materially, senior officials and government scientists in Ottawa envisioned and created a northern space in the name of science and strategic necessity. This history helps explain the power wielded by ideas to shape and perpetuate Cold War–era ideology in Canada. Where the Canadian government and northern science in the Cold War is concerned, ideas about progress and dominance outpaced the creation of adequate and objective administrative oversight structures. Knowledge production held value for government actors in search of solutions to military, political, and economic problems. In an atmosphere of intense anxiety, the potential benefits of science significantly outweighed the dangers of inactivity.

The circumstances, attitudes, decisions, and consequences of the military-sponsored Arctic research activities carried out in Canada during the first half of the Cold War period represent a field of study with critical implications for Canadian history, military history, and the history of science. Although DRNL was a relatively small facility, Fort Churchill supported a diverse and extensive scientific program that attracted government and academic researchers, high-ranking officials, and senior military planners from all three North Atlantic security partners. In the face of increasing scientific research activity in the Soviet Arctic, government officials and military leaders in Britain and the US looked favourably upon northern Canada as a space to advance Western military capabilities and knowledge. "As Caesar's legions did of old, so did the Canadian Army build its *castra* and leave it, a legacy to its country's future in the North," declared Fort Churchill's last commander, Colonel Strome Galloway, in his farewell piece written for *Canadian Army Journal*. "Here, on the rim of Canada's Arctic it has been proven once again that, '*Nothing has ever been made until the soldier has made safe the field where the building shall be built, and the soldier is the scaffolding until it has been built, and the soldier gets no reward but honour.*'"[21] Roads, housing, heating, medical buildings, recreational facilities, and research laboratories were all among the so-called southern amenities planted on the shore of Hudson Bay to create the garrison deemed necessary to prepare the Canadian armed services for the Arctic battlefield of the future.

Indeed, Fort Churchill and DRNL were inherently preparatory in purpose. They functioned as a safety mechanism for Canadians in the South. Northern Manitoba offered an ideal location for year-round military training and scientific research, and senior officials in Ottawa expended significant financial and diplomatic capital to support and promote Fort

Churchill as a unique Canadian resource. Canada's northern geography served as a natural laboratory for cold- and warm-weather experiments involving military personnel, civilians, animals, and the natural environment. Funding from the DRB and affiliated branches of the federal government enabled scientists to travel north and conduct research on behalf of the armed services or sponsoring departments. During the first two decades of the Cold War, senior military and defence officials in Ottawa required the assistance of government and civilian scientists to prepare the armed services in peacetime. Assessment of potential enemies is fundamental to military preparations, and the defence department funded experimental science involving military personnel and civilians to learn about Canada's northern climate and devise solutions to defend against the Soviet Union's Arctic warfighting capabilities. The military garrison also served as an access point to the Far North, providing the federal government with a tangible geostrategic asset to leverage Canada's sovereignty and security interests in the Arctic.

This approach to science also characterized the non-military research activities sponsored and carried out by the health and welfare department. To investigate and combat the health hazards of radioactive fallout in the North, medical authorities in Ottawa relied upon scientific colleagues and research networks established within and outside Canada. People, lands, and the environment in northern Canada and the Arctic represented a research subject, one that was bound to the governance structures and paternalistic responsibilities of federal oversight. The immediate and lasting consequences of Cold War research in the North, whether military or non-military, derived from a cultural and institutionalized belief in Canadian nordicity.

Yet Canada's northern identity was a facade, a superficial stereotype that appealed to narrow expectations of "Canadian." To be Canadian meant to live and understand the North – at least, that was the expectation of US and British officials who looked to Canada for Arctic expertise. Military science promised to assert Canadian dominance in the North, over both climate and geography. For the senior military and defence officials who championed Arctic research, the combination of strength, endurance, ingenuity, skill, and scientific expertise demonstrated and reinforced Canada's northern identity. But ideas about Canadian nordicity clashed with the natural environment, as demonstrated by science's failure to achieve state goals for military control in northern Canada and the Arctic. Scientists and soldiers alike faced hostile conditions that challenged and ultimately compromised the goals of Canada's Cold War Arctic research program, yielding results that had little applicability outside the spatial and perceptual confines of the imagined northern frontier.

Was any of it necessary? Did Wilfred Bigelow *need* to freeze and kill dogs for military research purposes? Did Malcolm Brown *need* to perform needle biopsies on Inuit and experiment on White medical students to study cold tolerance? Did C.R. Twinn *need* to spray chemical insecticides and attempt to eradicate biting insects in the North? Did Peter Bird *need* to collect specimens of human bone, organ, and tissue to investigate the threat of radioactive fallout to public health? The passage of time affords historians the privilege of hindsight, and it would be disingenuous and irresponsible to cast aspersions or blame. But let it be said that the scientists and senior government officials documented in this book assumed a wartime level of risk in peacetime. They approached the early Cold War period as if they were still at war, embracing a military world view towards professional work that explains (without justifying) the unintended and largely negative consequences of Canada's Arctic research program. Whether the experiments were necessary matters not; what matters is that the scientists and their government sponsors knew what they were doing and did it anyway.

Ultimately, Canadian scientists embraced military research during the early Cold War for a dual purpose: to advance scientific knowledge and their own professional careers. The sense of duty associated with contributing to security and national defence, although important in the face of real and escalating dangers, was seemingly tertiary to the lures of research funding and the opportunities available through military research. As Canada's former minister of science and one-time metallurgist for the armed services, William "Bill" Winegard, once expressed in an interview, the DRB was a large pile of money and hundreds of scientists across the country received federal funding to conduct military research, regardless of any personal interest in Canadian defence.[22] Winegard himself had little interest in supporting the armed services or contributing to national defence, but he understood the circumstances of the period and performed military research for the DRB in order to advance metallurgical work and create graduate student opportunities at his academic institution, the University of Toronto. Several other academics did the same, and while the entire defence science community aimed to protect military personnel from operational hazards, money and research opportunities were the principal motivators for undertaking research on behalf of the armed services.

Despite the unintended consequences of the diverse but interrelated scientific experiments discussed in the preceding chapters, the term "military research" should not carry negative connotations. In the early years of the Cold War, military research was common practice for scientists and engineers in Canada, Britain, and the US. The success of science in the Second World War, especially among the Allies, signalled the continued growth and progression of scientific and technological research in peacetime. The entire corps of Canadian scientists and engineers who devised and carried out the DRB's Arctic research program had conducted military research for the armed services during the war. These men lived, studied, taught, interacted, and worked at war. Conflict and external threats had shaped their world view. The experience of war remained with them, affecting how they understood and approached scientific research at the dawn of the Cold War.

As such, the role of scientists, engineers, doctors, and other so-called experts in facilitating the expansion of military activities and profiting from state-sponsored research needs further study, particularly in relation to public health concerns and the ethics of human experimentation. In failing to help realize state goals for security and national defence, the scientists and experimental research projects discussed in the preceding chapters laid bare the deficiencies of Western capability. But even with rigorous archival research and constant pursuit of access to information, the history of military-sponsored research in northern Canada and the Arctic during the Cold War will remain incomplete without ongoing contributions from Indigenous scholars and residents in the North. Cold War knowledge of science, technology, and the environment continues to affect people, lands, and resources in Canada today, and it is incumbent on historians, scholars, community leaders, and the public to gather and interpret new sources of information for current and future generations.

In the same vein, policy considerations concerning science, security, and governance in northern Canada and the Arctic should rely on records, stories, and voices of the past. As climate change and environmental degradation provide increasing access to the North, and as competition for resources grows among government and non-state actors, Ottawa will need to develop new policies and approaches to work with northern residents and help ensure and protect regional and national interests in northern Canada. From the late 1940s through the 1970s, policy-making for the North was too limited. Policies created to protect and promote Canada's northern security and defence empowered military leaders and scientists to authorize and carry out experimental research projects that had unintended consequences for soldiers and civilians alike, not to mention for the environment and Indigenous lands and resources. Effective policy-making requires open and collaborative dialogue among government and civilian partners. This lesson becomes poignantly clear when we examine the history of Cold War research in northern Canada and the Arctic.

Notes

Introduction: Scientists, War, and Canada's Northern Frontier

1 Library and Archives Canada (LAC), Record Group (RG) 128, vol. 237, file Hypothermia, W.G. Bigelow, J.E. McBirnie, and H.H. Karachi, *The Fatal Exposure of Unanaesthetized Dogs to Severe Cold: Observations on Body Temperature, Blood Sugar, Serum Magnesium and Serum Potassium*, DR Report No. 51 (Ottawa: Defence Research Board, Department of National Defence, February 1952), DRB Project D50-90-20-01, funded by DRB Extramural Research Grant No. 125.
2 For information about DRML, see LAC, RG 24, vol. 2529, file 801-100-M91 pt. 1, Annual Report, Defence Research Medical Laboratories, November 1950.
3 Bigelow, Callaghan, and Hopps, "General Hypothermia for Experimental Intracardiac Surgery."
4 Klingle, "The Multiple Lives of Marjorie."
5 Bigelow and McBirnie, "Further Experiences with Hypothermia."
6 LAC, RG 24, vol. 4119, file DRBS 3-50-43-2 pt. 1, Defence Research Board: Defence Medical Research Advisory Committee, Minutes of the Third Meeting, Toronto, 6 December 1952, 20.
7 Coates et al., *Arctic Front*, 5.
8 For information about the history of Canada's Mobile Striking Force, an air-transportable brigade, see Maloney, "The Mobile Striking Force"; and Raymond Stouffer, "Military Culture and the Mobile Striking Force," in *De-icing Required!*, ed. Lackenbauer, 58–70.
9 LAC, RG 85, vol. 299, file 1009-2[5], *Defence Research Northern Laboratory: Progress Report on Indoctrination Training for Military Operations in the North, DRNL Project Report No. 4* (Ottawa: DRB, Department of National Defence, Canada, 1954), 2.
10 For biographical information about Bigelow, see the Canadian Medical Hall of Fame, "Dr. Wilfred G. Bigelow," http://www.cdnmedhall.org/inductees/dr-wilfred-g-bigelow.

11 Webster and Bigelow, "Injuries Due to Cold, Frostbite, Immersion Foot, and Hypothermia."
12 I use the term peacetime in reference to the post-1945 period following the Second World War, recognizing fully that Canadians fought and died in conflicts that occurred during the Cold War.
13 Webster and Bigelow, "Injuries Due to Cold, Frostbite, Immersion Foot, and Hypothermia."
14 Bigelow, *Cold Hearts*, 47. For the specific study that Bigelow refers to in this quote, see Bigelow et al., "Oxygen Transport and Utilization."
15 Federal employees of the DRB swore an oath of secrecy to the Crown in writing. Non-employed researchers who received a financial grant from the DRB also swore an oath of secrecy, but only if they were the lead researcher or co-researcher on the grant. In rare instances, all researchers involved in the project, including research assistants and graduate students, swore an oath of secrecy. See LAC, RG 85, vol. 298, file 1009-2[2], Defence Research Board, Care and Communication of Classified Information, 10 January 1949.
16 Bigelow, *Cold Hearts*, 49.
17 Bothwell, *The Big Chill*, 13.
18 Whitaker and Marcuse, *Cold War Canada*, 27–30; Whitaker and Hewitt, *Canada and the Cold War*, 13–17; Dennis Molinaro, "How the Cold War Began ... with British Help."
19 United States, Department of the Army, *Army Information Digest*, "Cold War Terminology – US," reprinted in *Canadian Army Journal* 16, no. 2 (1962): 94.
20 Wolfe: *Freedom's Laboratory*, 3.
21 Laura McEnaney, *Civil Defense Begins at Home*, 6. See also Rohde, *Armed with Expertise*, 3.
22 Chastain and Lorek, eds., *Itineraries of Expertise*.
23 Matthias Heymann, "In Search of Control: Arctic Weather Stations in the Early Cold War," in *Exploring Greenland*, ed. Doel, Harper, and Heymann, 75–98; Farish, *The Contours of America's Cold War*.
24 Sverker Sörlin, "Ice Diplomacy and Climate Change: Hans Ahlmann and the Quest for a Nordic Region beyond Borders," in *Science, Geopolitics, and Culture*, ed. Sörlin.
25 Whitaker and Marcuse, *Cold War Canada*; Kinsman, Buse, and Steedman, eds., *Whose National Security?*; Cavell, ed., *Love, Hate, and Fear*; Iacovetta, *Gatekeepers*; Kinsman and Gentile, *The Canadian War on Queers*; Sethna and Hewitt, *Just Watch Us*.
26 Turner, "The Defence Research Board of Canada" (PhD diss.); Goodspeed, *A History of the Defence Research Board of Canada*.
27 Ridler, *Maestro of Science*.
28 Avery, *Pathogens for War*. See also Avery, *The Science of War*; and Donald Avery, "The Canadian Biological Weapons Program and the Tripartite Alliance," in *Deadly Cultures*, ed. Wheelis, Rózsa, and Dando, 84–107.

29 Godefroy, *Defence and Discovery*; Jones-Imhotep, *The Unreliable Nation*. For a detailed list of the internal institutional histories referenced here, see Turner, "The Defence Research Board of Canada," 4–6.
30 Dziuban, *United States Army in World War II*, 185–6.
31 Pennie, *Defence Research Northern Laboratory*.
32 Goodspeed, *A History of the Defence Research Board of Canada*, 175–88; Margaret A. Carroll, "Defence Forces Operations in Hudson Bay," in *Science, History, and Hudson Bay*, vol. 2, ed. Beals and Shenstone, 897–934; Iarocci, "Opening the North." Also, in *Lessons in Northern Operations*, Lackenbauer and Kikkert devote considerable attention to the history of Fort Churchill in their work on the Canadian Army and northern operations in the early post-war period.
33 Souchen, *War Junk*, 5.
34 Souchen calls on historians to consider "life cycles" in historical analysis. See Souchen, *War Junk*, 204.
35 Davis, "Grounds for Permanent War" (PhD diss.); Smith, *Toxic Exposures*, 32–41; McVety, *The Rinderpest Campaigns*), 164–6, 177–90.
36 Wiseman, "Canadian Scientists and Military Research."
37 For a useful source on the principle of functionalism, see Chapnick, "The Canadian Middle Power Myth."
38 Holmes, "Most Safely in the Middle," 367. See also Holmes, *The Shaping of Peace*, vol. 1. Another useful source on Canada and the concept of middle power is Soward, "On Becoming and Being a Middle Power."
39 For a more recent and detailed account of Canada's role as a middle power in international forums, see Chapnick, *The Middle Power Project*.
40 Campbell, *Unlikely Diplomats*, 3.
41 Nicol, "Reframing Sovereignty"; Shadian, *The Politics of Arctic Sovereignty*.
42 Dodds and Powell, "Polar Geopolitics," 4.
43 Faragher, ed., *Rereading Frederick Jackson Turner*.
44 A.R.M. Lower's earlier works examined the role of the forest in Canadian development, building on the staples thesis elaborated by Harold Innis: Lower, "The Trade in Square Timber," in University of Toronto Studies, History and Economics, *Contributions to Canadian Economics*, vol. 6, 40–61; Lower, *Settlement and the Forest Frontier in Eastern Canada*; Lower, *The North American Assault on the Canadian Forest*. Written during the Second World War, Lower's book *Colony to Nation* (1946) sought a basis for a sense of national community that might unite French and English Canadians.
45 Grant, "Myths of the North in Canadian Ethos."
46 Giehmann, *Writing the Northland*, 47.
47 Northrop Frye, "'Conclusion' to Literary History of Canada," in *Divisions on a Ground*, ed. Polk, 58–69; see also Frye, *The Bush Garden*.
48 Giehmann, *Writing the Northland*, 48.
49 Coleman, "Science and Symbol," 22.

50 The so-called strong hypothesis of linguistic relativity suggests that language (categories and usage) *determines* thought, whereas the weak hypothesis suggests that linguistic categories and usage only *influence* human thought and decisions. On the theory of linguistic relativity, see Laura M. Ahearn, *Living Language: An Introduction to Linguistic Anthropology* (Hoboken: John Wiley and Sons, 2011).

51 Sangster, *The Iconic North*, 3.

52 Peter Kikkert and P. Whitney Lackenbauer, "'Men of Frontier Experience': Yukoners, Frontier Masculinity, and the First World War," *Northern Review* 44 (2017): 209–42.

53 Farish, "Frontier Engineering"; Farish, *The Contours of America's Cold War*, 173–9.

54 Janice Cavell, "The Second Frontier: The North in English-Canadian Historical Writing," *Canadian Historical Review* 83, no. 3 (September 2002): 364.

55 Kerry M. Abel, "Quelques arpents de neige? Or, How Fares the Myth of the North?," *Underhill Review* (Fall 2008): 2.

56 Suzanne Zeller, *Inventing Canada: Early Victorian Science and the Idea of a Transcontinental Nation* (Toronto: University of Toronto Press, 1987), 179.

57 Jesse Thistle, *From the Ashes: My Story of Being Métis, Homeless, and Finding My Way* (Toronto: Simon and Schuster, 2019).

58 Section 91(24) of the British North America Act (1867) gave the federal government unilateral control over Indigenous nations and territories; this led, in part, to the creation and dispatch of the NWMP during the Red River Rebellion (1869–70) and the Northwest Rebellion (1885). That dispatch violated existing treaties between Indigenous nations and the British Crown. See Edward Butts, "North-West Mounted Police (NWMP)," *Canadian Encyclopedia*, https://www.thecanadianencyclopedia.ca/en/article/north-west-mounted-police.

59 James Daschuk, *Clearing the Plains: Disease, Politics of Starvation, and the Loss of Aboriginal Life* (Regina: University of Regina Press, 2013); John S. Milloy, *A National Crime: The Canadian Government and the Residential School System* (Winnipeg: University of Manitoba Press, 2017).

60 Brenda Macdougall, "Space and Place within Aboriginal Epistemological Traditions: Recent Trends in Historical Scholarship," *Canadian Historical Review* 98, no. 1 (2017): 64–82; Brittany Luby, "From Milk-Medicine to Public (Re)Education Programs: An Examination of Anishinabek Mothers' Responses to Hydroelectric Flooding in the Treaty #3 District, 1900–1975," *Canadian Bulletin of Medical History* 32, no. 2 (2015): 363–89; Lianne C. Leddy, "Intersections of Indigenous and Environmental History in Canada," *Canadian Historical Review* 98, no. 1 (2017): 83–95; Mary Jane Logan McCallum, "Starvation, Experimentation, Segregation, and Trauma: Words for Reading Indigenous Health History," *Canadian Historical Review* 98, no. 1 (2017): 96–113.

61 Macdougall, "Space and Place within Aboriginal Epistemological Traditions."
62 Jeffrey Schiffer (@jeffschiffer), "When Europeans first stumbled upon Turtle Island there were 40 Million #Indigenous peoples speaking 1800 different languages," *Twitter*, 1 July 2020, https://www.pinterest.ca/pin/jeffrey-schiffer-phd-on-twitter–228839224804586914/.
63 Leddy, *Serpent River Resurgence*, 10.
64 Leddy, *Serpent River Resurgence*, 9.
65 Emilie Cameron, *Far Off Metal River: Inuit Lands, Settler Stories, and the Making of the Contemporary Arctic* (Vancouver: UBC Press, 2015), 9.
66 Here I add science to what Rosanna White calls the settler-colonial "assemblage of Arctic imaginations." See White, "Ceremonies of Possession" (PhD diss.), 14.
67 Bravo and Sörlin, eds., *Narrating the Arctic*; Sörlin, "The Historiography of the Enigmatic North," 559; Andrew Stuhl, *Unfreezing the Arctic: Science, Colonialism, and the Transformation of Inuit Lands* (Chicago: University of Chicago Press, 2016), 10, 100–1; Hans M. Carlson, "'That's the Place Where I Was Born': History, Narrative Ecology, and Politics in Canada's North," in *Ice Blink*, ed. Bocking and Martine, 295–332.
68 Shelley Wright, *Our Ice Is Vanishing / Sikuvut Nunguliqtuq: A History of Inuit, Newcomers, and Climate Change* (Montreal and Kingston: McGill–Queen's University Press, 2014), 8.
69 Jones-Imhotep, *The Unreliable Nation*, 191.
70 Tina Adcock, "Many Tiny Traces: Northern Exploration and Antimodernism between the Wars," in *Ice Blink*, ed. Bocking and Martine, 131–77.
71 For instance, see Stephen Bocking, *Ecologists and Environmental Politics: A History of Contemporary Ecology* (New Haven: Yale University Press, 1997); and Bocking, *Nature's Experts: Science, Politics, and the Environment* (New Brunswick: Rutgers University Press, 2004).
72 Stephen Bocking, "Indigenous Knowledge and the History of Science, Race, and Colonial Authority in Northern Canada," in *Rethinking the Great White North*, ed. Baldwin, Cameron, and Kobayashi, 59.
73 Mary Jane McCallum, "This Last Frontier: 'Isolation' and Aboriginal Health," *Canadian Bulletin of Medical History* 22, no. 1 (2005): 103–20.
74 Julie Cruikshank, *Do Glaciers Listen? Local Knowledge, Colonial Encounters, and Social Imagination* (Vancouver: UBC Press, 2005). See also Bocking, "Indigenous Knowledge," 39–61.
75 "Terry Audla – His Story," IQQAUMAVARA Project, http://www.iqqaumavara.com/en/familles/entrevues/terry-audla-his-story.
76 The culling of vast Indigenous nations and their confinement on reserves, the establishment and enforcement of band governance, and missionaries' long-term attempts to Christianize Indigenous peoples also did great damage

during what amounted to the cultural genocide of Indigenous peoples in what is now Canada. See Canada, *Honouring the Truth, Reconciling for the Future: Summary of the Final Report of the Truth and Reconciliation Commission of Canada* (Ottawa: Truth and Reconciliation Commission of Canada: 2015).
77 Piper and Sandlos, "A Broken Frontier."
78 Bocking, "Indigenous Knowledge," 41.
79 Sörlin, "The Historiography of the Enigmatic North," 560.
80 Russell, *War and Nature*; Hamblin, *Poison in the Well*; Edgington, *Range Wars*; Reno, *Military Waste*.
81 Lackenbauer, *Battle Grounds*, 10.
82 Lackenbauer, *The Canadian Rangers*, 152.
83 Canada, "The Distant Early Warning Line: An Environmental Legacy Project," https://www.canada.ca/en/department-national-defence/corporate/video/other/distant-early-warning-line-an-environmental-legacy-project.html.
84 Parr, *Sensing Changes*; Souchen, *War Junk*.
85 Canadian War Museum, Hartland Molson Library Collection, Defence Research Board, *Heads of Establishments Conference* (Fort Churchill: Defence Research Northern Laboratory, 10–14 March 1954).
86 LAC, RG 24, vol. 2425, file Speeches – Reporting etc. 1947 – March 1953, vol. 1, Policy and Plans for Defence Research in Canada: A Preliminary Review by O.M. Solandt, Director General of Defence Research, Appendix A: Suggested Fields for Defence Research, Ottawa, May 1946, 12.
87 Lester Pearson, Memorandum from Under-Secretary of State for External Affairs to Secretary of State for External Affairs, 19 January 1948, in *Documents on Canadian External Relations*, vol. 14: *1948* (Ottawa: Department of Foreign Affairs and International Trade, 1994), 1514.
88 Lester B. Pearson, "Canada Looks 'Down North,'" *Foreign Affairs: An American Quarterly Review* (July 1946): 638–48.
89 Grace, *Canada and the Idea of North*, 9.
90 Pearson, Cabinet Memorandum, in *Documents on Canadian External Relations*, vol. 14, 1514.
91 For a detailed retrospective of the "Great White North" as an idea and historical concept, see Baldwin, Cameron, and Kobayashi, eds., *Rethinking the Great White North*.
92 Lackenbauer and Kikkert, *Lessons in Northern Operations*, vi.
93 I borrow the definitions here from historian P. Whitney Lackenbauer; see "From Polar Race to Polar Saga: An Integrated Strategy for Canada and the Circumpolar World," in *Canada and the Changing Arctic*, ed. Griffiths, Huebert, and Lackenbauer, 69–179; note on definitions provided on pages 71–2.
94 Canadian War Museum, Hartland Molson Library Collection, Defence Research Board, *Heads of Establishments Conference* (Fort Churchill: Defence Research Northern Laboratory, 10–14 March 1954).

95 The "Canadian North" and "northern Canada" include the territories and the northern regions of Quebec, Ontario, Manitoba, Saskatchewan, Alberta, and British Columbia; "the North" includes Yukon, the Northwest Territories, and Nunavut – commonly referred to as "North of 60."
96 The US National Archives and Records Administration (NARA), College Park, MD, RG 319, vol. 2761, file Precis on Canada's Arctic and Sub-Arctic North of the 60th Parallel, Ottawa: Department of Indian Affairs, January 1948, 32.
97 During the late 1940s and early 1950s, the territorial government for the NWT included Hugh L. Keenleyside (Commissioner); Roy A. Gibson (Deputy Commissioner); John G. McNiven, Louis de la C. Audette, Harold B. Godwin, James G. Wright; Stuart T. Wood; and Robert A. Hoey (Members of Council).
98 I borrow definitions for "sovereignty" and "security" from Canadian Arctic policy scholar Rob Huebert. See "Canadian Arctic Security: Shifting Challenges," in *International Relations and the Arctic*, ed. Murray and Nuttall, 132.

1. Fort Churchill and Defence Research Northern Laboratory

1 Galloway, "The Army Says Goodbye to 'The Shining Land,'" 40.
2 For a detailed account of the establishment of the JSES at Fort Churchill, see Iarocci, "Opening the North."
3 On the militarization of northern Canada and particularly the Canadian Northwest during the Second World War, see Coates and Morrison, *The Alaska Highway in World War II*. On the Canol Project, see Finnie, *CANOL*.
4 On the establishment of the Permanent Joint Board on Defence, see Sokolsky and Jockel, eds., *Fifty Years of Canada–United States Defense Cooperation*.
5 Margaret A. Carroll, "Defence Forces Operations in Hudson Bay," in *Science, History and Hudson Bay*, vol. 2, ed. Beals and Shenstone, 907.
6 James Reston, "Unified Arctic Defense Plan Proposed by U.S. to Canada: Joint Bases, Weather Stations in Far North, Coordinated Training and Equipping of Forces in Scheme Put to Ottawa," *New York Times*, 18 May 1946, 1; Jan Drent, "'A Good, Workable Little Fleet': Canadian Naval Policy, 1945–1950," in *A Nation's Navy*, ed. Hadley, Huebert, and Crickard, 205–35.
7 For information about the MSF, see Maloney, "The Mobile Striking Force," 75–88; Horn, *Bastard Sons*; Raymond Stouffer, "Military Culture and the Mobile Striking Force," in Lackenbauer, *De-icing required!*, 58–70; and Coates et al., *Arctic Front*, 65.
8 The concept of the Canadian Rangers emerged in 1947, writes P. Whitney Lackenbauer, when officials in Ottawa created "a military space for citizens who live in isolated coastal and northern communities and who would not otherwise be suitable for or interested in military service." See Lackenbauer, *The Canadian Rangers*, 7.

9 P. Whitney Lackenbauer, "At the Crossroads of Militarism and Modernization: Inuit–Military Relations in the Cold War Arctic," in *Roots of Entanglement*, ed. Rutherdale, Abel, and Lackenbauer, 137.
10 Major K.C. Eyre, "Tactics in the Snow: The Development of a Concept," *Canadian Defence Quarterly* 4, no. 4 (Spring 1975), 7–12.
11 Lackenbauer and Kikkert, *Lessons in Northern Operations*, vii–viii.
12 Eyre, "Custos Borealis" (PhD diss.), 150; Lackenbauer and Kikkert, *Lessons in Northern Operations*, viii.
13 Lackenbauer and Kikkert, *Lessons in Northern Operations*, viii.
14 Eyre, "Custos Borealis," 152; Lackenbauer and Peter Kikkert, *Lessons in Northern Operations*, viii.
15 Lackenbauer and Kikkert, *Lessons in Northern Operations*, viii–ix; Hugh Halliday, "Recapturing the North: Exercises 'Eskimo,' 'Polar Bear,' and 'Lemming' 1945," *Canadian Military History* 6, no. 2 (Autumn 1997): 29–38.
16 Coates et al., *Arctic Front*, 55; Lackenbauer and Kikkert, *Lessons in Northern Operations*, ix.
17 Lackenbauer and Kikkert, *Lessons in Northern Operations*, v.
18 Lackenbauer and Kikkert, *Lessons in Northern Operations*, vi.
19 Lackenbauer and Kikkert, *Lessons in Northern Operations*. See also Godefroy, *In Peace Prepared*, 91.
20 LAC, RG 24, vol. 7330, file DRBS-100-40/8 pt. 1, W.W. Goforth, "Defence Geography of Canada," 22 August 1945.
21 Matthew Evenden, "Mapping Cold War Canada: George Kimble's *Canadian Military Geography*, 1949," in *Method and Meaning in Canadian Environmental History*, ed. MacEachern and Turkel, 257.
22 Government of Canada, *Minutes of the First Meeting of the Cabinet Committee on Research and Defence, 4 December, 1945*.
23 In addition to winter warfare, the paragraph listed chemical warfare, radar, electronics, vehicle design, and protective clothing as example fields of Canadian speciality. See DHH, George Lindsey fonds 87/253, box 15, 85/334 file 104, Defence Research Estimates & Supplementary Information, "To the Cabinet Committee on Research for Defence: Post-War Policy for Scientific Research for Defence," 31 October 1945.
24 Foulkes had commanded the Second Canadian Division in Normandy and the First Canadian Corps in Italy during the war. His battlefield experience gave him a keen understanding of modern combat, and he was impressed with the tremendous influence of wartime science. Considerations for the application of military research in peacetime were at the front of his mind as the hostilities ended.
25 For a biography of Solandt, see Ridler, *Maestro of Science*. See also Law, Lindsey, and Grenville, eds. *Perspectives in Science and Technology*.

26 LAC, RG 24, vol. 2425, file Speeches – Reporting etc. 1947 – March 1953 vol. 1, A Five Year Plan for Defence Research and Development, 1950–1954, April 1950, 1.
27 Godefroy, *In Peace Prepared*, 91.
28 Galloway, "The Army Says Goodbye to 'The Shining Land,'" 41.
29 Document 879, DEA/7-DA (S), E.W.T. Gill, Memorandum for Cabinet Defence Committee, 29 July 1947. See Norman Hillmer and Donald Page, eds., *Documents on Canadian External Relations*, vol. 13: *1947* (Ottawa: External Affairs and International Trade Canada, 1993), 1501–2.
30 The majority of the work in 1947–8 was the responsibility of No. 18 Works Company, Royal Canadian Engineers, who completed the construction with the assistance of civilian labour.
31 Carroll, "Defence Forces Operations in Hudson Bay," in *Science, History, and Hudson Bay*, vol. 2, ed. Beals and Shenstone, 910.
32 General A.G.L. McNaughton, quoted in "Experiments at Churchill: Canada, U.S. Defence Chiefs Here for Talks," *Winnipeg Free Press*, 18 February 1947, 1.
33 "Soviet Accepts Invitation to Churchill," *Winnipeg Free Press*, 18 February 1947, 1.
34 Iarocci, "Opening the North," 79.
35 "IRON FIST: Soviet Paper Raps Canada, U.S. Defence," *Winnipeg Free Press*, 18 February 1947, 1.
36 Brooke Claxton, quoted in "Claxton Says Churchill Rumor Wrong," *Winnipeg Free Press*, 18 February 1947, 1.
37 For information about Musk Ox, see John Lauder (author), P. Whitney Lackenbauer, and Peter Kikkert, eds., *Tracks North: The Story of Exercise Muskox* (Arctic Operational History Series, No. 5, 2018); Thrasher, *Exercise Musk Ox:* (MA thesis); Lackenbauer and Kikkert, *Lessons in Northern Operations*, x–xii. See also the 1946 film "Expedition in the Arctic," which shows Domashev standing with a group of observers during Musk Ox. Chapter 4 discusses the exercise from the Canadian point of view in detail.
38 "Russian Tight-Lipped: Inspection Party off to Churchill after Spending Night in Winnipeg," *Winnipeg Free Press*, 27 February 1947, 1.
39 LAC, RG 24, vol. 7329, file DRBS-100-30/0, "Department of National Defence Agreement for Expenditure of U.S. Funds, Fort Churchill, Manitoba, Canada"; Appendix "D" to Long Term Plan for the Combined Experimental and Training Station Fort Churchill, 21 April 1947.
40 LAC, RG 24, vol. 7329, file DRBS-100-30/0, Long Term Plan for the Combined Experimental and Training Station Fort Churchill, 21 April 1947.
41 LAC, RG 24, vol. 7329, file DRBS-100-30/0, Joint Organization Order No: Fort Churchill.
42 At the end of the Cold War, Canada and the US jointly funded and constructed a fourth radar system called the North Warning System to replace the DEW Line. The stations transmit data south to Canadian and

US military command centres, enabling identification of incoming aircraft and intercontinental ballistic missiles.

43 O'Brian, *The Bomb in the Wilderness*, 69.
44 Langford and Langford, *A Cold War Tourist and His Camera*, 7.
45 O'Brian, *The Bomb in the Wilderness*, 71.
46 O'Brian, *The Bomb in the Wilderness*.
47 Ryan Dean and P. Whitey Lackenbauer, "A Northern Nuclear Nightmare? Operation Morning Light and the Recovery of Cosmos 954 in the Northwest Territories, 1978," in *Nuclear North*, ed. Colbourn and Andrews Sayle, 181–206.
48 LAC, RG 24, vol. 4117, file DRBS 2-1-172-10, Research and Development Board, Washington D.C., "Strategic Guidance," 7 December 1949.
49 On scientific intelligence in the DRB, see Wiseman, "The Origins and Early History."
50 LAC, RG 24, vol. 4117, file DRBS 2-1-172-10, Research and Development Board, Washington, D.C., "Strategic Guidance," 7 December 1949.
51 LAC, RG 24, vol. 4117, file DRBS 2-1-172-10, Research and Development Board, Washington, D.C., "Strategic Guidance," 7 December 1949.
52 Wiseman, "The Origins and Early History," 8.
53 LAC, RG 24, vol. 17601, file 004–100–74/34–3, DRB, Scientific Intelligence Division, Organization and Functions, 9 November 1949.
54 Webb, "'How to Raise a Curtain.'"
55 Keenleyside, *Memoirs*, vol. 2, 308.
56 Keenleyside, *Memoirs*, vol. 2.
57 Keenleyside, *Memoirs*, vol. 2.
58 Lize-Marié van der Watt, Peder Roberts, and Julia Lajus, "Institutions and the Changing Nature of Arctic Research during the Early Cold War," in *Cold Science*, ed. Bocking and Heidt, 205.
59 Grant, *Polar Imperative*, 279.
60 McCannon, "The Commissariat of Ice."
61 Keenleyside, *Memoirs*, vol. 2, 308–9.
62 For information about the Black Brant and Alouette programs, see Godefroy, *Defence and Discovery*.
63 Galloway, "The Army Says Goodbye to 'The Shining Land,'" 44.
64 CWM, HMLC, Defence Research Board, *Heads of Establishments Conference*. Transair Limited, a civilian airline, also operated daily scheduled flights to and from Winnipeg in the 1950s, save for weather delays and Sundays. The airline flew to and from Ottawa and Montreal once a week.
65 LAC, RG 24, vol. 20317, file 951.009 (D71) DRBS 100-40/18 vol. 1, "HQS 2001-1975/9 – Use of Fort Churchill by the Canadian Army," 10 June 1952, 2.
66 CWM, HMLC, Defence Research Board, *Heads of Establishments Conference*.
67 Galloway, "The Army Says Goodbye to 'The Shining Land,'" 53.
68 CWM, HMLC, Defence Research Board, *Heads of Establishments Conference*.

69 CWM, HMLC, Defence Research Board, *Heads of Establishments Conference*.
70 CWM, HMLC, Defence Research Board, *Heads of Establishments Conference*.
71 CWM, HMLC, Defence Research Board, *Heads of Establishments Conference*.
72 CWM, HMLC, Defence Research Board, *Heads of Establishments Conference*.
73 CWM, HMLC, Defence Research Board, *Heads of Establishments Conference*.
74 The Department of Transport Radio Division also operated a radio station at Churchill that received all meteorological data transmitted from weather stations and ships in the eastern Canadian Arctic, and personnel in the ionospheric station supported the sounding rocket program of the nearby Churchill Research Range in the late 1950s and early 1960s.
75 CWM, HMLC, Defence Research Board, *Heads of Establishments Conference*.
76 US Army, *An Introduction to Churchill, Fort Churchill and Surrounding Area*.
77 US Army, *An Introduction to Churchill, Fort Churchill and Surrounding Area*.
78 Galloway, "The Army Says Goodbye to 'The Shining Land,'" 42.
79 Pennie, "Defence Research Northern Laboratory."
80 CWM, HMLC, Defence Research Board, *Heads of Establishments Conference*.
81 Goodspeed's official history of the DRB states that Croal arrived in September 1947, but Croal himself wrote that he first arrived in August. See J.P. Croal, "First in the Field," in *Defence Research Northern Laboratory 1947–1965*, Report DR 179, ed. Pennie (Ottawa: Defence Research Board, 1966), 31–6; Goodspeed, *A History of the Defence Research Board of Canada*, 180; Carroll, "Defence Forces Operations in Hudson Bay," in *Science, History, and Hudson Bay*, vol. 2, ed. Beals and Shenstone, 911–12; Turner, "The Defence Research Board of Canada" (PhD diss.), 93–4.
82 J.P. Croal, "First in the Field," 31.
83 J.P. Croal, "First in the Field," 31–2.
84 In addition to Croal, the DRB staff at Fort Churchill included A.V. Hannam, Guy Marier, W. Beckel, and J.D. O'Connor. See Goodspeed, *A History of the Defence Research Board of Canada*, 180.
85 A.C. Jones, "Churchill Diary 1946–1949," in *Defence Research Northern Laboratory 1947–1965*, ed. Pennie, 12–30; Carroll, "Defence Forces Operations in Hudson Bay," in *Science, History, and Hudson Bay*, vol. 2, ed. Beals and Shenstone, 912.
86 Goodspeed, *A History of the Defence Research Board of Canada*, 182.
87 LAC, RG 24, vol. 7329, file DRBS-100-30/0, DRBS 52-750-1 (Arct), C.P. McNamara, 14 July 1948.
88 LAC, RG 24, vol. 7329, file DRBS-100-30/0, Joint Organization Order No 1: Fort Churchill, 9 May 1952.
89 Goodspeed, *A History of the Defence Research Board of Canada*, 180.
90 DHH, George Lindsey fonds (87/253), box 15, 85/334 file 107, "Copy No. 12 Department of National Defence Estimates 1949–50: Defence Research and Development," 194.

91 DHH, George Lindsey fonds (87/253), box 15, 85/334 file 107, "Copy No. 12 Department of National Defence Estimates 1949–50: Defence Research and Development," 195.
92 CWM, HMLC, Defence Research Board, *Heads of Establishments Conference.*
93 CWM, HMLC, Defence Research Board, *Heads of Establishments Conference.*
94 CWM, HMLC, Defence Research Board, *Heads of Establishments Conference.*
95 For a full list of DRNL superintendents and their biographical information, see Pennie, ed., *Defence Research Northern Laboratory.* See also Pennie, "Significant Contributions of Canadian Defence Science," in *Perspectives in Science and Technology,* 83.
96 Pennie, ed., *Defence Research Northern Laboratory 1947–1965,* 1.
97 LAC, RG 24, vol. 10341, file August 54 to March 55, "Pennie Gets New Post," *Windsor Star,* 28 August 1954; "Archie Pennie Named to High Defence Post," *Ottawa Journal,* 28 August 1954.
98 Pennie, "Defence Research Northern Laboratory," *Canadian Army Journal* 10, no. 1 (January 1956), 48.
99 On DRB research at Suffield, see Avery, *Pathogens for War,* 75–7. On research at Valcartier, see Directorate of Public Relations, National Defence, Ottawa, "New Anti-Tank Weapon is Developed for Canadian Army," *Canadian Army Journal* 9, no. 2 (April 1955): 46–8; Director of Artillery, Army Headquarters, Ottawa, "Guns and Guided Missiles," *Canadian Army Journal* 9, no. 2 (April 1955): 49–64; and Defence Research Board, "DRB Pioneers Missile Study," *Canadian Army Journal* 15, no. 3 (Summer 1961): 54.
100 Pennie, "Defence Research Northern Laboratory," 48.
101 LAC, RG 85, vol. 299, file 1009-2[5], *Defence Research Northern Laboratory: Progress Report on Indoctrination Training for Military Operations in the North, DRNL Project Report No. 4* (Ottawa: DRB, Department of National Defence, Canada, 1954), 2.
102 Goodspeed, *A History of the Defence Research Board of Canada,* 182.
103 Goodspeed, *A History of the Defence Research Board of Canada,* 182–3.
104 Pennie, "Significant Contributions of Canadian Defence Science," in *Perspectives in Science and Technology,* ed. Law, Lindsay, and Grenville, 82.
105 Carroll, "Defence Forces Operations in Hudson Bay," in *Science, History, and Hudson Bay,* vol. 2, ed. Beals and Shenstone, 923.
106 Coffey, *Debriefing of an Ice Floe.*
107 Carroll, "Defence Forces Operations in Hudson Bay," in *Science, History, and Hudson Bay,* vol. 2, ed. Beals and Shenstone, 922.
108 "Study More Defence in Far North," *Brandon Daily Sun,* 3 February 1953, 1.
109 "Big Atomic Cannon Tested at Churchill," *Winnipeg Free Press,* 18 February 1954, 1.
110 LAC, RG 24, vol. 7329, file DRBS-100-30/0, DRBS 52-0-240 (Arct), Visit to DRNL, 22 December 1954.

111 LAC, RG 24, vol. 2425, file Speeches – Reporting etc. 1947 – March 1953 vol. 1, Policy and Plans for Defence Research in Canada: A Preliminary Review by O.M. Solandt, Director General of Defence Research, Part II: A Suggested Policy to Guide External Relations on Defence Research and Development, Ottawa, May 1946, 5.
112 LAC, RG 24, vol. 4168, file 225-1-53-1 vol. 3, "Defence Research Board, Department of National Defence: Programme of Scientific Work for the Fiscal Year 1955–1956 – Prepared for the Advisory Panel of Scientific Policy of the Privy Council Committee on Scientific and Industrial Research," 10.
113 Iarocci, "Opening the North."
114 Goodspeed, *A History of the Defence Research Board of Canada*, 188.

2. Acclimatization, Cold Tolerance, and Biochemical Experimentation

1 LAC, RG 128, vol. 258, Malcolm Brown, Queen's University Arctic Expeditions 1947, 1948, 1949, 1950: Progress Report, December 1950. The Defence Research Board published the report in October 1951 as Dr. G. Malcolm Brown, *Progress Report on Clinical and Biochemical Studies of the Eskimo* (Ottawa: DRB, Department of National Defence, 1951).
2 For a brief description of DRB Grant No. 80, "Clinical and Biochemical Studies of the Eskimo," see National Archives and Records Administration (NARA), College Park, Maryland, RG 319, box 865, DRB: Annual Report on the Progress of Defence Medical Research, Report No. D.R., 15 December 1949.
3 This chapter uses the word *Inuit* to refer to persons indigenous to the Canadian North, who were the "test subjects" of the discussed acclimatization research. I use the term *Inuit* because the evidentiary record does not distinguish the "Eskimo" persons subjected to the research, except for one reference to "Iviliks." It is possible that the scientists conducted research on Indigenous persons other than Inuit.
4 While the December 1950 report detailed experiments on 282 "Eskimo test subjects," Brown co-published a medical research article in 1954 that detailed further experiments on an additional six adult "Eskimo" test subjects. See Baugh, Bird, Brown, et al., "Blood Volumes of Eskimos and White Men."
5 Brown, Bird, et al., "The Circulation in Cold Acclimatization."
6 Brown, Bird, et al., "The Circulation in Cold Acclimatization." See also Brown, Bird, et al., "Cold Acclimatization."
7 Researchers performed needle biopsies of the liver in two so-called test subjects in 1947; see "Northern Research Reports: Medicine," *Arctic* 1, no. 1 (1948): 65. See also in 1948: Brown, "Northern Research Reports: Medical Investigation at Southampton Island." The December 1950 report (see note 1) states that "punch biopsies have been carried out on ten subjects."

8 The word *Inuit* means "people" and is a plural reference to three or more people. The word *Inuk* is a singular reference to one person. For a more detailed explanation, see Marcus, *Relocating Eden*, xv–xvi.
9 LAC, RG 128, vol. 258, file Miscellaneous part 2, Recent Research on Cold Acclimatization and Other Arctic Medical Problems, G. Malcolm Brown, 14.
10 LAC, RG 128, vol. 258, file Miscellaneous part 2, Recent Research on Cold Acclimatization and Other Arctic Medical Problems, G. Malcolm Brown, 14.
11 Grygier, *A Long Way from Home*, 58.
12 For details on the Canadian Government Eastern Arctic Patrol of 1935, see Rabinowitch, "Clinical and Other Observations."
13 The Hudson's Bay Company ship *Nascopie* patrolled eastern Arctic waters in 1947, but the main patrol belonged to the *C.D. Howe*, which made its maiden voyage in 1950 and sustained operations until the patrol was discontinued in 1969; see Grygier, *A Long Was from Home*, 86–103.
14 Coates et al., *Arctic Front*.
15 Brown, "Cold Acclimatization in Eskimo."
16 For a comprehensive account of southern views about Indigenous health in the Canadian North, see Piper and Sandlos, "A Broken Frontier."
17 C.D. Naylor, "The CMA's First Code of Ethics: Medical Morality or Borrowed Ideology?" *Journal of Canadian Studies* 17, no. 4 (1982–3): 20.
18 For a brief history of the Nuremberg Code, see Weindling, "The Origins of Informed Consent." For specific reference to the formalization of ethical guidelines for medical research in Canada, see Cotton, Manning-Kroon, and McNally, "An Overview of the Law Regarding Informed Consent."
19 See Brown, "Northern Research Reports," 70. The experiences of Inuit who had contact with Brown and his group of medical researchers may be lost to history. Unfortunately, it seems those experiences went unrecorded and untold. I contacted and received no response from Aboriginal Affairs and Northern Development Canada, nor any from the Nunavut Food Security Coalition. The Government of Nunavut's Department of Health sent a kind response, recommending further contacts, but had no information pertaining to environmental or medical research in the 1940s and 1950s. Likewise, the friendly people at the Unikkaarvik Visitors Centre of Nunavut Tourism graciously provided a list of contacts, but also had no information. The Hamlet of Coral Harbour and the Hamlet of Igloolik also politely responded to my inquiries, but equally had no information. If Inuit experiences or stories have survived, the details may further elucidate what information we do have and/or bring to light important insights for further consideration. Such insights may point to what Mary-Ellen Kelm refers to as "medical pluralism," which recognizes the resistance of Indigenous peoples to colonial medicine. See Kelm, *Colonizing Bodies*, 153–72. Currently, it remains unclear whether Inuit

subjected to cold acclimatization research resisted or embraced medical treatment services. Available records state that the "Eskimos" were "very cooperative."

20 LAC, RG 128, vol. 258, Malcolm Brown, Queen's University Arctic Expeditions 1947, 1948, 1949, 1950: Progress Report, December 1950.
21 Brown, Bird, et al., "Cold Acclimatization," 259.
22 Stephen Bocking, "Indigenous Knowledge and the History of Science, Race, and Colonial Authority in Northern Canada," in *Rethinking the Great White North*, ed. Baldwin, Cameron, and Kobayashi, 46.
23 Mosby, "Administering Colonial Science."
24 P. Whitney Lackenbauer has challenged protagonist narrative structures that pit "Aboriginal peoples on one hand, and the 'Euro-Canadian' (non-Aboriginal) camp on the other." His extensive work on military–Aboriginal relations has shown that avoiding dichotomous language can serve well in assessing complex historical interactions; see Lackenbauer, "The Irony and the Tragedy of Negotiated Space."
25 Acclimatization research was not unique in this regard, as evidenced by the diverse range of other federal scientific research initiatives that became essential to Canada's Cold War agenda; see, for instance, Bocking, "Seeking the Arctic."
26 On the "colonial project," see Burnett, *Taking Medicine*, 7.
27 See, by date of publication, Lackenbauer and Farish, "The Cold War on Canadian Soil"; Kasurak, *A National Force*; and Godefroy, *In Peace Prepared*.
28 Daniel Heidt and P. Whitney Lackenbauer, "Sovereignty for Hire: Civilian Airlift Contractors and the Distant Early Warning (DEW) Line, 1954–1961," in Lackenbauer, *De-icing Required!*, 95–112.
29 LAC, RG 85, vol. 299, file 1009-2[5], *Annual Report on the Progress of Environmental Protection Research, December 1953: Report No. DR 80* (Ottawa: DRB, Department of National Defence, Canada, 1954), 1.
30 To be precise, the DRB's Environmental Protection Program coordinated and facilitated various scientific research projects involving scientists and engineers in the National Research Council, the Department of National Health and Welfare, the Department of Agriculture, and the Ontario Research Foundation, to name some of the main sponsors and participants.
31 LAC, RG 85, vol. 299, file 1009-2[3], DRBS 3-750-43-2, DRB: Minutes of the 2/51 Meeting of the Arctic Research Advisory Committee, 30 April 1951, "Activities of the Defence Research Northern Laboratory."
32 LAC, RG 85, vol. 299, file 1009-2[3], DRBS 3-750-43-2, DRB: Minutes of the 2/51 Meeting of the Arctic Research Advisory Committee, 30 April 1951, "Activities of the Defence Research Northern Laboratory."
33 Mary Jane McCallum, "This Last Frontier: 'Isolation' and Aboriginal Health," *Canadian Bulletin of Medical History* 22, no. 1 (2005): 103–20.

34 For details pertaining to the personal life and professional working career of Malcolm Brown, see LAC, RG 128, vol. 259, file Dr. G. Malcolm Brown – List of Publications, Curriculum Vitae, etc. (pt. 1), Curriculum Vitae – Dr. G. Malcolm Brown, 11 March 1978.
35 The personal papers of G. Malcolm Brown are located at LAC as part of RG 128 (Medical Research Council/Canadian Institutes of Health Research). The papers comprise seven volumes of documentation, nearly the entire contents of which were released to me, following review, under the Access to Information and Privacy Act. Volumes and files of interest are listed extensively in the notes of this chapter.
36 Alison Li, *J.B. Collip and the Development of Medical Research in Canada* (Montreal and Kingston: McGill–Queen's University Press, 2003), 158.
37 Li, *J.B. Collip*, 159.
38 Li, *J.B. Collip*, 159.
39 In the December 1950 progress report, Brown listed the following personnel: R.G. Sinclair, L. Bruce Cronk, G.C. Clarke, J.E. Green, John Page, J.E. Gibbons, D.L. (Don) Whittier, Frederick deSinner, J.D. Hatcher, T.J. (Thomas) Boag, L.C. Boag, Donald Delahaye, Morley G. Whillans, and Gordon Bird. Although they did not travel to the North, Dorothy Knapman, Eve Minovitch, Claire McAdam, Shirley Davy, and Mary M. Sleeth assisted biochemical work in Kingston. See LAC, RG 128, vol. 258, Malcolm Brown, Queen's University Arctic Expeditions 1947, 1948, 1949, 1950: Progress Report, December 1950.
40 "Flying to Arctic to Ascertain If Eskimo Immune to Cancer," *Toronto Daily Star*, 1 August 1947, 3.
41 "Medsmen Rub Noses with Eskimos During Cool Summer Study," *Queen's Journal*, 5 October 1948, 5, in *Queen's Journal*, vol. 76: 1948–9, Queen's University Archives.
42 Jordan wrote the introduction to Ronald Wild's 1955 book about the *Nascopie's* Captain Smellie: Wild, *Arctic Command*. See also Geller, "The 'True North' in Pictures?," 180.
43 "Medsmen Rub Noses with Eskimos," 5.
44 "Medsmen Rub Noses with Eskimos," 5.
45 LAC, RG 24, vol. 4129, file DRBS 4-78-53, vol. 1, Minutes of the First Meeting of the Panel on Arctic Medical Research, DRB, 16 December 1948.
46 LAC, RG 24, vol. 2529, file 801-100-M91, vol. 1, Minutes of the Third Meeting of the Medical Research Advisory Committee – Appendix "C": Arctic Medical Research Panel, 3 February 1950.
47 LAC, RG 24, vol. 2529, file 801-100-M91 pt. 1, DRB: Arctic Medical Research Panel, 12 January 1949.
48 LAC, RG 85, vol. 299, file 1009-2[3], DRBS 3-750-43-2 DRB: Minutes of the 1/51 Meeting of the Arctic Research Advisory Committee, Appendix "A" Confidential, 2.

49 For an outline of the security policy of the DRB, see LAC, RG 24, vol. 2529, file 801-100-M91 pt. 1, DRB: Care and Communication of Classified Information, 10 January 1949.
50 LAC, RG 24, vol. 2529, file 801-100-M91 pt. 1, DRB: Care and Communication of Classified Information, 10 January 1949.
51 See LAC, RG 24, vol. 2484, file HQS-726-40-17-11, Defence Research Northern Laboratory: Acclimatization Research Programme 1949–50, Fort Churchill. For more information about the DRB and northern military science, see Wiseman, "The Development of Cold War Soldiery."
52 NARA, RG 319, box 865, DRB: Annual Report on the Progress of Defence Medical Research, Report No. D.R. 15, December 1949.
53 NARA, RG 319, box 856, "DRB Second Semi-Annual Report: 1 October 1947–31 March 1948," 5 June 1948, 84.
54 NARA, RG 319, box 856, "DRB Third Semi-Annual Report: 1 April – 30 September 1948," 10 December 1948, 153–8. The five projects included Grants No. 79 (W.H. Johnson, "Physiology of Motion Sickness"), 80 (Malcolm Brown, "Clinical and Biochemical Studies on the Eskimo"), 81 (Louis-Paul Dugal, "Physiological Factors Involved in Resistance and Acclimatization to Cold Temperatures"), 82 (C.K. van Rooyen and L. McClelland, "Studies on the Inhibition of Virus Multiplication"), and 83 (J.W. Stevenson, "Studies of the Effects of the Toxin of Clostridium Botulinum Type A on the Transmission of Nerve Impulses").
55 NARA, RG 319, box 856, "DRB Third Semi-Annual Report: 1 April–30 September 1948," 10 December 1948, 153.
56 NARA, RG 319, box 856, "DRB Second Semi-Annual Report: 1 October 1947–31 March 1948," 5 June 1948, 90.
57 Ibid.
58 NARA, RG 319, box 856, "DRB Third Semi-Annual Report: 1 April–30 September 1948," 10 December 1948, 146.
59 "Northern Research Reports – Medicine: Queen's University Expedition to Southampton Island," 65.
60 "Northern Research Reports – Medicine: Queen's University Expedition to Southampton Island."
61 See Mosby, "Administering Colonial Science."
62 Brown, Bird, et al., "Cold Acclimatization," 259.
63 See LAC, RG 128, vol. 237, file Ascorbic acid – the response to cold and other Dugal work. For additional data charts, see Baugh, Bird, Brown, et al., "Blood Volumes of Eskimos and White Men"; Brown, Bird et al., "The Circulation in Cold Acclimatization"; and Brown, Bird, et al., "Cold Acclimatization."
64 Brown, Bird, et al., "Cold Acclimatization," 260.
65 LAC, RG 128, vol. 237, file Acclimatization, Alan C. Burton, Abstract of Discussion on Acclimatization to Cold. Burton founded the Department of

Biophysics at the University of Western Ontario and led it between 1948 and 1970. Considered a founding father of modern biophysics and a pioneer of interdisciplinary health research, he was a key scientist for the Canadian armed services. He worked on the design and development of protective clothing for the NRC during the Second World War and served on the DRB's Arctic Medical Research panel during the 1950s and 1960s. See Wiseman, "The Weather Factory."

66 Burton and Edholm, *Man in a Cold Environment*, xi.
67 Lex Schrag, "Canadian Lemmings Have Sense: Low-Slung Arctic Beastie Will Be Shaved for Test," *Globe and Mail*, 20 May 1952, 10.
68 Research involving lemmings continued for no fewer than three years at DRNL, where scientists expressed great interest in finding out how small rodents survive the rigours of the Arctic. "Indeed the sarcastic smile on the Eskimo faces reveals that our reputation of 'strange' white people is not changed when we ask them this question," wrote J.S. LeBlanc in a letter to the editor about reports of Arctic lemmings committing mass suicide. See Dr. J.S. LeBlanc, DRNL, Fort Churchill, Manitoba, "Letters to the Editors of the Journal: Ways of the Lemmings," *Ottawa Journal*, 15 February 1955.
69 LAC, RG 24, vol. 4117, file DRBS 2-150-302, Defence Research Northern Laboratory Fort Churchill, Manitoba: A Summary of Operational Research Studies with Human Resources Implications Conducted at DRNL during 1952–3, 8 January 1954, 5.
70 LAC, RG 24, vol. 4117, file DRBS 2-150-302, 6.
71 LAC, RG 24, vol. 4117, file DRBS 2-150-302, 8.
72 LAC, RG 24, vol. 4117, file DRBS 2-150-302, 13.
73 "Red Cross Blood Donor Clinic Helps Arctic Survival Research," *Winnipeg Free Press*, 28 August 1954, 3.
74 "Red Cross blood donor clinic helps Arctic survival research."
75 LAC, RG 24, vol. 10341, file August 54 to March 55, "Why Don't Eskimos Freeze? Blood Tests May Help 'Warm' Army," *Winnipeg Tribune*, 28 August 1954.
76 Eileen Jacob, "Fort Churchill Military Base Photos," ca. 2003. Accessed and viewed online; the website is no longer active.
77 DND generated a staggering $400,000 in revenue in 1962 alone, with some $22,000 outstanding and considered presumably uncollectable. See Fort Churchill Sub-Committee, Advisory Committee on Northern Development, DND Report on Proposed Assumption of Canadian Army and RCAF Responsibilities by other Government Departments, 31 October 1963, file G-1988-004: 8–10 (CVC/Churchill), Department of the Executive fonds, NWT Archives/Northwest Territories.
78 The word *Inuuk* refers to two people, whereas the word *Inuit* refers to three or more people.

79 "Study Immunity: Eskimos Handle Icy Metal," *Hamilton Spectator*, 21 December 1954.
80 "Study Immunity: Eskimos Handle Icy Metal."
81 The drugs were priscol, administered in tablet form, and tetraethyl ammonium chloride, which doctors injected into patients. Colonel Francis Pruitt, consultant on internal medicine for the US Army's Far East command, commented that the drugs were "very successful in giving relief." See LAC, RG 128, vol. 237, file Frostbite, Extract from "The Evening Star," Washington: Two New Drugs Used to Combat Frostbite, 30 November 1950.
82 For instance, see LAC, RG 128, vol. 237, file Frostbite, T.Ya Aryev (Leningrad), "On the Question of the Pathology and Clinical Treatment of General and Local Hypothermia," *Klinicheskaya Medicina* (USSR) 28, no. 3 (1950): 15–24, translated by E.R. Hope, Scientific Intelligence Division, Defence Research Board, 22 June 1950.
83 Lux, *Separate Beds*, 112.
84 Lux, *Separate Beds*.
85 LAC, RG 24, vol. 10341, file August 53 to August 54, "Medical Research Set in Far North," *Montreal Gazette*, 13 February 1954.
86 Western Archives, Western University, Scrap books containing clippings, etc., relating to the University of Western Ontario, call no. LE3 W522S82 v. 43 1955–6, "Reports Aiding Defence Presented at 'U' Talks," *London Free Press*, 14 October 1955.
87 For information about Burton's connection with the DRB, see LAC, RG 24-F-1, vol. 7553, file DRBC 9310-37 pt. 1, A.C. Burton, Progress Report of Project D 40-93-10-37, DRB Grant 341, Reaction of Men and Animals to Cold and Damp, 6 August 1953.
88 For further details and images of Eagan's experimental research at DRNL, see Wiseman, "Frontier Footage: Science and Colonial Attitudes on Film in Northern Canada, 1948–1954," in *Cold Science*, ed. Bocking and Heidt, 68–9.
89 Heggie, *Higher and Colder*, 148.
90 Farish, "The Lab and the Land."
91 Farish, "The Lab and the Land," 3.
92 Angela N.H. Creager, "Atomic Tracings: Radioisotopes in Biology and Medicine," in *Science and Technology in the Global Cold War*, ed. Oreskes and Krige, 39.
93 Creager, "Atomic Tracings," 19.
94 Committee on Evaluation of 1950s Air Force Human Health Testing in Alaska Using Radioactive Iodine[131]; Polar Research Board Commission on Geosciences, Environment, and Resources in cooperation with Board on Health Promotion and Disease Prevention Institute of Medicine; Board on Radiation Effects Research Commission of Life Sciences; National Research Council, *The Arctic Aeromedical Laboratory's Thyroid Function Study*.

95 Creager, "Atomic Tracings," 32–3.
96 For instance, see LAC, RG 128, vol. 238, file Thyroid Diseases, Henry M. Thomas Jr., "Effect of Thyroid Hormone on Circulation," *Journal of the American Medical Association* 163, no. 5 (1957): 337–41; and Hendrick, "Diagnosis and Management of Thyroids."
97 LAC, RG 128, vol. 259, file Dr. G. Malcolm Brown – List of Publications, Curriculum Vitae, etc. (pt. 1), Letter from Malcolm Brown to Dr. K.J.R. Wightman, 2 November 1954. Keith John Roy "Kager" Wightman succeeded Ray Farquharson as chair of the Department of Medicine at the University of Toronto in 1960. In the same year, he also became Physician-in-Chief at Toronto General Hospital, a position he retained until 1970. For biographical information about Wightman, see "Dr. Keith JR Wightman," Wightman-Berris Academy, University of Toronto, http://wbacademy.utoronto.ca/about-us/dr-keith-jr-wightman.
98 Mann wanted to co-publish Brown's results with his own from similar research conducted in the United States, but Brown refused because it would be "unsatisfactory" to have the data combined. See LAC, RG 128, vol. 259, file Dr. G. Malcolm Brown – List of Publications, Curriculum Vitae, etc. (pt. 1), letter from Malcolm Brown to Dr. George V. Mann, 2 March 1955.
99 LAC, RG 24, vol. 4129, file DRBS 4-78-53 pt. 1, Minutes of the First Meeting of the Panel on Arctic Medical Research, DRB, 16 December 1948.
100 LAC, RG 24, vol. 4129, file DRBS 4-78-53 part 2, DRB Panel on Arctic Medical Research: Minutes of the Tenth Meeting, 15 January 1954.
101 For instance, see LAC, RG 24, vol. 4206, file 270-0-89-6, Winter Exercise "Sun Dog One," June 1951.
102 LAC, RG 24, vol. 10341, file November 52 to August 53, Wellington Jeffers, "Finance at Large: Survival Suits Could Treble Arctic Population, Including Whites, Indians and Eskimo; Industry, Defense and Science May Find Other Answers," *Globe and Mail*, 17 December 1952, 20.
103 LAC, RG 24, vol. 10341, file November 52 to August 53, "The Well-Dressed Man in the North Wears a 'Survival Suit,' " *Ottawa Citizen*, reprinted in the *Hamilton Spectator*, 7 January 1953.
104 Gerald Waring, transcribed from *CBC News Roundup*, "Nylon Fur Coats Introduced," 3 June 1949, http://www.cbc.ca/archives/entry/nylon-fur-coats-introduced.
105 "It's Usually Rabbit – Brother Rabbit's Newest Rival: Canadian Nylon Fur," *Reader's Digest*, March 1952, 73.
106 Waring, transcribed from, *CBC News Roundup*, "Nylon Fur Coats Introduced."
107 LAC, RG 24, vol. 10341, file August 54 to March 55, "Defence Research Achievement: New Medical Tourniquet May Help Lower Deaths in Warfare," *Ottawa Journal*, 21 August 1954.
108 McNeill and Engelke, *The Great Acceleration*, 137.

109 "The Well-Dressed Man in the North Wears a 'Survival Suit.'"
110 Unnamed RCAF pilot, quoted in "Army Rejection of Nylon 'Fur' Draws 'Blast' from Arctic Vets," *Ottawa Journal*, 30 January 1953.
111 Unnamed member of the RCMP, quoted in "Army Rejection of Nylon 'Fur' Draws 'Blast' from Arctic Vets."
112 In the many published reports of the acclimatization research documented extensively in this chapter, the Canadian government goes unmentioned, save for the odd footnote that credits financial support from Grant DRB no. 80 and the Department of National Health and Welfare.
113 For an extensive study about the history of Inuit relocation, see Marcus, *Relocating Eden*.
114 These persons include, but are not limited to Omond Solandt, Hugh Keenleyside, and other leading government officials who supported cold acclimatization research.
115 Brown, "Cold Acclimatization in Eskimo," 344.
116 Brown, "Cold Acclimatization in Eskimo," 351.
117 Baugh, Bird, Brown, et al., "Blood Volumes of Eskimos and White Men," 354.

3. Entomology, Insect Control, and Biological Warfare

1 Russell, *Fighting Humans and Insects*, 156.
2 Langston, *Toxic Bodies*, 85.
3 Brown, *Plutopia*, 224.
4 Twinn, "Studies of the Biology and Control of Biting Flies."
5 Note that the original document refers to Goose Creek as "Goose River"; see LAC, RG 24, vol. 2484, file HQS-726-40-17-11, C.R. Twinn, Entomologist, Canadian Party, Biting Fly Survey and Control Report, Churchill, Manitoba, 1947: Progress Summary – Week Ending 19 July 1947.
6 For information about the widespread use of dichlorodiphenyltrichloroethane (DDT) during and after the Second World War, see Davis, *Banned*; and Dunlap, ed., *DDT: Classic Texts*.
7 LAC, RG 24, vol. 10341, file March 1947 to November 1952, Cyril Bassett, "Science: Our Armed Forces' $35 Million Fourth Arm," *The Financial Post* (Toronto) 46, no. 7, 16 February 1952.
8 For information about the extent of the scientific efforts to survey and control biting insects in the early post-war period, see T.N. Freeman, "Some Problems of Insect Biology"; Twinn, "Review of Recent Progress"; and Freeman, "The Canadian Northern Insect Survey."
9 Freeman, "The Canadian Northern Insect Survey."
10 LAC, RG 24, vol. 2484, file HQS-726-40-17-11, DRBS 135-120-267, R.G. MacNeill, Secretary, Defence Research Board to David Johnson, Secretary, Canadian Section, Permanent Joint Board on Defence, Ottawa, 29 January 1948.

11 Twinn, "Studies of the Biology and Control of Biting Flies," 19.
12 Twinn, "Studies of the Biology and Control of Biting Flies," 21.
13 Jennifer Bonnell, "Early Insecticide Controversy and Beekeeper Advocacy in the Great Lakes Region," *Environmental History* 26 (2021): 79–101.
14 Kuhlberg, *Killing Bugs for Business and Beauty*, 4–5.
15 LAC, RG 24, vol. 2484, file HQS-726-40-17-11, W.L. Coke, Brigadier, Director General of Medical Services to D.G.D.R. [Director General Defence Research, O.M. Solandt], Memorandum, Department of National Defence, Army, Mosquito and Black Fly Control, Ottawa, 5 January 1948.
16 LAC, RG 24, vol. 2484, file HQS-726-40-17-11, 5 January 1948.
17 Smith, *Toxic Exposures*, 34; Avery, *Pathogens for War*, 261.
18 LAC, RG 24, vol. 2484, file HQS-726-40-17-11, DRBS 270-120-301, E.L. Davies, Vice-Director General for Chairman, Defence Research Board, [Omond Solandt] to DGMS [Director General of Medical Services, Brigadier W.L. Coke], re: Mosquito and Black Fly Control, 8 March 1948.
19 LAC, RG 24, vol. 2484, file HQS-726-40-17-11, FCGS 4-8-1-11, D.C. Williams, Proposed Plan for a Study of the Psychological Effects of Insect Pests on Army Personnel, 18 March 1948.
20 LAC, RG 24, vol. 2484, file HQS-726-40-17-11, DRBS 135-120-267-1-20, E.L. Davies, Vice-Director General for Chairman, Defence Research Board, [Omond Solandt] to C.G.S. [Chief of the General Staff, Lieutenant-General Charles Foulkes], re: Biting Flies: Acclimatization of Personnel, Ottawa, 30 March 1948.
21 For the official military report of the research, see Freeman, *Entomological Research in Northern Canada*; see also Twinn, *Entomological Research in Northern Canada*.
22 LAC, RG 24 R 112-6655-4-F, vols. 31013 to 31016, DRB, *Annual Report on the Progress of Environmental Protection Research*, Report No. DR. 25, 30 April 1950.
23 "Groups Start Fight against Insects in Northern Canada: Eight Field Parties Carry Out 'Operation Insect' for Board," *Brandon Sun*, 17 June 1948.
24 "Exercise Fishing Engaged 'Small Fry,' Camp Borden: In the Air Force," *Barrie Examiner*, 1 June 1950, 14.
25 LAC, RG 24, vol. 4118, file DRBS 2-935-172, Defence Research Board, Scientific Intelligence Division, B.W. in the Northern half of the Northern Hemisphere: Possible use of insects as B.W. vectors, Scientific Intelligence aspects, 4 August 1949, 1.
26 See Avery, *Pathogens for War*.
27 On the long history of insect vectors and war, see McNeill, *Mosquito Empires*; Roger Webber, *Disease Selection*.
28 LAC, RG 24, vol. 4118, file DRBS 2-935-172, Defence Research Board, Scientific Intelligence Division, B.W. in the Northern half of the Northern Hemisphere: Possible use of insects as B.W. vectors, Scientific Intelligence aspects, 4 August 1949, 1.

29 Avery, *Pathogens for War*, 70–1.
30 Linn, *Elvis's Army*, 194.
31 Avery, *Pathogens for War*, 70–1.
32 For a history of the Atomic Energy Project at Chalk River, see Bothwell, *Nucleus*.
33 For information about Chalk River and the widespread use of isotopes in post-war Canada, see Wellington Jeffries, "Finance at Large: Use of Isotopes in Applying Atomic Energy to Wide Industrial Uses in Canada Grows Rapidly in Steel, Oil, Paper, Pulp and Other Industries," *Globe and Mail*, 27 February 1953, 20.
34 LAC, RG 24, vol. 4116, file 1-0-262-2 to 1-205-81, *Radioisotopes Price List No. 2: Effective on Purchase Orders received after November 1, 1948* (Isotopes Branch, National Research Council, Atomic Energy Project, Chalk River, Ontario), Application for Radioactive Isotope, 10 June 1949. For a detailed description of the research, see Appendix A of the same file, titled "Radioactive Mosquito Project for Defence Research Board, to Be Conducted by Dr. D.W. Jenkins, Medical Division, United States War Department."
35 LAC, RG 24, vol. 4116, file 1-0-262-2 to 1-205-81, *Radioisotopes Price List No. 2: Effective on Purchase Orders received after November 1, 1948*; see Appendix A, "Radioactive Mosquito Project for Defence Research Board."
36 Jenkins, "A Field Method of Marking Arctic Mosquitoes."
37 Bugher and Taylor, "Radiophosphorus and Radiostrontium in Mosquitoes."
38 "Soviet Organ See Confusion in U.S.," *New York Times*, 13 April 1951, 6; *The Problem of Chemical and Biological Warfare*, vol. 1: *The Rise of CB Weapons* (Stockholm: Stockholm International Peace Research Institute, 1971), 224.
39 For a detailed history of biological and chemical weapons simulation research that was conducted in the US and Canada during the early Cold War, see Lisa Martino-Taylor, *Behind the Fog: How the US Cold War Radiological Weapons Program Exposed Innocent Americans* (New York: Routledge, 2018).
40 Harvey Hickey, "Benefits from Isotopes' Use Seen Worth All Atomic Cost," *Globe and Mail*, 5 March 1953, 1.
41 C.J. Mackenzie was not the first Canadian official to extol the benefits of peaceful atomic research, of course. The special committee appointed to examine government operations in the field of atomic energy garnered extensive attention on Parliament Hill. See Government of Canada, House of Commons Debates, 21st Parliament, 7th Session, vol. 2, "Atomic Energy: Appointment of Committee to Examine into Operations of Control Board," 17 February 1953, 2026–33.
42 Bruce-Chwatt, "Radioisotopes for Research," 492.
43 "Half-Life: Radioactivity," *Encyclopaedia Britannica*, https://www.britannica.com/science/half-life-radioactivity.

44 See, for instance, Bruce-Chwatt, "Radioisotopes for Research," 495.
45 LAC, RG 24, vol. 4118, file DRBS 2-935-172, A.A. James, Medical Liaison Officer, Canadian Joint Staff, Washington, to A.J.G., Director, Scientific Intelligence, Defence Research Board, 23 August 1949.
46 LAC, RG 24, vol. 4118, file DRBS 2-935-172, A.A. James, Medical Liaison Officer, Canadian Joint Staff, Washington, to A.J.G., Director, Scientific Intelligence, Defence Research Board, 30 August 1949.
47 LAC, RG 24, vol. 4118, file DRBS 2-935-172, Defence Research Board, Scientific Intelligence Division, B.W. in the Northern half of the Northern Hemisphere: Possible use of insects as B.W. vectors, Scientific Intelligence aspects, 4 August 1949, 1.
48 LAC, RG 24, vol. 17601, file 004-100-74/34-3, DRBS 2-0-172-3, "Defence Research Board Scientific Intelligence Division: Organization and Functions," 9 November 1949. The other projects listed in Appendix D included reports on nerve gases, jets and rockets, radar, the possibilities of wind-borne attack across the Polar Basin, and the organization of Russian science and defence research.
49 LAC, RG 24, vol. 2425, file Speeches – Reporting etc. 1947 – March 1953, vol. 1, J.C. Arnell, Defence Research Chemical Laboratories, Ottawa, "Research in Defence," lecture delivered at various local sections of the Chemical Institute of Canada in the Spring of 1953, 2.
50 A.A. Kingscote, quoted in *Report of the Ontario Veterinary College 1947*, 47.
51 LAC, RG 24, vol. 10341, file March 1947 to November 1952, Cyril Bassett, "Science: Our Armed Forces' $35 Million Fourth Arm," *Financial Post* (Toronto) 46, no. 7, 16 February 1952.
52 LAC, RG 24, vol. 7392, file DRBS 170-80/S6, vol. 1, Defence Research Board: Statement of Awards of Grants-in-Aid of Research, 6 March 1956; see page 3 – Grant No. 75 Kingsctoe A.A., for the period 1948–51 and page 8 – Grant No. 254 (6801–11) Kingscote, A.A., for the period 1951–7, excluding the years 1953–5.
53 A.A. Kingscote, quoted in *Report of the Ontario Veterinary College 1948*, 52.
54 *Report for the Year 1953–54: The Ontario Veterinary College Guelph, Ontario*, 9.
55 LAC, RG 24, vol. 4230, file DRBS 2-1-87, vol. 1, Schedule "A," Defence Research Board: Accountable Grants in Air for Research, Serial DRB – 75, 3.
56 LAC, RG 24-C-1-c, vol. 35893, file 2001-91/S6, Report on Exercise "Shoo Fly" 20 July 1950–2 August 1950, 10.
57 For a full description of Exercise Deer Fly, see the operational report printed in Lackenbauer and Kikkert, *Lessons in Northern Operations*, 126–35.
58 The report also indicated that soldiers adapted to avoid suffering from local insect life, but at no time did the biting flies appear to limit the efficiency of the fighting troops; see LAC, RG 24-C-1-c, vol. 35873, file 2001-91/D5 pt. 2, Canadian Army Report: Sub-unit Arctic Training Programme, "Deer Fly" Series, Fort Churchill, Summer 1952, 8.

59 LAC, RG 24, vol. 10341, file November 52 to August 53, "Black and White and Bugs," *Ottawa Journal,* 28 May 1953.
60 LAC, RG 24-C-1-c, vol. 35873, file 2001-91/D5 pt. 2, Canadian Army Report: Sub-unit Arctic Training Programme, "Deer Fly" Series, Fort Churchill, Summer 1952, 5.
61 LAC, RG 24, vol. 2425, file Speeches – Reporting etc. 1947–March 1953, vol. 1, [Omond Solandt], Annual Report of the Chairman, Defence Research Board, pt. 1: Defence Research, October 1952, 6.
62 *Defence Research Board Newsletter* 1, no. 7 (July 1955).
63 NARA, College Park, MD, RG 319, vol. 857, *Annual Report on the Progress of Environmental Protection Research, 1 May 1949–30 April 1950* (Ottawa: DRB Report No. DR. 25, April 1950), 24.
64 Iarocci, "Opening the North," 90.
65 NARA, RG 319, vol. 857, *Annual Report on the Progress of Environmental Protection Research, 1 May 1949–30 April 1950* (Ottawa: DRB Report No. DR. 25, April 1950), 24.
66 Iarocci, "Opening the North," 91.
67 Parr, *Sensing Changes,* 32.
68 Parr, *Sensing Changes,* 28.
69 Cecil, *Herbicidal Warfare,* 48.
70 National Defence and the Canadian Armed Forces, "The Use of Herbicides at CFB Gagetown from 1952 to Present Day: Agent Orange Ex-gratia Payment," http://www.forces.gc.ca/en/about-reports-pubs/herbicides-gagetown.page.
71 Department of National Defence, Canada, "ARCHIVED – The Use of Herbicides at CFB Gagetown from 1952 to Present Day," 6 September 2013, https://www.canada.ca/en/department-national-defence/corporate/reports-publications/health/use-of-herbicides-at-cfb-gagetown-from-1952-to-present-day.html.
72 Kinkela, *DDT and the American Century,* 206.
73 Kinkela, *DDT and the American Century,* 105.
74 "Worse and Worse," *Globe and Mail,* 23 August 1950, 6.
75 "Summer Orchestra," *Globe and Mail,* 15 June 1954, 6.
76 LAC, RG 24-G-18-3, vol. 10, file SES-S 1800-1 pt. 1, G.R. Vavasour, Head, BW and CW Research Section, Directorate of Atomic Research, Defence Research Board, "BW Attack Against North America," 30 November 1962.
77 LAC, RG24-F-1, vol. 7538, file DRBC-6800-7 pt. 2, "Bite in It," 18 June 1958.
78 Dave McIntosh, "Strange Bugs May Hitchhike on Jet Planes," *Simcoe Reformer,* 14 April 1961, 9.
79 "Defence Board Produces: World's First Muskeg Map," *Brandon Sun,* 26 April 1961, 15.
80 Churchill Public Archives, Churchill, Manitoba, box 991.1-991.2.4, file 991.2.4, "DDT Discontinued in Canada's Nat'l Parks: Churchill Not Affected?" *Taiga Times* (Churchill) no. 23, vol. 4, 20 July 1968.

81 LAC, RG 24, vol. 7981, file 2-6678-2 pt. 3, Air Defence Command Headquarters in St. Hubert, Quebec, Operation Order 7/64 Airspray, Annex B: Restricted Publicity, 23 April 1964.
82 To be clear, Schenk expressed concern about the air spray program for reasons of sport fishing. He seemed less concerned about the negative ecological or environmental impacts of DDT in general. See LAC, RG 24, vol. 7981, file 2-6678-2 pt. 3, Carl F. Schenk, Biologist, Ontario Water Resources Commission to Mr. A.E. Winmill, Technical Officer 7, Canadian Forces Headquarters, Department of National Defence, 1 April 1965.
83 LAC, RG 24, vol. 7981, file 2-6678-2 pt. 3, A.E. Winmill, Technical Officer 7, Canadian Forces Headquarters, Department of National Defence to Carl F. Schenk, Biologist, Ontario Water Resources Commission, 12 April 1965.
84 LAC, RG 24, vol. 7981, file 2-6678-2 part 5, Raymond Miller, Regional Medical Health Officer, Moose Jaw, Saskatchewan to Squadron Leader Burden, Canadian Forces Base, Moose Jaw, Saskatchewan, 29 August 1966.
85 Churchill Public Archives, Churchill, Manitoba, box 991.1-991.2.4, file 991.2.4, "DDT Discontinued in Canada's Nat'l Parks: Churchill Not Affected?" *Taiga Times* (Churchill) no. 23, vol. 4, 20 July 1968.
86 "CDA Learning All Possible about Biology in Arctic," *Simcoe Reformer*, 9 May 1969, 9.
87 In her research on the history and regulation of the hormone disruptor diethylstilbestrol (DES), Nancy Langston suggests that the human body represents a dynamic system of "material and cultural feedbacks" that is a capable indicator for environmental change and degradation. Government regulation of chemicals is weak when restrictive paradigms of the natural world present humans as alien to their habitat, the author argues. See Langston, *Toxic Bodies*, 136.
88 Sandlos and Keeling, "Toxic Legacies," 9. See also Schlosberg and Carruthers, "Indigenous Struggles"; and Sandlos and Keeling, "Aboriginal Communities."

4. The Changing Science of Arctic Warfare

1 See Greg Donaghy, ed., *Documents on Canadian External Relations*, vol. 16: *1950* (Ottawa: Department of Foreign Affairs and International Trade Canada, 1996), document 40, chapter 2, Korean Conflict, pt. 1: Creation of United Nations' Unified Command, "Secretary of State for External Affairs to Permanent Delegate to United Nations," Ottawa, 12 July 1950, http://epe.lac-bac.gc.ca/100/206/301/faitc-aecic/history/2013-05-03/www.international.gc.ca/department/history-histoire/dcer/details-en.asp@intRefid=7060.

2 LAC, RG 24, vol. 2425, file O.M. Solandt Speeches and Reports, Omond Solandt, "An Address to the Manitoba Chamber of Miners," Winnipeg, 20 October 1950.
3 For instance, authorities in the DRB sent operational research analysts to Korea to examine the fighting tactics and weapons effectiveness of the coalition forces. An international scientific presence strengthened Canada's stature among allies and showed officials in Ottawa exactly where the DRB could continue to make an important contribution to the Western cause. See LAC, RG 24, vol. 4206, file DBRS 270-180-105-1, vol. 1, W.L. Archer, "Canadian Army Operational Research Establishment: Notes on Operational Experiences in Korea," 21 November 1951; and LAC, RG 24, vol. 4206, file DBRS 270-180-105-1, vol. 1, A.R. Menzies, Head of Canadian Liaison Mission in Japan, "Letter No. 564 – From the Canadian Liaison Mission, Tokyo, Japan to the Under-Secretary of State for External Affairs – Subject: Return from Korea of Dr. W.L. Archer, Defence Research Board." See also A.R. Menzies, "Canadian Brigade for Korea – Command and Support Problems," Letter No. 239, 10 March 1951, in *Documents on Canadian External Relations*, vol. 17: *1951*, ed. Greg Donaghy (Ottawa: Department of Foreign Affairs and International Trade Canada, 1996), 178–80.
4 LAC, RG 24, vol. 2425, file Speeches – Reporting etc 1947 – March 1953, vol. 1, Edmond Cloutier, *Canada's Defence Programme 1951–52* (Ottawa: Printer to the King's most Excellent Majesty Controllers of Stationery, 1951), 5.
5 LAC, RG 24, vol. 2425, file Speeches – Reporting etc 1947 – March 1953, vol. 1, O.M. Solandt, "Fifth Annual Birthday Address," in *5 Anniversary: Defence Research Board Ottawa* (Ottawa: Defence Research Board, 1952), 17.
6 CWM, George Metcalf Archival Collection, box 58 A1 294, file 58 A1 294.11, Arctic Aeromedical Laboratory, "Review of Research on Military Problems in Cold Regions: Technical Documentary Report AAL-TDR-64-28," December 1964.
7 Bidd, ed., *The NFB Film Guide*, 508. For a historical analysis of the film *Vigil in the North*, see Wiseman, "Frontier Footage," in *Cold Science*, ed. Bocking and Heidt, 61–74.
8 Starting at Fort Churchill, the Musk Ox convoy travelled along the shore of Hudson Bay to Eskimo Point (Arviat) and then northwest via Baker Lake and Perry River to Cambridge Bay. The Moving Force rested and explored the area for ten days before continuing to Coppermine, Port Radium, Tulita (Fort Norman), Fort Simpson, and to the Alaska Highway at Fort Nelson. From Nelson, the Moving Force began travelling south along roads to Edmonton, but the convoy stirred up dust storms on the highway that led to several breakdowns. The trek ended at Grande Prairie, Alberta, where

all personnel and vehicles boarded a train to their final destination. After eighty-one days on the trail, the Moving Force reached Edmonton on 6 May 1946, just one day behind schedule. For a detailed timeline and analysis of Exercise Musk Ox, see John Lauder (author), P. Whitney Lackenbauer, and Peter Kikkert, eds., *Tracks North: The Story of Exercise Muskox* (Arctic Operational History Series, No. 5, 2018); Lackenbauer and Kikkert, *Lessons in Northern Operations*, x–xii; Thrasher, "Exercise Musk Ox"; G.W. Rowley, "Exercise Muskox," *Geographical Journal* 109, nos. 4–6 (1947): 175–85; J.T. Wilson, "Exercise Musk-Ox, 1946," *Polar Record* 5, nos. 33–4 (1947): 14–27; and Hugh A. Halliday, "Exercise 'Musk Ox': Asserting Sovereignty 'North of 60,'" *Canadian Military History* 7, no. 4 (2012): 37–44.

9 Lackenbauer and Kikkert, *Lessons in Northern Operations*, x. Lackenbauer and Kikkert credit Kenneth Eyre with making this observation. See Eyre, "Custos Borealis" (PhD diss.), 157.
10 Lackenbauer and Kikkert, *Lessons in Northern Operations*.
11 Lackenbauer and Kikkert, *Lessons in Northern Operations*, xi.
12 University of Toronto Archives and Records Management Services, Omond McKillop Solandt fonds, box B93-0041/034, file Hattersley-Smith, O.M. Solandt, quoted in G. Hattersley-Smith, *North of Latitude Eighty: The Defence Research Board in Ellesmere Island* (Ottawa: Defence Research Board, 1973), 2.
13 H.H. Watson, "Army Operational Research Aims and Methodology," *Canadian Army Journal* 15, no. 3 (Summer 1961): 22–9.
14 Watson, "Army Operational Research Aims," 22.
15 Rearmament for the Korean conflict saw the Canadian defence budget grow to $1.9 billion by 1953, reaching ten times the budget of 1947. Over the same period, between 1950 and 1953, the Canadian armed services grew from 47,000 to 104,000 personnel. See Morton, *A Military History of Canada*, 238.
16 Pennie, "Defence Research Northern Laboratory," 50.
17 See also Wiseman, "The Development of Cold War Soldiery," 127–55; and Wiseman, "Frontier Footage," in *Cold Science*, ed. Bocking and Heidt, 61–74.
18 A.M. Pennie, "Defence Research Northern Laboratory," *Canadian Army Journal* 10, no. 1 (January 1956): 48–9.
19 Galloway, "The Army Says Goodbye to 'The Shining Land,'" 46.
20 On the Canadian Joint Air Training Centre at Rivers, see Heide, "Stations of the RCAF."
21 LAC, RG 85, vol. 299, file 1009-2[5], *Defence Research Northern Laboratory: Progress Report on Indoctrination Training for Military Operations in the North, DRNL Project Report No. 4* (Ottawa: DRB, Department of National Defence, Canada, 1954), 2.
22 Pennie, "Significant Contributions of Canadian Defence Science," in *Perspectives in Science and Technology*, edited by Law, Lindsey, and Grenville, 83.

23 "Science Serves a Banquet: Paratroops' Appetites Probed," *Winnipeg Free Press*, 10 December 1954, 18.
24 Pennie, "Defence Research Northern Laboratory," 49. For information about the Canadian Joint Air Training Centre, Rivers, see Major J.S. Hitsman, "Medical Problems of Paratroop Training," *Canadian Army Journal* 4, no. 1 (April 1950): 9–18.
25 General Staff Branch, Headquarters Western Command, Edmonton, "Exercise Bulldog III," *Canadian Army Journal* 9, no. 2 (April 1955): 10–19.
26 For a detailed account of Exercise Bull Dog III and the participation of the Yellowknife Rangers, see Lackenbauer, *The Canadians Rangers*, 152–4.
27 Pennie, "Significant Contributions of Canadian Defence Science," in *Perspectives in Science and Technology*, edited by Law, Lindsey, and Grenville, 83.
28 Pennie, "Significant Contributions of Canadian Defence Science," in *Perspectives in Science and Technology*, edited by Law, Lindsey, and Grenville.
29 George R. Lindsey, "Management of Science in the Defence Research Board," in *Perspectives in Science and Technology*, edited by Law, Lindsey, and Grenville, 93.
30 Lindsey, "Management of Science in the Defence Research Board," in *Perspectives in Science and Technology*, edited by Law, Lindsey, and Grenville, 94.
31 Goodspeed, *A History of the Defence Research Board of Canada*, 90.
32 Authorities in the DRB denied publication requests in limited situations only, usually to protect Ottawa from public scrutiny. In the mid-1950s, for instance, McGill psychologist Donald Hebb made several requests that he be able to share his DRB-sponsored research into sensory deprivation with public audiences. Scientists and government officials in Ottawa consistently denied Hebb's request, citing national security and the potential for misinterpretation as reasons to uphold the veil of secrecy.
33 Pennie, "Defence Research Northern Laboratory," 52.
34 Directorate of Public Relations (Army), "Winter Training at Churchill," *Canadian Army Journal* 10, no. 2 (April 1956): 47.
35 Directorate of Public Relations (Army), "Winter Training at Churchill," *Canadian Army Journal* 10, no. 2 (April 1956): 47.
36 A product of the DRB's Canadian Armament Research and Development Establishment at Valcartier, Quebec, the Heller was the first complete weapon, ammunition, and fire control system fully designed, developed, and manufactured in Canada. See Directorate of Public Relations, National Defence, Ottawa, "New Anti-Tank Weapon Is Developed for Canadian Army," *Canadian Army Journal* 9, no. 2 (April 1955): 46–8.
37 The participating units included the Royal Canadian Regiment, the Princess Patricia's Canadian Light Infantry, and the Royal 22ᵉ Régiment. For a

complete and detailed description of the winter warfare training course, including a selection of photographs taken near Fort Churchill, see Sargent-Major F.J. Way, "Mobile Striking Force Winter Training: Training at Fort Churchill," *Canadian Army Journal* 11, no. 2 (April 1957): 20–9.
38 Major J.W.P. Bryan, "Mobile Striking Force Winter Training: Training in Kirkland Lake Area," *Canadian Army Journal* 11, no. 2 (April 1957): 29–30.
39 Bryan, "Mobile Striking Force Winter Training," 30.
40 Pennie, "Defence Research Northern Laboratory," 53.
41 Galloway, "The Army Says Goodbye to 'The Shining Land,'" 51.
42 Pennie, "Defence Research Northern Laboratory," 53–4.
43 Directorate of Public Relations (Army), Ottawa, "Winter Training at Churchill," *Canadian Army Journal* 10, no. 2 (April 1956): 41.
44 Directorate of Public Relations (Army), "Winter Training at Churchill," *Canadian Army Journal* 10, no. 2 (April 1956): 42–5.
45 "Match Soviets in Arctic Race, Scientists Urge," *Globe and Mail*, 25 January 1958, 1.
46 *Defence Research Board Newsletter* 1, no. 6 (June 1955), n.p.
47 *Defence Research Board Newsletter* 1, no. 11 (November 1955), n.p.
48 The total trip covered nearly 10,000 miles of flight over sub-Arctic tundra, Arctic islands, and parts of Greenland. For a full and detailed record of the trip, see *Defence Research Board Newsletter* 2, no. 5 (May 1956).
49 Information Canada, *Defence Research Board*, 17.
50 *Defence Research Board Newsletter* 2, no. 3 (March 1956), n.p.
51 *Defence Research Board Newsletter* 2, no. 3.
52 *Defence Research Board Newsletter* 2, no. 3.
53 Galloway, "The Army Says Goodbye to 'The Shining Land,'" 46.
54 Galloway, "The Army Says Goodbye to 'The Shining Land,'" 47.
55 Galloway, "The Army Says Goodbye to 'The Shining Land,'" 46.
56 Galloway, "The Army Says Goodbye to 'The Shining Land,'" 51.
57 Galloway, "The Army Says Goodbye to 'The Shining Land,'" 47.
58 Galloway, "The Army Says Goodbye to 'The Shining Land.'"
59 Galloway, "The Army Says Goodbye to 'The Shining Land,'" 47.
60 The responsibility for military training at or near Fort Churchill in the 1950s and 1960s was not exclusively Canadian, however. The US Army Arctic Test Center carried out several training courses on location, allowing Canadian Army personnel serving at the garrison opportunities to participate and further their military qualifications and expertise. For more information, see Directorate of Supplies and Transport, Army Headquarters, Ottawa, "U.S. Army Commends Canadian," *Canadian Army Journal* 14, no. 3 (Summer 1960): 92–3.
61 Pope, "Canada's Defence Research Board," 29.
62 Galloway, "The Army Says Goodbye to 'The Shining Land,'" 40.

63 Noble, "Reflections on Fort Churchill," 23.
64 Galloway, "The Army Says Goodbye to 'The Shining Land,'" 40.
65 Galloway, "The Army Says Goodbye to 'The Shining Land.'"
66 Galloway, "The Army Says Goodbye to 'The Shining Land,'" 50.
67 Galloway, "The Army Says Goodbye to 'The Shining Land,'" 52.
68 Pennie, ed., *Defence Research Northern Laboratory*, 2.
69 For a detailed historical account of the Canadian economy in this period, see Muirhead, *Dancing Around the Elephant.*
70 Iarocci, "Opening the North," 74.
71 Muirhead, *Dancing Around the Elephant*, 17–18.

5. Operation Hazen and the International Geophysical Year

1 LAC, RG 24, vol. 10341, file November 52 to August 53, "Defence Research Board Chalks-Up Noted Record: 1952 Highlighted by Canadian Role in British Atomic Blast," *Montreal Star*, 3 January 1953.
2 It is unclear whether Tuktuk is a reference to the community of Tuktoyaktuk in the Northwest Territories or to a different coastal location near the present-day Tuktut Nogait National Park.
3 "Snow Geese Breed on Remote Island in Arctic Circle," *Barrie Examiner*, 2 October 1953, 19.
4 LAC, RG 24, vol. 10341, file November 52 to August 53, "Ottawan Spending Summer on Remote Arctic Island," *Ottawa Citizen*, 30 May 1953, 2.
5 Manning, Höhn, and Macpherson described the plant and animal life of Banks Island in an article published in the 1956 bulletin of the National Museum. See T.H. Manning, E.O. Höhn, and A.H. Macpherson, "The Birds of Banks Island," *National Museum of Canada Bulletin* 143 (1956): 182–97.
6 LAC, RG 24, vol. 10341, file August 53 to August 54, Frank Swanson, "Get Data on Bleak Island in Arctic," *Ottawa Citizen*, 24 August 1953.
7 Della Dora, "From the Radio Shack to the Cosmos," 126.
8 Howkins, *The Polar Regions*, 137.
9 On *Sputnik* and the IGY, see Robert W. Smith, "A Setting for the International Geophysical Year," in *Reconsidering Sputnik*, ed. Logsdon, Launius, and Smith.
10 Smith, "A Setting for the International Geophysical Year."
11 T. Harwood, "The International Geophysical Year," in *The Unbelievable Land*, ed. Norman Smith, 114–18; Duffin, *Stanley's Dream*, 13.
12 Doel, "Constituting the Postwar Earth Sciences."
13 Powell, "Science, Sovereignty, and Nation," 618.
14 On the US government and the IGY, see Belanger, *Deep Freeze*. While a Canadian equivalent of Belanger's study has yet to be written, geographer Richard C. Powell from the University of Liverpool wrote a succinct

summary of the Canadian experience with the IGY; see Powell, "Science, Sovereignty, and Nation." On the Canadian experience with the IGY, see also Doyle and Skog, eds., *The International Geophysical Year*, 32–5.
15 Turner, "Politics and Defence Research," 52.
16 Taylor, "Exploring Northern Skies."
17 G. Hattersley-Smith, "Operation Hazen," in *The Unbelievable Land*, ed. Smith, 119–24; Lotz, *The Best Journey in the World*.
18 Andrew Stuhl, *Unfreezing the Arctic: Science, Colonialism, and the Transformation of Inuit Lands* (Chicago: University of Chicago Press, 2016), 96.
19 Stuhl, *Unfreezing the Arctic*, 99–102.
20 See discussion and notes about Croal's permafrost research in chapter 1.
21 Stuhl, *Unfreezing the Arctic*, 100.
22 Stuhl, *Unfreezing the Arctic*, 101.
23 For a detailed analysis of the DEW Line and military infrastructure in the Canadian Arctic during the Cold War, see Lackenbauer and Farish, *The Distant Early Warning (DEW) Line*.
24 Heidt and Lackenbauer, *The Joint Arctic Weather Stations*, 4–5.
25 P. Whitney Lackenbauer, "At the Crossroads of Militarism and Modernization: Inuit-Military Relations in the Cold War Arctic," in *Roots of Entanglement*, ed. Rutherdale, Abel, and Lackenbauer, 127.
26 Loo, *Moved by the State*, 38.
27 Stacey A. Fritz, "DEW Line Passage: Tracing the Legacies of Arctic Militarization," (PhD diss., University of Alaska Fairbanks, 2010), 4.
28 Matthew Heymann et al., "Exploring Greenland: Science and Technology in Cold War Settings," *Scientia Canadensis* 33, no. 2 (2010): 11–42.
29 Martin-Nielsen, "The Other Cold Wa,"; Howkins, *The Polar Regions*, 133–4.
30 Heymann et al., "Exploring Greenland," 11; Howkins, *The Polar Regions*, 133.
31 Howkins, *The Polar Regions*, 133.
32 LAC, RG 24, vol. 10341, file August 53 to August 54, Department of Defense, Office of Public Information, Washington, US–Canadian Group to Study Western Arctic Ocean Area, 2 April 1954.
33 LAC, RG 24, vol. 10341, file August 53 to August 54, Department of Defense, Office of Public Information, Washington, 2 April 1954.
34 LAC, RG 24, vol. 10341, file August 53 to August 54, Department of Defense, Office of Public Information, Washington, 2 April 1954.
35 LAC, RG 24, vol. 10341, file August 53 to August 54, William Cameron, quoted in "Canadian Will Head Expedition to Arctic," *Evening Telegram* (St. John's), 19 July 1954, 5.
36 LAC, RG 24, vol. 10341, file August 53 to August 54, "Canadian Will Head Expedition to Arctic," *Evening Telegram* (St. John's), 19 July 1954, 5.
37 LAC, RG 24, vol. 10341, file August 53 to August 54, "For Research: H.M.C.S. 'Labrador' Is Newest Arctic Vessel," *Western Star* (Corner Brook), 21 July 1954.

38 J.P. Croal and J.R. Lotz, "Armed Forces Serve Science," *Canadian Army Journal* 15, no. 2 (Spring 1961): 19.
39 Kobalenko, "Geoffrey Hattersley-Smith," 489.
40 For biographical details on Hattersley-Smith, see Kobalenko, "Geoffrey Hattersley-Smith."
41 For a general interpretive overview of ice islands in the context of the early post-war period, see Koenig et al., "Arctic Ice Islands"; and Smith, *Ice Islands in Arctic Waters*, 1.
42 Lajeunesse, *Lock, Stock, and Icebergs*, 49.
43 Smith, *Ice Islands in Arctic Waters*, 1–3.
44 Lajeunesse, *Lock, Stock, and Icebergs*, 49. The United States also launched a project to discover and track ice islands drifting in the Arctic Ocean. See "Arctic ice islands: They Make Stable Platforms for Observing Weather Phenomena," *New York Times*, 9 November 1952, E9.
45 Lajeunesse, *Lock, Stock, and Icebergs*, 49.
46 Lajeunesse, *Lock, Stock, and Icebergs*, 50.
47 Lajeunesse, *Lock, Stock, and Icebergs*, 51.
48 LAC, RG 25, vol. 4, file 9057-40, Memorandum, 20 May 1954.
49 LAC, RG 25, vol. 4, file 9057-40, T.A. Harwood, "Probable Russian Activity in the Canadian Section of the Arctic Ocean," 5 August 1954.
50 LAC, RG 25, vol. 1, file 50211-40, O.M. Solandt, "Russian Activities on the Canadian Side of the Pole," 12 August 1954.
51 DHH, 2002/17, Joint Staff fonds, "The Sector Theory and Floating Ice Islands in the Arctic," 30 August 1954.
52 University of Toronto Archives and Records Management Services (UTA), Omond McKillop Solandt fonds, box B93-0041/034, file Hattersley-Smith, G. Hattersley-Smith, *North of Latitude Eighty: The Defence Research Board in Ellesmere Island* (Ottawa: Defence Research Board, 1973), 102.
53 Croal and Lotz, "Armed Forces Serve Science," 20.
54 LAC, RG 24, vol. 10341, file November 52 to August 53, "Ottawa Men Recover Peary's Arctic Records," *Ottawa Journal*, 4 June 1953.
55 LAC, RG 24, vol. 10341, file November 52 to August 53, "Canadian Scientists Find Peary Records in Arctic," *Toronto Daily Star*, 4 June 1953.
56 Hattersley-Smith had hoped to return the piece of flag to Peary's widow, who made the ensign for her husband. Peary left six pieces of the flag at various locations marking his travels in the Arctic. With three other pieces previously found and returned, Hattersley-Smith wanted to do the same with the fourth. See LAC, RG 24, vol. 10341, file November 52 to August 53, George Bain, "Relics of Explorers: Two Scientists Bring Back Piece of Peary's Flag," *Globe and Mail*, 27 August 1953; "Relics of Polar Expeditions Brought Back by Canadians," *Montreal Daily Star*, 27 August 1953.
57 Croal and Lotz, "Armed Forces Serve Science," 19.

58 Croal and Lotz, "Armed Forces Serve Science," 27.
59 Croal and Lotz, "Armed Forces Serve Science," 20.
60 Croal and Lotz, "Armed Forces Serve Science," 20–1.
61 Croal and Lotz, "Armed Forces Serve Science," 19–20.
62 Croal and Lotz, "Armed Forces Serve Science," 22.
63 Croal and Lotz, "Armed Forces Serve Science," 23.
64 Croal and Lotz, "Armed Forces Serve Science," 27.
65 Croal and Lotz, "Armed Forces Serve Science," 27.
66 Directorate of Public Relations, Army Headquarters, Ottawa, "Army Surveyors in Far North," *Canadian Army Journal* 15, no. 3 (Summer 1961): 64.
67 Directorate of Public Relations, Army Headquarters, "Army Surveyors in Far North."
68 Directorate of Public Relations, Army Headquarters, "Army Surveyors in Far North."
69 Dave McIntosh, "Northern Evidence: Even Eskimos Get Cold," *Gazette* (Montreal), 4 October 1958.
70 UTA, Omond McKillop Solandt fonds, box B93-0041/034, file Hattersley-Smith, G. Hattersley-Smith, *North of Latitude Eighty: The Defence Research Board in Ellesmere Island* (Ottawa: Defence Research Board, 1973), 99.
71 In addition to being presented in academic publications, the research activities and findings of Operation Hazen received wide dissemination and attracted notable publicity in the press.
72 UTA, Omond McKillop Solandt fonds, box B93-0041/034, file Hattersley-Smith, G. Hattersley-Smith, *North of Latitude Eighty*, 100.
73 UTA, Omond McKillop Solandt fonds, box B93-0041/034, *North of Latitude Eighty*, 103.
74 UTA, Omond McKillop Solandt fonds, box B93-0041/034, *North of Latitude Eighty*.
75 UTA, Omond McKillop Solandt fonds, box B93-0041/034, *North of Latitude Eighty*, 105.
76 UTA, Omond McKillop Solandt fonds, box B93-0041/034, *North of Latitude Eighty*, 106.
77 Dalhousie University Archives, Halifax (DUA), Dalhousie University Reference Collection (MS-1-Ref), box 135, file 135.4 Public Relations Scrapbook 1–31 January 1957, "Defence Research Board Year-End Roundup 1956," *Daily News* (St. John's), 7 January 1957.
78 On the history of radio propagation and ionosphere research in northern Canada, see Jones-Imhotep, *The Unreliable Nation*.
79 LAC, RG 24, vol. 10341, file August 53 to August 54, Leslie Wilson, "DRB Bending an Ear to 'Chatter' from Space," *Ottawa Journal*, 14 August 1954.
80 Simon Singh, *Big Bang: The Origin of the Universe* (New York: Harper Perennial, 2005), 402–8.

81 LAC, RG 24, vol. 2425, file Speeches – Reporting etc. 1947–March 1953, vol. 1, Policy and Plans for Defence Research in Canada: A Preliminary Review by O.M. Solandt, Director General of Defence Research, Ottawa, May 1946, 15.
82 LAC, RG 24, vol. 2425, file Speeches – Reporting etc. 1947–March 195,3, vol. 1, Policy and Plans for Defence Research in Canada: A Preliminary Review by O.M. Solandt, Director General of Defence Research, Appendix A: Suggested Fields for Defence Research, Ottawa, May 1946, 12.
83 For information about the JANET system, see Jones-Imhotep, *The Unreliable Nation*, 153, 173–7.
84 LAC, RG 24, vol. 10341, file November 52 to August 53, "Defence Researchers Solve Tough Problems in Military Science," *Saint John Telegraph-Journal*, 3 December 1952.
85 Jones-Imhotep, *The Unreliable Nation*.
86 LAC, RG 24, vol. 10341, file August 53 to August 54, Leslie Wilson, "DRB Bending an Ear to 'Chatter' from Space," *Ottawa Journal*, 14 August 1954.
87 DUA, MS-1-Ref, box 135, file 135.4 Public Relations Scrapbook 1–31 January 1957, "Defence Research Board Year-End Roundup 1956," *Daily News* (St. John's), 7 January 1957.
88 Defence Research Board, "Canada to Enter Space Race," *Canadian Army Journal* 14, no. 3 (Summer 1960): 61.
89 Galloway, "The Army Says Goodbye to 'The Shining Land,'" 45.
90 Galloway, "The Army Says Goodbye to 'The Shining Land.'"
91 Galloway, "The Army Says Goodbye to 'The Shining Land.'"
92 Doyle and Skog, eds., *The International Geophysical Year*, 33.
93 Doyle and Skog, eds., *The International Geophysical Year*.
94 Directorate of Public Relations (Army), Ottawa, "Winter Training at Churchill," *Canadian Army Journal* 10, no. 2 (April 1956): 47.
95 Jones-Imhotep, "Laboratory Cultures," 17–18.
96 Galloway, "The Army Says Goodbye to 'The Shining Land,'" 40.
97 Galloway, "The Army Says Goodbye to 'The Shining Land.'"
98 Doyle and Skog, eds., *The International Geophysical Year*, 33.
99 "International Collaboration in Space Science Exemplified by Top Side Sounder," *Roundel* 12, no. 6 (1960): 23.
100 Defence Research Board, "Canada to Enter Space Race," *Canadian Army Journal* 14, no. 3 (Summer 1960): 60.
101 For full details on Canada's military space program, see Godefroy, *Defence and Discovery*; information on the Alouette program is covered in chapter 4, 95–120.
102 Rand, "Falling Cosmos," 78.
103 Fred Cleverley, "North Base Waits: Fort Churchill Ready for Nuclear Tests," *Winnipeg Free Press*, 6 February 1963, 1.

104 Herzberg, Kehrt, and Torma, "Exploring Ice and Snow in the Cold War," in *Ice and Snow in the Cold War*, ed. Herzberg, Kehrt, and Torma, 3.
105 Herzberg, Kehrt, and Torma, "Exploring Ice and Snow in the Cold War."
106 Harper, *Make It Rain*.
107 Churchill Public Archives, Churchill, Manitoba, box 991.1-991.2.4, file 991.2.4, "The Hudson Bay Plan – A Canadian Alternative to the Bering Straits Proposal," *Taiga Times* (Churchill), no. 25, vol. 4, 20 July 1968.
108 P.M. Borisov, "Can we Control the Arctic Climate?" *Bulletin of the Atomic Scientists* (March 1969): 43–8.
109 Churchill Public Archives, Churchill, Manitoba, box 991.1-991.2.4, file 991.2.4, "Diefenbaker Advocates Look at Warming the Bay," *Taiga Times* (Churchill), nos. 33 and 34, 1968.
110 Lajeunesse, *Lock, Stock, and Icebergs*, 48; Coates et al., *Arctic Front*, 55.
111 Farish, *The Contours of America's Cold War*, 174.

6. Nuclear Fallout and the Northern Radiation Study

1 For information and historical analysis about *The Beaver*, see Sangster, *The Iconic North*, ch. 3, "*The Beaver*: Northern Indigenous Life in Popular Education," 69–103.
2 William O. Pruitt Jr., "Is A-Test Fallout Poisoning Caribou – and the Eskimos Who Eat Them?," *Whitehorse Star*, 14 January 1963; reprinted from William O. Pruitt Jr., "A New 'Caribou Problem,'" *The Beaver* (Winter 1962), 24–5.
3 Pruitt Jr., "Is A-Test Fallout Poisoning Caribou."
4 For a full biography of Otto Schaefer, see Hankins, *Sunrise over Pangnirtung*. A brief biographical sketch of the life and work of Schaefer is also available through the University of Alberta's Education and Research Archive, Canadian Circumpolar Institute, Dr. Otto Schaefer Collection, Lauren Wheeler, "Dr. Otto Schaefer Biography," 3 August 2011.
5 Myra Rutherdale, "Alaska Highway Nurses and DEW Line Doctors: Medical Encounters in Northern Canadian Indigenous Communities," in *Roots of Entanglement*, ed. Rutherdale, Abel, and Lackenbauer, 159–77 at 168.
6 LAC, RG 29, vol. 2886, file 851-1-27 pt. 1, Otto Schaefer, M.D. to Dr. P.G. Mar, Radio Chemical Laboratory, Radiation Protection Division, Department of National Health and Welfare, 18 January 1963.
7 LAC, RG 29, vol. 2886, file 851-1-27 pt. 1, Otto Schaefer, M.D. to Dr. P.G. Mar, 18 January 1963.
8 In January 1950, senior officials in the Industrial Health Division of the Department of National Health and Welfare created a Health Radiation Laboratory to assess the public health risks from exposure to medical X-rays and radioactive isotopes. Later renamed the Radiation Protection

Division, the laboratory became one of three bureaux of the new Environmental Health Directorate established in 1969. In its current form, the Radiation Protection Bureau protects Canadians from the dangers associated with radioactivity by monitoring radioactivity levels in the air and water, measuring and recording radiation exposure of workers, evaluating radioisotope license applications, and evaluating consumer products that emit radiation under the authority of the Radiation Emitting Devices Act of 1972. For a brief overview of the origins of the Radiation Protection Division, see Government of Canada, "Canada's Programme for Radiation Protection: An Address by Mr. J. Waldo Monteith, Minister of National Health and Welfare, at the Jubilee Meeting of the Canadian Public Health Association, Montreal, Quebec, on June 1, 1959" (Ottawa: Information Division, Department of External Affairs, 1959).
9 Graham Farmelo, "The Cold War's Effect on Science and Scientists," in *Out of the Cold*, ed. Fitzgerald, 92.
10 For instance, see, by order of publication date, Spiers, "Radioactivity in Man and His Environment"; Solon et al., "Investigations of Natural Environmental Radiation"; United Nations, *Report of the Scientific Committee*.
11 Luedee, "Locating the Boundaries of the Nuclear North," 69.
12 Luedee, "Locating the Boundaries of the Nuclear North," 89.
13 Higuchi, *Political Fallout*, 23.
14 Higuchi, *Political Fallout*, 23.
15 Higuchi, *Political Fallout*, 23–5.
16 Higuchi, *Political Fallout*, 140.
17 Higuchi, *Political Fallout*, 150.
18 Heidt, "I Think That Would Be the End of Canada."
19 Heidt, "I Think That Would Be the End of Canada,'" 352.
20 Higuchi, *Political Fallout*, 150.
21 Higuchi, *Political Fallout*, 168.
22 Higuchi, *Political Fallout*, 169.
23 Higuchi, *Political Fallout*, 167.
24 Serebryanny, "The Colonization and Peoples," 305.
25 Howkins, *The Polar Regions*, 132; Higuchi, *Political Fallout*, 167.
26 Jacobs, *Nuclear Bodies*, 159.
27 Jacobs, *Nuclear Bodies*, 159–60.
28 Emmerson, *The Future History of the Arctic*, 114.
29 Higuchi, *Political Fallout*, 170.
30 "Big Blast Could Raze Ontario Industry Heart," *Globe and Mail*, 18 October 1961, 9.
31 Higuchi, *Political Fallout*, 170.
32 Higuchi, *Political Fallout*.
33 Howkins, *The Polar Regions*, 132–3.

34 Doel, "Constituting the Postwar Earth Sciences"; Doel, "Cold Conflict: The Pentagon's Fascination with the Arctic (and Climate Change) in the Early Cold War," in *LASHIPA*, ed. Hacqueboord, 147–60; Doel, Zeller, and Wråkberg, "Science, Environmental Knowledge."
35 Farish, "The Lab and the Land."
36 Howkins, *The Polar Regions*, 135.
37 Farmelo, "The Cold War's Effect," in *Out of the Cold*, ed. Fitzgerald, 91.
38 Farmelo, "The Cold War's Effect," 91.
39 Farmelo, "The Cold War's Effect," 93.
40 Gregg Herken, *Brotherhood of the Bomb: The Tangled Lives and Loyalties of Robert Oppenheimer, Ernest Lawrence, and Edward Teller* (New York: Henry Holt, 2002).
41 Farmelo, "The Cold War's Effect," in *Out of the Cold*, ed. Fitzgerald, 93.
42 See, in order of publication date, Bothwell, *Nucleus*, 139–42; Simpson, *NATO and the Bomb*, 41–9; and Avery, *Pathogens for War*, 92.
43 C.P. McNamara and W.G. Penney, "The Technical Feasibility of Establishing an Atomic Weapons Proving Ground in the Churchill Area" (Ottawa: Defence Research Board; London: Ministry of Supply, ca. 1949–50). Document available online through Government of Canada, Defence Research Reports, https://cradpdf.drdc-rddc.gc.ca/PDFS/unc14/p518842.pdf.
44 For a full description of the top-secret report and its declassification in the post–Cold War years, see Clearwater and O'Brien, "O Lucky Canada."
45 Maloney, *Learning to Love the Bomb*, 82.
46 John Clearwater and David O'Brien infer that Canadian approval for the final go-ahead would have been a mere formality, whereas historian Sean Maloney suggests that the British went to Australia because the Canadians disagreed over using a new blast site for each detonation. See Clearwater and O'Brien, "O Lucky Canada," 63; Maloney, *Learning to Love the Bomb*, 82. See also James Eayrs, *In Defence of Canada: Growing Up Allied* (Toronto: University of Toronto Press, 1980), 242–3. In an interview with Eayrs, the DRB's founding chair Omond Solandt claimed that "the Canadian attitude toward the whole idea of participation in [a] nuclear weapon program was so negative that it was not pursued beyond the informal discussion stage."
47 McNamara and Penney, "The Technical Feasibility," 8.
48 McNamara and Penney, "The Technical Feasibility," 4.
49 In mid-1954, a Civil Defence committee including DRB scientific advisor Dr. E.E. Massey publicly disclosed information about the possible effects of a hydrogen bomb attack on Canada to prevent speculation and encourage Canadians to support emergency preparedness. See Burtch, *Give Me Shelter*, 84.
50 John Clearwater and David O'Brien have suggested that Soviet spies operated in the area, negating any possibility that the secret could have been kept from the Kremlin. See Clearwater and O'Brien, "O Lucky Canada," 64.

51 Clearwater and O'Brien, "O Lucky Canada," 64–5.
52 For information about the ecology of northern Manitoba and the Churchill region of Hudson Bay, see Brandson, *Churchill Hudson Bay*, 167–205.
53 Clearwater and O'Brien, "O Lucky Canada," 65.
54 Clearwater and O'Brien, "O Lucky Canada."
55 Gorham, "A Comparison of Lower and Higher Plants," 327–9.
56 Participants at the First Expert Meeting on the Radioactivity Investigations in Scandinavia discussed current research and the known extent of the problem. The Second Northern Meeting on Radioactive Food Chains was held in Helsinki during the first three days of April in 1962. See Viereck, *Radioactivity Report*, vol. 4, 6.
57 Officially "held in abeyance" by the Atomic Energy Commission, Project Chariot has never been formally cancelled. See O'Neil, *The Firecracker Boys*.
58 US Department of Energy, "Hanford Site: Understand the Past," https://www.hanford.gov/page.cfm/understandPAST.
59 Hanson, Whicker, and Dahl, "Iodine-131."
60 Viereck, *Radioactivity Report*, vol. 4.
61 Linus Pauling, quoted in Peter Worthington, "Dubious National Distinction: Canada – Land of Hottest Fallout," *Brandon Sun*, 14 May 1963.
62 "The Nobel Prize in Chemistry 1954: Linus Pauling," https://www.nobelprize.org/prizes/chemistry/1954/pauling/facts.
63 "Treaty Banning Nuclear Weapon Tests in the Atmosphere, in Outer Space and Under Water," United Nations Treaty Collection, https://treaties.un.org/pages/showDetails.aspx?objid=08000002801313d9.
64 Lauriston Taylor, quoted in Worthington, "Dubious National Distinction."
65 John E. Bird, "Fallout Hazard Studied," *Brandon Sun*, 9 December 1963, 11.
66 Worthington, "Dubious National Distinction."
67 A.H. Booth, quoted in "Animals Are 'Hot Spots': Radioactivity Test of Caribou Bones," *Lethbridge Herald*, 10 May 1963, 10; Arch MacKenzie, "Test for Fallout on Caribou Bone," *Brandon Sun*, 15 May 1963, 23.
68 LAC, RG 29, vol. 2886, file 851-1-27 pt. 1, A.H. Booth, Senior Scientific Officer, Radiation Protection Division to Mr. G.W. Elliott, President, Baker Lake Resident's Association, Baker Lake, NWT, 28 March 1963.
69 "High Fallout Rate in Eskimo Diets," *Winnipeg Free Press*, 15 May 1964, 47.
70 The term nanocurie expresses radiation units and is equal to 1000 micro-micro curies. A curie is the amount of radiation released by one gram of radium.
71 Government of Canada, "Canada's Programme for Radiation Protection," 3.
72 LAC, RG 29, vol. 2886, file 851-1-27 pt. 1, P.M. Bird, Chief, Radiation Protection Division, [Department of National Health and Welfare] to Dr. Otto Schaefer, Charles Camsell Hospital in Edmonton, Alberta, 6 February 1963.

73 Lux, *Separate Beds*, 4.
74 Lux, *Separate Beds*, 122.
75 LAC, RG 29, vol. 2886, file 851-1-27 pt. 1, Dr. P.M. Bird, Radiation Protection Division to Dr. P.E. Moore, Director, Medical Services, 21 February 1963.
76 Bird wrote a nine-page report in February 1963 summarizing the issue of northern fallout and proposing a large-scale systematic study; see LAC, RG 29, vol. 2886, file 851-1-27, pt. 1, P.M. Bird, *On the Problem of Radioactive Fallout Levels in Canada's North* (Ottawa: Radiation Protection Division, Department of National Health and Welfare, 1963).
77 LAC, RG 29, vol. 2886, file 851-1-27, pt. 1, Zone Superintendent, Yukon Zone to Regional Superintendent, Foothills Region, 25 March 1963.
78 LAC, RG 29, vol. 2886, file 851-1-27, pt. 1, Autopsies Performed on Eskimos, Central Region 1961–2, list sent from O.J. Rath, Regional Superintendent, Medical Services, Central Region to Director, Medical Services, Department of National Health and Welfare, 28 June 1963.
79 LAC, RG 29, vol. 2886, file 851-1-27, pt. 1, P.M. Bird, *On the Problem of Radioactive Fallout Levels*, 6.
80 Meren, "'Commend Me the Yak,'" 343–70 at 360.
81 Michael Best, "Eskimos May Get Yaks from India in Scheme to Relieve Austerity," *Winnipeg Free Press*, 27 August 1954, 7.
82 Meren, "'Commend Me the Yak.'"
83 LAC, RG 29, vol. 2886, file 851-1-27, pt. 1, P.M. Bird, *On the Problem of Radioactive Fallout Levels*, 9.
84 LAC, RG 29, vol. 2886, file 851-1-27, pt. 1, P.E. Moore, Director, Medical Services to Colonel H.M. Jones, Director, Indian Affairs Branch, Department of Citizenship and Immigration; same letter addressed to Dr. B.G. Sivertz, Director, Northern Administration Branch, Department of Northern Affairs and National Resources, 14 March 1963.
85 Bird, "Radiation Protection in Canada, Part III," 1115.
86 LAC, RG 29, vol. 2886, file 851-1-27, pt. 1, Fallout Levels in the Canadian North: Whole Body Counting – January–July 1965 Summary, July 1965.
87 Brown, *Plutopia*, 251.
88 LAC, RG 29, vol. 2886, file 851-1-27, pt. 1, H.A. Proctor to Regional Superintendent, Eastern Region, Medical Services re: Caesium 137 Levels in Northern Residents, 2 April 1965.
89 LAC, RG 29, vol. 2886, file 851-1-27, pt. 1, P.M. Bird to Dr. H.B. Brett, Zone Superintendent, MacKenzie Zone, Medical Services, Department of National Health and Welfare, 2 April 1965.
90 Brown, *Plutopia*, 248.
91 US Department of Defense, *An Appraisal of the Health and Nutritional States*.
92 Between November 1959 and December 1960, surgeons obtained bones from children and adults ranging in age from four months to sixty-two years. See Schulert, "Strontium-90 in Alaska," 146–8.

93 Bird, "Radiation Protection in Canada, Part III," 1116.
94 Curiously, Bird made no mention of Indigenous peoples in his article. When describing the source of the human bone samples, he opted for a general reference to northern residents.
95 LAC, RG 29, vol. 2886, file 851-1-27, pt. 1, Dr. O. Schaefer to Dr. O.C. Gray, Medical Superintendent, Charles Camsell Hospital, 26 April 1963.
96 "Eskimos Get Radioactive Cesium Doses," *Brandon Sun*, 5 September 1964, 1.
97 William Dean Howe (Hamilton South), quoted in House of Commons Debates, 26th Parliament, 2nd Session, vol. 7, 4 September 1964, 7659.
98 Hon. Judy V. LaMarsh (Minister of National Health and Welfare), quoted in House of Commons Debates, 26th Parliament, 2nd Session, vol. 7, 4 September 1964, 7659.
99 Government of Canada, *Data from Radiation Protection Programs*, vols. 1–6; see also *Annual Report on the Radioactive Fallout Study Program*, published consecutively from 1957 onward during the lifespan of the research program.
100 Hon. Judy V. LaMarsh (Minister of National Health and Welfare), quoted in House of Commons Debates, 26th Parliament, 2nd Session, vol. 7, 4 September 1964, 7659.
101 Declassified records from the Radiation Protection Division indicate that chemists in Ottawa received no fewer than six bone samples between March and November 1965. The records also indicate the receipt of four samples of rib, but it is unclear whether the word "rib" refers to rib bone. If the rib samples recorded were samples of rib bone, chemists in Ottawa received no fewer than ten human bone samples obtained through autopsies conducted at hospitals in northern Canada. However, if the samples of rib did not "qualify" as samples of rib bone, the actual number of human bone samples obtained and analysed for the study may have been six. Note that openly published records refer to "rib" as rib bone; see Government of Canada, *Data from Radiation Protection Programs*, vol. 1, No. 2 (Ottawa: Radiation Protection Division, Department of National Health and Welfare, 1963), 21. The last of the archival documents indicating the receipt of human bone did not specify the number of samples received, meaning the total number may have exceeded the estimated ten documented here; see LAC, RG 29, vol. 2886, file 851-1-27 pt. 1, Bone and Soft Tissue Samples Received as part of Special Northern Study (January–15 July 1965); E.R. Samuels, Chemist-in-Charge, Analytical Laboratory, Radiochemistry Section, [Radiation Protection Division, Department of National Health and Welfare] to Dr. Wilbush, Inuvik General Hospital, 22 November 1965.
102 LAC, RG 29, vol. 2886, file 851-1-27, pt. 1, Bone and Soft Tissue Samples Received as part of Special Northern Study (January–15 July 1965).
103 LAC, RG 29, vol. 2886, file 851-1-27, pt. 1, Bone and Soft Tissue Samples (January–15 July 1965).

104 LAC, RG 29, vol. 2886, file 851-1-27, pt. 1, E.R. Samuels, Chemist-in-Charge, Analytical Laboratory, Radiochemistry Section, [Radiation Protection Division, Department of National Health and Welfare] to Dr. Wilbush, Inuvik General Hospital, 22 November 1965.
105 LAC, RG 29, vol. 2886, file 851-1-27, pt. 1, H.A. Procter, Radiation Protection Division to Regional Superintendents, Foothills, Central and Eastern Regions, Medical Services, 10 September 1965.
106 LAC, RG 29, vol. 2886, file 851-1-27, pt. 1, H.A. Procter to Regional Superintendent, Eastern Region, Medical Services, 10 September 1965.
107 LAC, RG 29, vol. 2886, file 851-1-27, pt. 1, H.A. Procter to Regional Superintendent, 10 September 1965.
108 Kaufert and O'Neil, "Cooptation and Control"; reference borrowed from Dyck and Lux, "Population Control in the 'Global North'?"
109 Dyck and Lux, "Population Control in the 'Global North'?," 495.
110 LAC, RG 29, vol. 2886, file 851-1-27, pt. 1, W.V. Mayneord, The Institute of Cancer Research: Royal Cancer Hospital to Dr. [P.M.] Bird, Radiation Protection Division, Department of National Health and Welfare, 29 May 1964.
111 LAC, RG 29, vol. 2886, file 851-1-27, pt. 1, P.E. Moore to Regional Superintendents, Foothills, Central and Eastern Regions, Medical Services re: Radioactive Assay Program Yukon and Northwest Territories, 29 December 1964.
112 Brookfield, *Cold War Comforts*, 72.
113 Primary teeth develop over a period of six to eight years, resulting in a lag between collection and analysis. The process worked nonetheless, according to the authorities involved.
114 LAC, RG 29, vol. 2886, file 851-1-27 pt. 1, D.J. Yeo, Director, Metropolitan Health Committee, Dental Health Services, Vancouver to Mrs. Keith Bramber [*sic*], box 663, Whitehorse, Yukon, 30 January 1963; see the attached research description of the study concerning radioactivity of dental tissues.
115 LAC, RG 29, vol. 2886, file 851-1-27, pt. 1, Bone and Soft Tissue Samples Received as part of Special Northern Study (January–15 July 1965).
116 LAC, RG 29, vol. 2886, file 851-1-27, pt. 1, E.R. Samuels, Chemist-in-Charge, Analytical Laboratory, Radiochemistry Section, [Radiation Protection Division] to Dr. G. Gray, Charles Camsell Hospital, Edmonton, 20 May 1965.
117 Fifteen is an aggregated number based on data compiled from receipts issued by the Radiation Protection Division; see LAC, RG 29, vol. 2886, file 851-1-27, pt. 1.
118 LAC, RG 29, vol. 2886, file 851-1-27, pt. 1, J.H. Wiebe to Dr. V.K. Mohindra, Physicist-in-Charge, Analytical Laboratory, Physics Section, Radiation Protection Division, 16 August 1965.
119 LAC, RG 29, vol. 2886, file 851-1-27, pt. 1, Joseph Sternberg, Professor, Faculty of Medicine, University of Montreal to Dr Jean F. Webb, Chief,

Child and Maternal Health Division, Department of National Health and Welfare, 2 January 1965.
120 For instance, see Sternberg, "Placental Transfers"; and sternberg, "radiation and Pregnancy."
121 The RPD's special northern study also included a systematic investigation of the dietary habits of selected northern residents.

Conclusion: Reflections on Northern Canada, Military Research, and the Cold War

1 Wiseman, "Frontier Footage," in *Cold Science*, ed. Bocking and Heidt, 61–74.
2 Grant, "Myths of the North"; Grant, "Northern Nationalists: Visions of a New North, 1940–1950," in *For Purposes of Dominion*, ed. Coates and Morrison.
3 Jones-Imhotep, *The Unreliable Nation*, 33.
4 White, "Ceremonies of Possession" (PhD diss.), 16.
5 Although the analysis in this book ends at 1970, the DRB continued to fund and conduct Arctic-related research at climate-simulation laboratories in Toronto and other urban centres in southern Canada until mid-decade.
6 See, for instance, McNeil, *Something New Under the Sun*; and McNeil and Unger, eds., *Environmental Histories of the Cold War*.
7 Howkins, *The Polar Regions*, 124.
8 Keenleyside, quoted in Grant, *Sovereignty or Security?*, 191.
9 Historian Paul Robinson has argued that intelligence does not improve policy-making, an assertion derived from a close study of Canadian foreign intelligence. See Robinson, "The Viability of a Canadian Foreign Intelligence Service."
10 Blair Fraser, "The Truth about Our Arctic Defense: We Have None," *Maclean's Magazine*, 15 November 1954, 54.
11 See Goodspeed, *A History of the Defence Research Board of Canada*, 229–31; "Human Problems in War Studied by New Division," *Globe and Mail*, 12 June 1952, 2.
12 On the fear and fatigue associated with military service in the North, watch Nicholas Balla, *Vigil in the North*, archival film 106B 0154 088 (Ottawa: National Film Board of Canada, 1954). For a written description of the film, see Bidd, ed., *The NFB Film Guide*, 508.
13 For a detailed description of Zubek's career and work in sensory deprivation, including images of experiments conducted at the University of Manitoba, see Cecil Rosner, "Isolation: A Canadian Professor's Research into Sensory Deprivation and Its Connection to Disturbing New Methods of Interrogation," *Canada's History* (August–September 2010): 29–37. See also John P. Zubek, "Behavioral and Physiological Effects of Prolonged

Sensory and Perceptual Deprivation: A Review," in *Man in Isolation and Confinement*, ed. John E. Rasmussen (Chicago: Aldine, 1973), 9–83.
14 Wiseman, "Canadian Scientists and Military Research."
15 Macdonald, ed. *The Arctic Frontier*.
16 Howkins, *The Polar Regions*, 126.
17 To read more about the militarization of Canada in the Cold War, see Lackenbauer and Farish, "The Cold War on Canadian Soil."
18 Parr, *Sensing Changes*.
19 More precisely, see the drastic financial cuts and bureaucratic restructuring of the Department of National Defence brought on in part by the findings and recommendations of the Royal Commission on Government Organization (Glassco Commission).
20 See, for instance, Korsmo and Graham, "Research in the North American North"; Farish, "Frontier Engineering"; and Bocking, "Seeking the Arctic," 61–74.
21 Galloway, "The Army Says Goodbye to 'The Shining Land,'" 53.
22 Wiseman, "Canadian Scientists and Military Research," 458.

Bibliography

Archival Sources

Canadian Broadcasting Corporation Archives

Canadian War Museum (CWM)
  George Metcalf Archival Collection
  Hartland Molson Library Collection
  Oral History Project

Churchill Public Archives, Churchill Public Library

Dalhousie University Archives
  Dalhousie University Reference Collection

Directorate of History and Heritage (DHH), Department of National Defence
  George Lindsey fonds
  Joint Staff fonds

Laurier Military History Archive, Laurier Centre for the Study of Canada
  George Lindsey fonds

Library and Archives Canada (LAC)
  Government Record Groups
    Advisory Committee on Northern Development, RG 112
    Department of External Affairs, RG 25
    Department of National Defence, RG 24
    Geological Survey of Canada, RG 45
    Indian Affairs and Northern Development, RG 22
    Medical Research Council, RG 128

National Health and Welfare, RG 29
Northern Affairs Program, RG 85
Privy Council Office, RG 2
Manuscript Groups and Private Papers
Arctic Institute of North America fonds, MG28-I79
G. Malcolm Brown
James Patrick Croal fonds, MG31-G34

National Archives and Records Administration (NARA), College Park
Government Record Groups
Army Staff, RG 319
Defense Intelligence Agency, RG 373

Northwest Territories Archives, Yellowknife
Department of the Executive fonds

Queen's University Archives
Queen's Journal

University of Alberta, Education and Research Archive
Canadian Circumpolar Institute, Dr. Otto Schaefer Collection

University of Saskatchewan Archives and Special Collections
J.W.T. Spinks fonds

University of Toronto Archives and Records Management Services
Omond McKillop Solandt fonds

Western University Archives and Special Collections
University of Western Ontario Scrapbooks

Government Publications

Beals, C.S., and D.A. Shenstone. Eds. *Science, History, and Hudson Bay*, vols. 1 and 2. Ottawa: Department of Energy, Mines and Resources, Queen's Printer, 1968.

Bigelow, W.G., J.C. Callaghan, and John A. Hopps. *Radio-Frequency Rewarming in Resuscitation from Severe Hypothermia*. Ottawa: Defence Research Board, 1952.

Bird, P.M., A.H. Booth, and P.G. Mar. *Annual Report for 1959 on the Radioactive Fallout Study Program*. Ottawa: Radiation Protection Division, Department of National Health and Welfare, 1960.

Booth, A.H., ed. *Annual Report for 1961 on the Radioactive Fallout Study Program*. Ottawa: Radiation Protection Division, Department of National Health and Welfare, 1962.

Bryan, Major J.W.P. "Mobile Striking Force Winter Training: Training in Kirkland Lake Area." *Canadian Army Journal* 11, no. 2 (April 1957): 29–30.

Brown, G. Malcolm. *Progress Report on Clinical and Biochemical Studies of the Eskimo.* Ottawa: DRB, Department of National Defence, 1951.

Canada. Atomic Energy of Canada Limited. Chalk River.
- Cabinet Committee on Research and Defence. Ottawa.
- Cabinet Defence Committee. Ottawa.
- Chiefs of Staff Committee. Ottawa.
- Defence Research Board. Ottawa.
- Department of Agriculture. Ottawa.
- Department of External Affairs. Ottawa.
- Department of Foreign Affairs, Trade and Development. Ottawa.
- Department of National Defence. Ottawa.
- Department of National Health and Welfare. Ottawa.
- Department of Northern Affairs and National Resources. Ottawa.
- *Documents on Canadian External Relations*, vol. 13: *1947.* Ottawa: External Affairs and International Trade Canada, 1993.
- *Documents on Canadian External Relations*, vol. 14: *1948.* Ottawa: Department of Foreign Affairs and International Trade Canada, 1994.
- *Documents on Canadian External Relations*, vol. 16: *1950.* Ottawa: Department of Foreign Affairs and International Trade Canada, 1996.
- *Documents on Canadian External Relations*, vol. 17: *1951.* Ottawa: Department of Foreign Affairs and International Trade Canada, 1996.
- *Documents on Canadian External Relations: The Arctic 1874–1949.* Ottawa: Global Affairs Canada, 2016.
- *Honouring the Truth, Reconciling for the Future: Summary of the Final Report of the Truth and Reconciliation Commission of Canada.* Ottawa: Truth and Reconciliation Commission of Canada, 2015.
- House of Commons Canada, *Debates.* Ottawa.
- House of Commons Canada, *Special Studies, Committee Reports.* Ottawa.
- Joint Intelligence Committee. Ottawa.
- Monteith, J. Waldo. "Canada's Programme for Radiation Protection: An address by Mr. J. Waldo Monteith, Minister of National Health and Welfare, at the Jubilee Meeting of the Canadian Public Health Association, Montreal, Quebec, on June 1, 1959." Ottawa: Information Division, Department of External Affairs, 1959.
- National Research Council. Ottawa.
- *Preliminary Report on the Measurements of Radioactive Strontium in Canadian Milk Powder Samples.* Ottawa: Radiation Services, Department of National Health and Welfare, 1958.
- Privy Council Committee on Scientific and Industrial Research. Ottawa.
- Privy Council Office. Ottawa.

- *Report of the Ontario Veterinary College 1947.* Toronto: Ontario Department of Agriculture, 1948.
- *Report of the Ontario Veterinary College 1948.* Toronto: Ontario Department of Agriculture, 1949.
- *Report for the Year 1953–54: The Ontario Veterinary College Guelph, Ontario.* Toronto: Ontario Department of Agriculture, 1949.
- Senate. Committee Reports. Ottawa.
- Statistics Canada, *Historical Statistics of Canada.*

Claxton, Brooke. *Canada's Defence: Information on Canada's Defence Achievements and Organization.* Ottawa: Department of National Defence, 1947.

Coffey, M.F. *Debriefing of an Ice Floe, Technical Paper 56/08519.* Ottawa: Directorate of Information Services, Defence Research Board, Department of National Defence.

Data from Radiation Protection Programs, vols. 1–6. Ottawa: Radiation Protection Division, Department of National Health and Welfare, 1963–8.

Davies, G.K. *An Annotated Bibliography of Unclassified Reports by Defence Research Northern Laboratory, 1947–1965.* Defence Scientific Information Service (DSIS) Report No. B–13. Ottawa: Defence Research Board, Department of National Defence, 1969.

Defence Research Board, Department of National Defence. *Annual Report on the Progress of Environmental Protection Research, December 1953: Report No. DR 80.* Ottawa: Defence Research Board, 1954.

- "Canada to Enter Space Race." *Canadian Army Journal* 14, no. 3 (Summer 1960): 60–1.
- *The Defence Research Board and the Defence Scientific Service.* Ottawa: Queen's Printer, 1953.
- *Defence Research Board Newsletter.* Ottawa: Defence Research Board.
- *Defence Research Board: The First Twenty-Five Years.* Ottawa: Department of National Defence, 1972.
- *Defence Research Northern Laboratory: Progress Report on Indoctrination Training for Military Operations in the North, DRNL Project Report No. 4.* Ottawa: Defence Research Board, 1954.

Directorate of Public Relations (Army). "Winter Training at Churchill." *Canadian Army Journal* 10, no. 2 (April 1956): 47–5.

Dziuban, Stanley W. *United States Army in World War II: Special Studies – Military Relations between the United States and Canada, 1939–1945.* Washington, D.C.: Office of the Chief of Military History, Department of Army, 1959.

Freeman, T.N. *Entomological Research in Northern Canada Progress Report 1948, Part I: The Northern Insect Survey.* Ottawa: Defence Research Board, Department of National Defence, 1948.

General Staff Branch, Headquarters Western Command, Edmonton. "Exercise Bulldog III." *Canadian Army Journal* 9, no. 2 (April 1955): 10–19.

Goodspeed, D.J. *A History of the Defence Research Board of Canada.* Ottawa: Queen's Printer, 1958.
Grummitt, W.E., A.P. James, and H.B. Newcombe. *The Analysis of the Strontium-90 Levels in Canadian Milk up to 1958, CRC-850.* Chalk River: Atomic Energy of Canada Limited, 1959.
Hattersley-Smith, G. *North of Latitude Eighty: The Defence Research Board in Ellesmere Island.* Ottawa: Defence Research Board, 1973.
Hitsman, Major J.S. "Medical Problems of Paratroop Training." *Canadian Army Journal* 4, no. 1 (April 1950): 9–18.
McNamara, C.P., and W.G. Penney. "The Technical Feasibility of Establishing an Atomic Weapons Proving Ground in the Churchill Area." Ottawa: Defence Research Board; London: Ministry of Supply, ca. 1947–50.
Pennie, A.M. *Defence Research Northern Laboratory 1947–1965,* Report DR 179. Ottawa: Defence Research Board, 1966.
Smith, Gordon. *Ice Islands in Arctic Waters.* Ottawa: Department of Indian and Northern Affairs, 1980.
Twinn, C.R. *Entomological Research in Northern Canada Progress Report 1948, Part II: Joint United-States-Canadian Biting Fly Survey and Experimental Control Programme Carried Out at Churchill, Manitoba, in 1948.* Ottawa: Defence Research Board, Department of National Defence, 1948.
United Kingdom. Ministry of Supply. London.
United Nations. *Report of the Scientific Committee on the Effects of Atomic Radiation, A/5216.* GAOR. 17th sess., suppl. no. 16, 1962.
– "Treaty Banning Nuclear Weapon Tests in the Atmosphere, in Outer Space and Under Water." United Nations Treaty Collection.
United States. *An Appraisal of the Health and Nutritional States of the Eskimo: Report of the Interdepartmental Committee on Nutrition for National Defense.* Fairbanks: Arctic Health Research Center, 1959.
– Army. *An Introduction to Churchill, Fort Churchill and Surrounding Area.* 7099th ASU, 1st Arctic Test Detachment, US Army, n.d.
– Department of Defense. Virginia.
– Department of Energy. Washington, D.C.
– Department of Fish and Game. Juneau.
– Department of the Army. "Army R&D Chief Entertains Quadripartite Group." *Army Research and Development* 4, no. 4 (April 1963).
– National Research Council. *The Arctic Aeromedical Laboratory's Thyroid Function Study: A Radiological Risk and Ethical Analysis.* Washington, D.C.: National Academy Press, 1996.
Viereck, Leslie A. *Radioactivity Report,* vol. 4: *Annual Project Segment Report, Federal Aid in Wildlife Restoration Project W-6-R-4, Work Plan L.* Juneau: Alaska Department of Fish and Game, 1964.
Watson, H.H. "Army Operational Research Aims and Methodology." *Canadian Army Journal* 15, no. 3 (Summer 1961): 22–9.

Way, Sargent-Major F.J., Directorate of Public Relations (Army). "Mobile Striking Force Winter Training: Training at Fort Churchill." *Canadian Army Journal* 11, no. 2 (April 1957): 20–9.

Film

Balla, Nicholas. *Vigil in the North.* National Film Board of Canada, Ottawa, 1954.

Internet Sites and Repositories

Canada Declassified, University of Toronto: https://declassified.library.utoronto.ca.
Canadian Broadcasting Corporation Archives: https://www.cbc.ca/archives.
Canadian Foreign Intelligence History Project, Carleton University: https://carleton.ca/csids/canadian-foreign-intelligence-history-project.

Interviews and Oral Testimony

Erickson, William, 3 August 2016.
Jacob, Eileen, Internet, ca. 2003.
Law, Cecil Ernest, CWM Oral History Project, 6 August 2008.
Pennie, Archie, CWM Oral History Project, 16 January 2009.
Winegard, William C., 5 April 2017.

Newspapers and Periodicals

Army Information Digest (US Army)
Beaver, The
Canadian Press
Churchill Observer
Daily News (St. John's, Newfoundland)
Globe and Mail
Hamilton Spectator
Lethbridge Herald
Maclean's Magazine
Montreal Gazette
New York Times
Ottawa Citizen
Ottawa Journal
Queen's Journal (Kingston, Ontario)
Roundel (RCAF)
Saint John Telegraph-Journal

Simcoe Reformer
Taiga Times (Churchill, Manitoba)
Toronto Daily Star
Whitehorse Star
Winnipeg Free Press

Secondary Sources

Abel, Kerry M. "Quelques arpents de neige? Or, How Fares the Myth of the North?" *Underhill Review* (Fall 2008): 1–4.
Ahearn, Laura M. *Living Language: An Introduction to Linguistic Anthropology.* Hoboken: John Wiley and Sons, 2011.
Avery, Donald. *Pathogens for War: Biological Weapons, Canadian Life Scientists, and North American Biodefence.* Toronto: University of Toronto Press, 2013.
– *The Science of War: Canadian Scientists and Allied Military Technology during the Second World War.* Toronto: University of Toronto Press, 1998.
Baldwin, Andrew, Laura Cameron, and Audrey Kobayashi, eds. *Rethinking the Great White North: Race, Nature, and the Historical Geographies of Whiteness in Canada.* Vancouver: UBC Press, 2011.
Baugh, C.W., G.S. Bird, G.M. Brown, et al. "Blood Volumes of Eskimos and White Men Before and During Acute Cold Stress." *Journal of Physiology* 140, no. 3 (1958): 354.
Belanger, Dian Olson. *Deep Freeze: The United States, the International Geophysical Year, and the Origins of Antarctica's Age of Science.* Boulder: University Press of Colorado, 2011.
Bidd, Donald W., ed. *The NFB Film Guide: The Productions of the National Film Board of Canada from 1939 to 1989.* Ottawa: National Film Board of Canada, 1991.
Bigelow, W.G. *Cold Hearts: The Story of Hypothermia and the Pacemaker in Heart Surgery.* Toronto: McClelland and Stewart, 1984.
Bigelow, W.G., et al. "Oxygen Transport and Utilization in Dogs at Low Body Temperatures." *American Journal of Physiology* 160, no. 1 (January 1950): 125–37.
Bigelow, W.G., J.C. Callaghan, and J.A. Hopps. "General Hypothermia for Experimental Intracardiac Surgery: The Use of Electrophrenic Respirations, an Artificial Pacemaker for Cardiac Standstill, and Radio-Frequency Rewarming in General Hypothermia." *Annals of Surgery* 132, no. 3 (September 1950): 531–7.
Bigelow, W.G., and J.E. McBirnie. "Further Experiences with Hypothermia for Intracardiac Surgery in Monkeys and Groundhogs." *Annals of Surgery* 137, no. 3 (March 1953): 361–5.
Bird, P.M. "Radiation Protection in Canada, Part III: The Role of the Radiation Protection Division in Safeguarding the Health of the Public." *Canadian Medical Association Journal* 90, no. 19 (May 1964): 1114–20.

Bocking, Stephen. *Ecologists and Environmental Politics: A History of Contemporary Ecology.* New Haven: Yale University Press, 1997.
- *Nature's Experts: Science, Politics, and the Environment.* New Brunswick: Rutgers University Press, 2004.
- "Seeking the Arctic: Science and Perceptions of Northern Canada." *Dalhousie Review* 90, no. 1 (Spring 2010): 61–74.

Bocking, Stephen, and Daniel Heidt, eds. *Cold Science: Environmental Knowledge in the North American Arctic during the Cold War.* New York: Routledge, 2019.

Bocking, Stephen, and Brad Martine. *Ice Blink: Navigating Northern Environmental History.* Calgary: University of Calgary Press, 2017.

Bonnell, Jennifer. "Early Insecticide Controversy and Beekeeper Advocacy in the Great Lakes Region." *Environmental History* 26 (2021): 79–101.

Borisov, P.M. "Can We Control the Arctic Climate?" *Bulletin of the Atomic Scientists* (March 1969): 43–8.

Bothwell, Robert. *The Big Chill: Canada and the Cold War.* Toronto: Irwin, 1998.
- *Nucleus: The History of Atomic Energy of Canada Limited.* Toronto: University of Toronto Press, 1988.

Brandson, Lorraine E. *Churchill Hudson Bay: A Guide to Natural and Cultural Heritage.* Churchill: Churchill Eskimo Museum, 2012.

Bravo, Michael, and Sverker Sörlin, eds. *Narrating the Arctic: A Cultural History of Nordic Scientific Practices.* Canton: Science History Publications, 2002.

Brookfield, Tarah. *Cold War Comforts: Canadian Women, Child Safety, and Global Insecurity.* Waterloo: Wilfrid Laurier University Press, 2012.

Brown, G. Malcolm. "Cold Acclimatization in Eskimo." *Arctic* 7, no. 3 and 4 (1954): 351.
- "Northern Research Reports: Medical Investigation at Southampton Island." *Arctic* 2, no. 1 (1949): 70–1.

Brown, G. Malcolm, G.S. Bird, et al. "Cold Acclimatization." *Canadian Medical Association Journal* 70, no. 3 (1954): 259.
- "The Circulation in Cold Acclimatization." *Circulation: Journal of the American Heart Association* (1954): 813–22.

Brown, Kate. *Plutopia: Nuclear Families, Atomic Cities, and the Great Soviet and American Plutonium Disasters.* New York: Oxford University Press, 2013.

Bruce-Chwatt, Leonard J. "Radioisotopes for Research on and Control of Mosquitos." *Bulletin of the World Health Organization* 15, nos. 3–5 (1956): 491–511.

Bugher, John C., and Marjorie Taylor. "Radiophosphorus and Radiostrontium in Mosquitoes. Preliminary Report." *Science* 110, no. 2849 (August 1949): 146–7.

Burnett, Kristin. *Taking Medicine: Women's Healing Work and Colonial Contact in Southern Alberta, 1880–1930.* Vancouver: UBC Press, 2010.

Burtch, Andrew. *Give Me Shelter: The Failure of Canada's Cold War Civil Defence.* Vancouver: UBC Press, 2012.

Burton, Alan C., and Otto G. Edholm. *Man in a Cold Environment: Physiological and Psychological Effects of Exposure to Low Temperatures.* London: Edward Arnold, 1955.
Cameron, Emilie. *Far Off Metal River: Inuit Lands, Settler Stories, and the Making of the Contemporary Arctic.* Vancouver: UBC Press, 2015.
Campbell, Isabel. *Unlikely Diplomats: The Canadian Brigade in Germany, 1951–64.* Vancouver: UBC Press, 2013.
Carson, Rachel. *Silent Spring: The Classic That Launched the Environmental Movement.* New York: Mariner Books, 2002.
Cavell, Janice. "The Second Frontier: The North in English-Canadian Historical Writing." *Canadian Historical Review* 83, no. 3 (September 2002): 364.
Cavell, Richard, ed. *Love, Hate, and Fear in Canada's Cold War.* Toronto: University of Toronto Press, 2004.
Cecil, Paul Frederick. *Herbicidal Warfare: The Ranch Hand Project in Vietnam.* New York: Praeger, 1986.
Chapnick, Adam. "The Canadian Middle Power Myth." *International Journal* 55, no. 2 (Spring 2000): 188–206.
– *The Middle Power Project: Canada and the Founding of the United Nations.* Vancouver: UBC Press, 2006.
Chastain, Andra B., and Timothy W. Lorek, eds. *Itineraries of Expertise: Science, Technology, and the Environment in Latin America.* Pittsburgh: University of Pittsburgh Press, 2020.
Clearwater, John, and David O'Brien. "O Lucky Canada." *Bulletin of Atomic Sciences* 59, no. 4 (July–August 2003): 60–5.
Coates, Ken S., et al. *Arctic Front: Defending Canada in the Far North.* Toronto: Thomas Allen, 2008.
Coates, Ken, and William R. Morrison. *The Alaska Highway in World War II: The US Army of Occupation in Canada's Northwest.* Norman: University of Oklahoma Press, 1992.
–, eds. *For Purposes of Dominion: Essays in Honor of Morris Zaslow.* Concord: Captus University, 1989.
Colbourn, Susan, and Timothy Andrews Sayle, eds. *The Nuclear North: Histories of Canada in the Atomic Age.* Vancouver: UBC Press, 2019.
Coleman, William. "Science and Symbol in the Turner Frontier Hypothesis." *American Historical Review* 72, no. 1 (1966): 22–49.
Cotton, Barb, Andrea Manning-Kroon, and William E. McNally. "An Overview of the Law Regarding Informed Consent." *The Barrister* 72, no. 10 (2004).
Croal, J.P., and J.R. Lotz. "Armed Forces Serve Science." *Canadian Army Journal* 15, no. 2 (Spring 1961): 19–27.
Cruikshank, Julie. *Do Glaciers Listen? Local Knowledge, Colonial Encounters, and Social Imagination.* Vancouver: UBC Press, 2005.

Daschuk, James. *Clearing the Plains: Disease, Politics of Starvation, and the Loss of Aboriginal Life.* Regina: University of Regina Press, 2013.

Davis, Frederick Rowe. *Banned: A History of Pesticides and the Science of Toxicology.* New Haven: Yale University Press, 2014.

della Dora, Veronica. "From the Radio Shack to the Cosmos: Listening to Sputnik during the International Geophysical Year (1957–1958)." *Isis* 114, no. 1 (2023): 123–49.

Dodds, Klaus, and Richard Powell. "Polar Geopolitics: New Researchers on the Polar Regions," *Polar Journal* 3, no. 1 (2013): 1–8.

Doel, Ronald E. "Constituting the Postwar Earth Sciences: The Military's Influence on the Environmental Sciences in the USA after 1945." *Social Studies of Science* 33 (2003): 635–66.

Doel, Ronald E., Kristine C. Harper, and Matthias Heymann, eds. *Exploring Greenland: Cold War Science and Technology on Ice.* New York: Palgrave Macmillan, 2016.

Doel, Ronald E., Suzanne Zeller, and Urban Wråkberg. "Science, Environmental Knowledge, and the New Arctic." *Journal of Historical Geography* 44 (2014): 2–14.

Doyle, Stephen E., and A. Ingemar Skog, eds. *The International Geophysical Year: Initiating International Scientific Space Co-operation.* Paris: International Astronautical Federation, 2012.

Duffin, Jacalyn. *Stanley's Dream: The Medical Expedition to Easter Island.* Montreal and Kingston: McGill–Queen's University Press, 2020.

Dunlap, Thomas R., ed. *DDT, Silent Spring, and the Rise of Environmentalism: Classic Texts.* Seattle: University of Washington Press, 2015.

Dyck, Erika, and Maureen Lux. "Population Control in the 'Global North'?: Canada's Response to Indigenous Reproductive Rights and Neo-Eugenics." *Canadian Historical Review* 97, no. 4 (December 2016): 481–512.

Eayrs, James. *In Defence of Canada: Growing Up Allied.* Toronto: University of Toronto Press, 1980.

Edgington, Ryan H. *Range Wars: The Environmental Contest for White Sands Missile Range.* Lincoln: University of Nebraska Press, 2014.

Emmerson, Charles. *The Future History of the Arctic.* London: Bodley Head, 2010.

Eyre, Major K.C. "Tactics in the Snow: The Development of a Concept." *Canadian Defence Quarterly* 4, no. 4 (Spring 1975): 7–12.

Faragher, John Mack, ed. *Rereading Frederick Jackson Turner: "The Significance of the Frontier in American History" and Other Essays.* New Haven: Yale University Press, 1999.

Farish, Matthew. *The Contours of America's Cold War.* Minneapolis: University of Minnesota Press, 2010.

– "Frontier Engineering: From the Globe to the Body in the Cold War Artic." *Canadian Geographer* 50, no. 2 (2006): 177–96.

- "The Lab and the Land: Overcoming the Arctic in Cold War Alaska." *Isis* 104, no. 1 (March 2013): 1–29.
Finnie, Richard. *CANOL: The Sub-Arctic Pipeline and Refinery Project Constructed by Bechtel-Price-Callahan for the Corps of Engineers United States Army 1942–1944*. San Francisco: Ryder and Ingram, 1945.
Fitzgerald, Michael R., ed. *Out of the Cold: The Cold War and Its Legacy*. New York and London: Bloomsbury Academic, 2013.
Freeman, T.N. "The Canadian Northern Insect Survey, 1947–57." *Polar Record* 9, no. 61 (1959): 299–307.
- "Some Problems of Insect Biology in the Canadian Arctic." *Arctic* 5, no. 3 (1952): 175–7.
Frye, Northrop. *The Bush Garden: Essays on the Canadian Imagination*. Toronto: House of Anansi Press, 1995.
Galloway, Colonel Strome. "The Army Says Goodbye to 'The Shining Land.'" *Canadian Army Journal* 18, no. 1 (1964): 40–53.
Geller, Peter. "The 'True North' in Pictures? Photographic Representation in the Hudson's Bay Company's *The Beaver* Magazine, 1920–1945." *Archivaria* 36 (1993): 166–88.
Giehmann, Barbara Stefanie. *Writing the Northland: Jack London's and Robert W. Service's Imaginary Geography*. Würzburg: Königshausen & Neumann, 2011.
Godefroy, Andrew B. *Defence and Discovery: Canada's Military Space Program, 1945–74*. Vancouver: UBC Press, 2011.
- *In Peace Prepared: Innovation and Adaptation in Canada's Cold War Army*. Vancouver: UBC Press, 2014.
Gorham, Eville. "A Comparison of Lower and Higher Plants as Accumulations of Radioactive Fall-Out." *Canadian Journal of Botany* 37 (1959): 327–9.
Grace, Sherrill. *Canada and the Idea of North*. Montreal and Kingston: McGill–Queen's University Press, 2007.
Grant, Shelagh D. "Myths of the North in Canadian Ethos," *Northern Review* 3, no. 4 (1989): 15–41.
- *Polar Imperative: A History of Arctic Sovereignty in North America*. Vancouver: Douglas and McIntyre, 2011.
- *Sovereignty or Security? Government Policy in the Canadian North, 1936–1950*. Vancouver: UBC Press, 1988.
Griffiths, Franklyn, Rob Huebert, and P. Whitney Lackenbauer, eds. *Canada and the Changing Arctic: Sovereignty, Security, and Stewardship*. Waterloo: Wilfrid Laurier University Press, 2011.
Grygier, Pat Sandiford. *A Long Way from Home: The Tuberculosis Epidemic among the Inuit*. Montreal and Kingston: McGill–Queen's University Press, 1997.
Hacquebord, Louwrens. *LASHIPA: History of Large Scale Resource Exploitation in Polar Areas*. Eelde: Barkhuis, 2012.

Hadley, Michael L., Rob Huebert, and Fred W. Crickard, eds. *A Nation's Navy: In Quest of Canadian Naval Identity.* Montreal and Kingston: McGill–Queen's University Press, 1996.

Halliday, Hugh A. "Exercise 'Musk Ox': Asserting Sovereignty 'North of 60.'" *Canadian Military History* 7, no. 4 (2012): 37–44.

– "Recapturing the North: Exercises 'Eskimo,' 'Polar Bear,' and 'Lemming' 1945." *Canadian Military History* 6, no. 2 (Autumn 1997): 29–38.

Hamblin, Jacob Darwin. *Poison in the Well: Radioactive Waste in the Oceans at the Dawn of the Nuclear Age.* New Brunswick: Rutgers University Press, 2009.

Hankins, Gerald W. *Sunrise over Pangnirtung: The Story of Otto Schaefer, M.D.* Calgary: Arctic Institute of North America, 2000.

Hanson, W.C., F.W. Whicker, and A.H. Dahl. "Iodine-131 in the Thyroids of North American Deer and Caribou: Comparison after Nuclear Tests." *Science* 140, no. 3568 (1963): 801–2.

Harper, Kristine C. *Make It Rain: State Control of the Atmosphere in Twentieth-Century America.* Chicago: University of Chicago Press, 2017.

Heggie, Vanessa. *Higher and Colder: A History of Extreme Physiology and Exploration.* Chicago: University of Chicago Press, 2019.

Heide, C.L. "Stations of the RCAF: CJATC Rivers." *Roundel* 13, no. 8 (1961): 10–15.

Heidt, Daniel. "I Think That Would Be the End of Canada": Howard Green, the Nuclear Test Ban, and Interest-Based Foreign Policy, 1946–1963." *American Review of Canadian Studies* 42, no. 3 (2012): 343–69.

Heidt, Daniel, and P. Whitney Lackenbauer. *The Joint Arctic Weather Stations: Science and Sovereignty in the High Arctic, 1946–1972.* Calgary: University of Calgary Press, 2022.

Hendrick, James W. "Diagnosis and Management of Thyroids." *Journal of the American Medical Association* 164, no. 2 (1957): 127–33.

Herken, Gregg. *Brotherhood of the Bomb: The Tangled Lives and Loyalties of Robert Oppenheimer, Ernest Lawrence, and Edward Teller.* New York: Henry Holt, 2002.

Herzberg, Julia, Christian Kehrt, and Franziska Torma, eds. *Ice and Snow in the Cold War: Histories of Extreme Climatic Environments.* New York: Berghahn Books, 2018.

Heymann, Matthew. et al. "Exploring Greenland: Science and Technology in Cold War Settings." *Scientia Canadensis* 33, no. 2 (2010): 11–42.

Higuchi, Toshihiro. *Political Fallout: Nuclear Weapons Testing and the Making of a Global Environmental Crisis.* Stanford: Stanford University Press, 2020.

Holmes, John W. "Most Safely in the Middle." *International Journal* 32, no. 2 (Spring, 1984): 366–88.

– *The Shaping of Peace: Canada and the Search for World Order, 1943–1957*, vol. 1. Toronto: University of Toronto Press, 1979.

Horn, Bernd. *Bastard Sons: An Examination of Canada's Airborne Experience 1941–1995.* St. Catharines: Vanwell, 2001.

Howkins, Adrian. *The Polar Regions: An Environmental History*. Cambridge: Polity Press, 2016.

Iacovetta, Franca. *Gatekeepers: Reshaping Immigrant Lives in Cold War Canada*. Toronto: Between the Lines, 2006.

Iarocci, Andrew. "Opening the North: Technology and Training at the Fort Churchill Joint Services Experimental Testing Station, 1946–64." *Canadian Army Journal* 10, no. 4 (Winter 2008): 74–95.

Jacobs, Robert A. *Nuclear Bodies: The Global Hibakusha*. New Haven: Yale University Press, 2022.

Jenkins, Dale W. "A Field Method of Marking Arctic Mosquitoes with Radiophosphorus." *Journal of Economic Entomology* 42, no. 6 (December 1949): 988–9.

Jones-Imhotep, Edward. "Laboratory Cultures." *Scientia Canadensis* 28 (2005): 7–26

– *The Unreliable Nation: Hostile Nature and Technological Failure in the Cold War*. Cambridge, MA: MIT Press, 2017.

Kasurak, Peter. *A National Force: The Evolution of Canada's Army, 1950–2000*. Vancouver: UBC Press, 2013.

Kaufert, Patricia A., and John D. O'Neil. "Cooptation and Control: The Reconstruction of Inuit Birth." *Medical Anthropology Quarterly* 4, no. 4 (1990): 438–9.

Keenleyside, Hugh L. *Memoirs of Hugh L. Keenleyside*, vol. 2: *On the Bridge of Time*. Toronto: McClelland and Stewart, 1982.

Kelm, Mary-Ellen. *Colonizing Bodies: Aboriginal Health and Healing in British Columbia, 1900–50*. Vancouver: UBC Press, 1999.

Kikkert, Peter, and P. Whitney Lackenbauer. "'Men of Frontier Experience': Yukoners, Frontier Masculinity, and the First World War." *Northern Review* 44 (2017): 209–42.

Kinkela, David. *DDT and the American Century: Global Health, Environmental Politics, and the Pesticide That Changed the World*. Chapel Hill: University of North Carolina Press. 2011.

Kinsman, Gary, Dieter K. Buse, and Mercedes Steedman, eds. *Whose National Security? Canadian State Surveillance and the Creation of Enemies*. Toronto: Between the Lines, 2000.

Kinsman, Gary, and Patrizia Gentile. *The Canadian War on Queers: National Security as Sexual Regulation*. Vancouver: UBC Press, 2010.

Klingle, Matthew. "The Multiple Lives of Marjorie: The Dogs of Toronto and the Co-Discovery of Insulin." *Environmental History* 23, no. 2 (2018): 368–82.

Kobalenko, Jerry. "Geoffrey Hattersley-Smith (1923 – 2012)." *Arctic* 65, no. 4 (December 2012): 488–91.

Koenig, L.S., K.R. Greenaway, Moira Dunbar, and G. Hattersley-Smith. "Arctic Ice Islands." *Arctic* 5, No. 2 (July 1952): 67–103.

Korsmo, Fael L., and Amanda Graham. "Research in the North American North: Action and Reaction." *Arctic* 55, no. 4 (December 2002): 319–25.

Kuhlberg, Mark. *Killing Bugs for Business and Beauty: Canada's Aerial War against Forest Pests, 1913–1930*. Toronto: University of Toronto Press, 2022.

Lackenbauer, P. Whitney. *Battle Grounds: The Canadian Military and Aboriginal Lands*. Vancouver: UBC Press, 2007.

– *The Canadian Rangers: A Living History*. Vancouver: UBC Press, 2013.

– *De-Icing Required!: The Historical Dimension of the Canadian Air Force's Experience in the Arctic*. Ottawa: National Defence and the Canadian Forces, 2012.

– "The Irony and the Tragedy of Negotiated Space: A Case Study on Narrative Form and Aboriginal-Government Relations during the Second World War." *Journal of the Canadian Historical Association* 15, no. 1 (2004): 177–206.

Lackenbauer, P. Whitney, and Matthew Farish. "The Cold War on Canadian Soil: Militarizing a Northern Environment." *Environmental History* 12, no. 4 (2007): 920–50.

–, eds. *The Distant Early Warning (DEW) Line Coordinating Committee Minutes and Progress Reports, 1955–63*. Documents on Canadian Arctic Sovereignty and Security, vol. 15, 2019.

Lackenbauer, P. Whitney, and Daniel Heidt, eds. *The Advisory Committee on Northern Development: Context and Meeting Minutes, 1948–66*. Documents on Canadian Arctic Sovereignty and Security, Volume 4, 2015.

Lackenbauer, P. Whitney, and Peter Kikkert. *Lessons in Northern Operations: Canadian Army Documents, 1945–56*. Documents on Canadian Arctic Sovereignty and Security, vol. 7, 2016.Lajeunesse, Adam. *Lock, Stock, and Icebergs: A History of Canada's Arctic Maritime Sovereignty*. Vancouver: UBC Press, 2016.

Langford, Martha, and John Langford. *A Cold War Tourist and His Camera*. Montreal and Kingston: McGill–Queen's University Press, 2011.

Langston, Nancy. *Toxic Bodies: Hormone Disruptors and the Legacy of DES*. New Haven: Yale University Press, 2010.

Lauder, John (author), P. Whitney Lackenbauer, and Peter Kikkert, eds. *Tracks North: The Story of Exercise Muskox*. Arctic Operational History Series no. 5, 2018.

Law, C.E., G.R. Lindsey, and D.M. Grenville, eds. *Perspectives in Science and Technology: The Legacy of Omond Solandt*. Kingston: Queen's Quarterly, 1994.

Leddy, Lianne C. "Intersections of Indigenous and Environmental History in Canada." *Canadian Historical Review* 98, no. 1 (2017): 83–95.

– *Serpent River Resurgence: Confronting Uranium Mining at Elliot Lake*. Toronto: University of Toronto Press, 2022.

Li, Alison. *J.B. Collip and the Development of Medical Research in Canada*. Montreal and Kingston: McGill-Queen's University Press, 2003.

Linn, Brian McAllister. *Elvis's Army: Cold War GIs and the Atomic Battlefield*. Cambridge, MA: Harvard University Press, 2016.

Logsdon, John M., Roger D. Launius, and Robert William Smith, eds. *Reconsidering Sputnik: Forty Years Since the Soviet Satellite.* Amsterdam: Psychology Press, 2000.

Loo, Tina. *Moved by the State: Forced Relocation and Making a Good Life in Postwar Canada.* Vancouver: UBC Press, 2019.

Lotz, Jim. *The Best Journey in the World: Adventures in Canada's High Arctic.* Lawrencetown Beach: Pottersfield Press, 2006.

Lower, Arthur R.M. *Colony to Nation: A History of Canada.* Toronto: Longmans, 1946.

– *The North American Assault on the Canadian Forest: A History of the Lumber Trade between Canada and the United States.* Toronto: Ryerson, 1938.

– *Settlement and the Forest Frontier in Eastern Canada.* Toronto: Macmillan, 1936.

Luby, Brittany. "From Milk-Medicine to Public (Re)Education Programs: An Examination of Anishinabek Mothers' Responses to Hydroelectric Flooding in the Treaty #3 District, 1900–1975." *Canadian Bulletin of Medical History* 32, no. 2 (2015): 363–89.

Luedee, Jonathan. "Locating the Boundaries of the Nuclear North: Arctic Biology, Contaminated Caribou, and the Problem of the Threshold." *Journal of the History of Biology* 54 (2021): 67–93.

Lux, Maureen K. *Separate Beds: A History of Indian Hospitals in Canada, 1920s–1980s.* Toronto: University of Toronto Press, 2016.

Macdonald, Ronald St. John, ed. *The Arctic Frontier.* Toronto: University of Toronto Press, 1966.

Macdougall, Brenda. "Space and Place within Aboriginal Epistemological Traditions: Recent Trends in Historical Scholarship." *Canadian Historical Review* 98, no. 1 (2017): 64–82.

MacEachern, Alan, and William J. Turkel, eds. *Method and Meaning in Canadian Environmental History.* Toronto: Nelson, 2009.

Maloney, Sean. *Learning to Love the Bomb: Canada's Nuclear Weapons during the Cold War.* Sterling: Potomac Books, 2007.

– "The Mobile Striking Force and Continental Defence 1948–1955." *Canadian Military Journal* 2, no. 2 (1993): 75–88.

Manning, T.H., E.O. Höhn, and A.H. Macpherson. "The Birds of Banks Island." *National Museum of Canada Bulletin* 143 (1956): 182–97.

Marcus, Alan Rudolph. *Relocating Eden: The Image and Politics of Inuit Exile in the Canadian Arctic.* Hanover: University Press of New England, 1995.

Martin-Nielsen, Janet. "The Other Cold War: The United States and Greenland's Ice Sheet Environment, 1948–1966." *Journal of Historical Geography* 38, no. 1 (2012): 69–80.

Martino-Taylor, Lisa. *Behind the Fog: How the US Cold War Radiological Weapons Program Exposed Innocent Americans.* New York: Routledge, 2018.

McCallum, Mary Jane Logan. "Starvation, Experimentation, Segregation, and Trauma: Words for Reading Indigenous Health History." *Canadian Historical Review* 98, no. 1 (2017): 96–113.

McCallum, Mary Jane. "This Last Frontier: 'Isolation' and Aboriginal Health." *Canadian Bulletin of Medical History* 22, no. 1 (2005): 103–20.

McCannon, John. "The Commissariat of Ice: The Main Administration of the Northern Sea Route (GUSMP) and Stalinist Exploitation of the Arctic, 1932–1939." *Journal of Slavic Military Studies* 20, no. 3 (2007): 393–419.

McEnaney, Laura. *Civil Defense Begins at Home: Militarization Meets Everyday Life in the Fifties*. Princeton: Princeton University Press, 2000.

McNeill, John Robert. *Mosquito Empires: Ecology and War in the Greater Caribbean, 1620–1914*. New York: Cambridge University Press, 2010.

– *Something New Under the Sun: An Environmental History of the Twentieth Century*. New York: W.W. Norton, 2000.

McNeill, John Robert, and Peter Engelke. *The Great Acceleration: An Environmental History of the Anthropocene Since 1945*. Cambridge, MA: Belknap Press of Harvard University Press, 2014.

McNeill, John Robert, and Corinna R. Unger, eds. *Environmental Histories of the Cold War*. Cambridge: Cambridge University Press, 2010.

McVety, Amanda Kay. *The Rinderpest Campaigns: A Virus, Its Vaccines, and Global Development in the Twentieth Century*. Cambridge: Cambridge University Press, 2018.

Meren, David. "'Commend Me the Yak': The Colombo Plan, the Inuit of Ungava, and 'Developing' Canada's North." *Histoire sociale/Social History* 50, no. 102 (November 2017): 343–70.

Milloy, John S. *A National Crime: The Canadian Government and the Residential School System*. Winnipeg: University of Manitoba Press, 2017.

Molinaro, Dennis. "How the Cold War Began ... with British Help: The Gouzenko Affair Revisited." *Labour/Le Travail* 79 (Spring 2017): 143–55.

Morton, Desmond. *A Military History of Canada*, 4th ed. Toronto: McClelland and Stewart, 1999.

Mosby, Ian. "Administering Colonial Science: Nutrition Research and Human Biomedical Experimentation in Aboriginal Communities and Residential Schools, 1942–1952." *Histoire sociale/Social History* 46, no. 91 (2013): 145–72.

Muirhead, Bruce. *Dancing Around the Elephant: Creating a Prosperous Canada in an Era of American Dominance, 1957–1973*. Toronto: University of Toronto Press, 2007.

Murray, Robert W., and Anita Dey Nuttall, eds. *International Relations and the Arctic: Understanding Policy and Governance*. Amherst: Cambria Press, 2014.

Naylor, C.D. "The CMA's First Code of Ethics: Medical Morality or Borrowed Ideology?" *Journal of Canadian Studies* 17, no. 4 (1982–3): 20–32.

Nicol, Heather N. "Reframing Sovereignty: Indigenous Peoples and the Arctic States." *Political Geography* 29, no. 2 (February 2010): 78–80.

Noble, H.R.R. "Reflections on Fort Churchill." *Roundel* 16, no. 8 (1964): 23–5.
Norman Smith, I., ed. *The Unbelievable Land: 29 Experts Bring Us Closer to the Arctic*. Ottawa: Queen's Printer, 1964.
"Northern Research Reports: Medicine." *Arctic* 1, no. 1 (1948): 65.
O'Brian, John. *The Bomb in the Wilderness: Photography and the Nuclear Era in Canada*. Vancouver: UBC Press, 2020.
O'Neil, Dan. *The Firecracker Boys: H-Bombs, Inupiat Eskimos, and the Roots of the Environmental Movement*. New York: St. Martin's Press, 1994.
Oreskes, Naomi, and John Krige, eds. *Science and Technology in the Global Cold War*. Cambridge, MA: MIT Press, 2014.
Parr, Joy. *Sensing Changes: Technologies, Environments, and the Everyday, 1953–2003*. Vancouver: UBC Press, 2010.
Pearson, Lester B. "Canada Looks 'Down North.'" *Foreign Affairs: An American Quarterly Review* (July 1946): 638–48.
Pennie, A.M. "Defence Research Northern Laboratory." *Canadian Army Journal* (January 1956): 1–8.
Piper, Liza, and John Sandlos. "A Broken Frontier: Ecological Imperialism in the Canadian North." *Environmental History* 12, no. 4 (October 2007): 759–95.
Polk, James, ed. *Divisions on a Ground: Essays on Canadian Culture*. Toronto: House of Anansi Press, 1982.
Pope, C.A. "Canada's Defence Research Board," *Roundel* 9, no. 1 (1957): 27–30.
Powell, Richard C. "Science, Sovereignty, and Nation: Canada and the Legacy of the International Geophysical Year, 1957–1958." *Journal of Historical Geography* 34 (2008): 618–38.
Problem of Chemical and Biological Warfare, The, vol. 1: *The Rise of CB Weapons*. Stockholm: Stockholm International Peace Research Institute, 1971.
Rabinowitch, I.M. "Clinical and Other Observations on Canadian Eskimos in the Eastern Arctic." *Canadian Medical Association Journal* 34, no. 5 (1936): 487–501.
Rand, Lisa Ruth. "Falling Cosmos: Nuclear Reentry and the Environmental History of Earth Orbit." *Environmental History* 24 (2019): 78–103.
Rasmussen, John E., ed. *Man in Isolation and Confinement*. Chicago: Aldine, 1973.
Reno, Joshua O. *Military Waste: The Unexpected Consequences of Permanent War Readiness*. Berkeley: University of California Press, 2020.
Ridler, Jason Sean. *Maestro of Science: Omond Mckillop Solandt and Government Science in War and Hostile Peace, 1939–1956*. Toronto: University of Toronto Press, 2015.
Robinson, Paul. "The Viability of a Canadian Foreign Intelligence Service." *International Journal* 64, no. 3 (2009): 703–16.
Rohde, Joy. *Armed with Expertise: The Militarization of American Social Research during the Cold War*. Ithaca: Cornell University Press, 2013.

Rosner, Cecil. "Isolation: A Canadian Professor's Research into Sensory Deprivation and Its Connection to Disturbing New Methods of Interrogation." *Canada's History* (August–September 2010): 29–37.

Rowley, G.W. "Exercise Muskox." *Geographical Journal* 109, nos. 4–6 (1947): 175–85.

Russell, Edmund. *Fighting Humans and Insects with Chemicals from World War I to Silent Spring*. Lawrence: University Press of Kansas, 2001.

– *War and Nature: Fighting Humans and Insects with Chemicals from World War I to Silent Spring*. Cambridge: Cambridge University Press, 2001.

Rutherdale, Myra, Kerry Abel, and P. Whitney Lakenbauer, eds. *Roots of Entanglement: Essays in the History of Native–Newcomer Relations*. Toronto: University of Toronto Press, 2018.

Sandlos, John, and Arn Keeling. "Aboriginal Communities, Traditional Knowledge, and the Environmental Legacies of Extractive Development in Canada." *Extractive Industries and Society* 3, no. 2 (2015): 278–87.

– "Toxic Legacies, Slow Violence, and Environmental Injustice at Giant Mine, Northwest Territories." *Northern Review* 42 (2016): 7–21.

Sangster, Joan. *The Iconic North: Cultural Constructions of Aboriginal Life in Postwar Canada*. Vancouver: UBC Press, 2016.

Schlosberg, David, and David Carruthers. "Indigenous Struggles, Environmental Justice, and Community Capabilities." *Global Environmental Politics* 10, no. 4 (November 2010): 12–35.

Schulert, Arthur R. "Strontium-90 in Alaska." *Science* 136, no. 3511 (April 1962): 146–8.

Serebryanny, Leonid. "The Colonization and Peoples of Novaya Zemlya Then and Now." *Journal of Nationalism and Ethnicity* 25, no. 2 (1997): 301–9.

Sethna, Cristabelle, and Steve Hewitt. *Just Watch Us: RCMP Surveillance of the Women's Liberation Movement in Cold War Canada*. Montreal and Kingston: McGill–Queen's University Press, 2018.

Shadian, Jessica M. *The Politics of Arctic Sovereignty: Oil, Ice, and Inuit Governance*. London: Routledge, 2014.

Simpson, Erika. *NATO and the Bomb: Canadian Defenders Confront Critics*. Montreal and Kingston: McGill–Queen's University Press, 2001.

Singh, Simon. *Big Bang: The Origin of the Universe*. New York: Harper Perennial, 2005.

Smith, Susan L. *Toxic Exposures: Mustard Gas and the Health Consequences of World War II in the United States*. New Brunswick: Rutgers University Press, 2017.

Sokolsky, Joel J., and Joseph T. Jockel, eds. *Fifty Years of Canada–United States Defense Cooperation: The Road from Ogdensburg*. Lewiston: Edwin Mellen, 1992.

Solon, L.R., et al. "Investigations of Natural Environmental Radiation." *Science* 131, no. 3404 (1960): 903–6.

Sörlin, Sverker. "The Historiography of the Enigmatic North." *Canadian Historical Review* 95, no. 4 (December 2014): 555–66.

–, ed. *Science, Geopolitics, and Culture in the Polar Region: Norden beyond Borders.* Farnham: Ashgate, 2013.

Souchen, Alex. *War Junk: Munitions Disposal and Postwar Reconstruction in Canada.* Vancouver: UBC Press, 2020.

Soward, F.H. "On Becoming and Being a Middle Power: The Canadian Experience." *Pacific Historical Review* 32, no. 2 (May 1963): 111–36.

Spiers, F.W. "Radioactivity in Man and His Environment, Presidential Address." *British Journal of Radiology* 29, no. 344 (1956): 409–17.

Sternberg, Joseph. "Placental Transfers: Modern Methods of Study." *American Journal of Obstetrics and Gynecology* 84, no. 11, pt. 2 (December 1962): 1731–48.

– "Radiation and Pregnancy." *Canadian Medical Association Journal* 109, no. 1 (July 1973): 51–7.

Stuhl, Andrew. *Unfreezing the Arctic: Science, Colonialism, and the Transformation of Inuit Lands.* Chicago: University of Chicago Press, 2016.

Taylor, C.J. "Exploring Northern Skies: The Churchill Research Range." *Manitoba History* 44 (Autumn/Winter 2002–033): 2–9.

Thistle, Jesse. *From the Ashes: My Story of Being Métis, Homeless, and Finding My Way.* Toronto: Simon and Schuster, 2019.

Turner, Jonathan. "Politics and Defence Research in the Cold War." *Scientia Canadensis* 35, no. 1–2 (2012): 39–63.

Twinn, C.R. "Review of Recent Progress in Mosquito Studies in Canada." *Mosquito News* (December 1955): 195–203.

– "Studies of the Biology and Control of Biting Flies in Northern Canada." *Arctic* 3, no. 1 (April 1950): 14–26.

University of Toronto Studies, History and Economics. *Contributions to Canadian Economics*, vol. 6. Toronto: University of Toronto Press, 1933.

Webb, Brandon. "'How to Raise a Curtain': Security, Surveillance, and Mobility in Canada's Cold War–Era Exchanges, 1955–65." *Cold War History* 21, no. 2 (2020): 215–33.

Webber, Roger. *Disease Selection: The Way Disease Changed the World.* Wallingford: CABI, 2015.

Webster, D.R., and W.G. Bigelow. "Injuries Due to Cold, Frostbite, Immersion Foot, and Hypothermia." *Canadian Medical Association Journal*, Civil Defence Issue 67 (December 1952): 534–8.

Weindling, Paul. "The Origins of Informed Consent: The International Scientific Commission on Medical War Crimes, and the Nuremburg Code." *Bulletin of the History of Medicine* 75, no. 1 (2001): 37–71.

Wheelis, Mark, Lajos Rózsa, and Malcolm Dando, eds. *Deadly Cultures: Biological Weapons Since 1945.* Cambridge, MA: Harvard University Press, 2006.

Whitaker, Reg, and Steve Hewitt. *Canada and the Cold War.* Toronto: James Lorimer, 2003.

Whitaker, Reg, and Gary Marcuse. *Cold War Canada: The Making of a National Insecurity State, 1945–1957*. Toronto: University of Toronto Press, 1994.

Wild, Roland. *Arctic Command: The Story of Smellie of the Nascopie*. Toronto: Ryerson Press, 1955.

Wilson, J.T. "Exercise Musk-Ox, 1946." *Polar Record* 5, nos. 33–4 (1947): 14–27.

Wiseman, Matthew S. "Canadian Scientists and Military Research in the Cold War, 1947–60." *Canadian Historical Review* 100, no. 3 (September 2019): 439–63.

– "The Development of Cold War Soldiery: Acclimatisation Research and Military Indoctrination in the Canadian Arctic, 1947–1953." *Canadian Military History* 24, no. 2 (Summer–Autumn 2015): 127–55.

– "The Origins and Early History of Canada's Cold War Scientific Intelligence, 1946–65." *International Journal* 77, no. 3 (2022): 7–25.

– "The Weather Factory: Alan C. Burton and Military Research at the University of Western Ontario, 1945–70." *Scientia Canadensis* 45, no. 1 (2023): 1–22.

Wolfe, Audra J. *Freedom's Laboratory: The Cold War Struggle for the Soul of Science*. Baltimore: Johns Hopkins University Press, 2018.

Wright, Shelley. *Our Ice Is Vanishing / Sikuvut Nunguliqtuq: A History of Inuit, Newcomers, and Climate Change*. Montreal and Kingston: McGill–Queen's University Press, 2014.

Zeller, Suzanne. *Inventing Canada: Early Victorian Science and the Idea of a Transcontinental Nation*. Toronto: University of Toronto Press, 1987.

Unpublished Theses

Davis, Brandon C. "Grounds for Permanent War: Land Appropriation, Exceptional Powers, and the Mid-Century Militarization of Western North American Environments." PhD diss., University of British Columbia, 2017.

Eyre, K.C. "Custos Borealis: The Military in the Canadian North." PhD diss., University of London – King's College, 1981.

Fritz, Stacey A. "DEW Line Passage: Tracing the Legacies of Arctic Militarization." PhD diss., University of Alaska Fairbanks, 2010.

Thrasher, Kevin Mendel. "Exercise Musk Ox: Lost Opportunities." MA thesis, Carleton University, 1998.

Turner, Jonathan. "The Defence Research Board of Canada, 1947–1974." PhD diss., University of Toronto, 2012.

White, Rossana. "Ceremonies of Possession: Performing Sovereignty in the Canadian Arctic." PhD diss., Royal Holloway, University of London, 2019.

Index

access to information, 27, 201, 215, 222, 238n35
acclimatization: vs. adaption, 81, 88; in animals, 81–2; biological understanding of, 68–71, 73, 78, 89–90, 94; and calorimeter research, 87; and classified research, 93–4; to cold, 67, 70, 74, 78, 94; and consent, 70–1, 80–1; and DRB research, 69, 72, 77, 81; and DRNL research, 59, 78, 81–2, 87; to high altitude, 87; and hyperthyroidism, 88; and Indigenous Alaskans, 88; and Indigenous soldiers, 82–3; international research on, 86–90; and Inuit, 67, 69, 78–81, 93–5; military applications of, 72, 77, 79, 88, 95; and radiation experiments, 88–9; and settler colonialism, 71, 94–5; theories of, 74, 80, 93; vs. tolerance, 81; to tropical environments, 81, 87–8, 97; and White medical students, 67, 80, 93; and White soldiers, 87, 90–1. *See also* Brown, G. Malcolm; human experiments
Admiralty (UK), 51

Advisory Committee on Northern Development, 76
Aerobee (rocket launch tower), 49
Africa, 9, 135
Ahlmann, Hans, 12
Air Defence Command, 119
aircraft: anti-aircraft missile, 170; B-52 bomber, 189; Canso, 162; cold-weather trials of, 28, 112, 118, 130, 165; commercial, 55; Dakota, 98–9, 146, 160–1; DDT spraying, 97–9, 103; Flying Boxcar, 161; Hercules, 162; landings on sea ice, 156–7; long-range bombers, 47; Martin Marietta, 174; search-and-rescue, 159; signals, 168; Soviet, 158; striking, 152; transport, 37, 39, 132, 146, 156, 214. *See also* Mobile Striking Force (MSF); Royal Canadian Air Force (RCAF)
Aklavik, Northwest Territories, 178
Aksayook, Etuangat, 179
Alaska: Alaska Highway, 37, 102, 249n8; Alaska–Canada border, 153; Alaska–Yukon border, 156; Arctic Aeromedical Laboratory, 88, 124; Distant Early Warning Line, 151; as a frontier, 20; and ice

Index

Alaska (*continued*)
  islands, 156; Klondike–Alaska Gold Rush, 19; Ladd Air Force Base, 124; militarization of, 88, 185; nuclear fallout over, 178, 189–90; permafrost in, 150; radiochemical analysis of residents in, 180, 190–1; research in, 89, 124; and the Second World War, 88; Sr-90 in, 197. *See also* University of Alaska–Fairbanks
Alberta, 48, 61, 101, 125, 138, 194, 203, 229n95
Alert, Nunavut, 151, 159
Aleutian Islands, Japanese invasion of, 37
Alexander, Scott, 91, 93
Alouette (satellite), 52, 171–2
Alter, Northwest Territories, 148
Ambleside, UK, 189
American Medical Association, 70
Amund Ringnes Island, Nunavut, 163
Amundsen Gulf, 153
Anaktuvuk Pass, Alaska, 192
Anglo-American relations, 18, 58, 61, 119, 140, 155, 183–4, 186–9, 220
animals: and acclimatization to cold, 78, 81; and colonization, 22, 34; and DDT, 112, 119–20; as disease vectors, 104; and DRB, 7; and DRML, 4; experimental trials with, 3–7, 11, 81, 115; and hypothermia research, 3–4, 220; and insulin trials, 4; and nuclear fallout, 179–80, 194; and radiation experiments, 190, 198. *See also* Bigelow, Wilfred G.
Antarctic Station McMurdo, 174
Antarctica: Falkland Islands Dependencies Survey in, 156; military research in, 174; nuclear power plant in, 174
anti-submarine warfare, 154
anti-war protesters, 117, 119
archaeology, 150, 155
Archangelsk, Russia, 185
Arctic: Canadian Archipelago, 33, 146, 153, 155–6, 161, 166, 176, 214; Circle, 9, 18, 34; colonization in the, 20–2, 27, 94, 181; definition of the, 33–5; Far North, 9, 19–20, 38, 68, 146–7, 152–3, 161, 168, 220; Novaya Zemlya archipelago, 183–4; nuclear tests in the, 183–5; nuclear-powered ships in the, 184; nuclear-powered submarines in the, 148, 184; Ocean, 37, 125, 138, 154–9, 175 184–5; research, 47, 51–2, 57–9, 62–5, 73–8, 94, 105, 125–6, 135–50, 157, 164–9, 207–22; science, 34, 66, 78; security, 35, 158; sovereignty, 32, 35; Soviet, 136, 183–4; strategic threat in the, 9, 50, 137, 144, 211; warfare, 9, 14, 29, 39–43, 46–7, 50–9, 64–5, 90, 111, 123, 132–3, 210–12. *See also* frontier; militarization; nuclear fallout
Arctic Aeromedical Laboratory (US), 88, 124
Arctic, Desert and Tropic Information Center (US), 51
Arctic Frontier, The (Macdonald), 216
Arctic Health Research Center, 197
Arctic Ice and Permafrost Project (US), 150
Arctic Institute of North America (AINA), 51–2, 67, 216
Arctic Research Institute (Soviet Union), 51
Arctic Research Laboratory (US), 150
Armstrong, Terence, 155
Army Snow, Ice and Permafrost Research Establishment (US), 165
Arviat, Nunavut. *See* Eskimo Point
Asia, 9

Associate Committee on Aviation Medical Research (ACAMR). *See* National Research Council (NRC)
Associate Committee on Medical Research (ACMR). *See* National Research Council (NRC)
astronomy, 167
astrophysics, 60
Atka (ship), 162
Atlantic Ocean, 30, 37, 169, 186, 189, 204
atmospheric studies, 147–9, 168–72, 177, 214. *See also* aurora borealis; nuclear fallout
atomic: bombing of Hiroshima and Nagasaki, 8; cannon, 64; energy, 89, 106–8, 181–2, 186, 189; and northern Canada, 5, 185–9; research, 89, 106–8; warfare, 10, 105, 183, 210; weapons, 6, 49, 64, 117, 138, 173, 181, 186–7, 190–2; World Conference against Atomic and Hydrogen Bombs, 191. *See also* radiation
Atomic Energy Act (US), 182
Atomic Energy Commission (AEC), 89, 181, 192
Atomic Energy of Canada Limited, 108
Atomic Energy Research Establishment (UK), 186
Atomic Knights, 186
Atomic Weapons Research Establishment (UK), 187
aurora borealis, 148–9, 167–72, 208–9, 214
Australia, 135, 138, 187, 260n46
aviation, 38, 55. *See also* Royal Canadian Air Force (RCAF)

Baffin Island, Nunavut, 102
Baker Lake, Nunavut, 49, 63, 191, 194, 205, 249n8
Banff, Scotland, 61
Banks Island, Northwest Territories, 146–7, 153
Banting, Frederick: Banting Institute, 89; and insulin trials, 4
Barlow, J.S., 81–2
Barret, H.M., 13
Bathurst Island, Nunavut, 163, 194
Beaufort Sea, 146, 153, 156, 159
Beckel, William, 114
beekeeping, 99
Bell Telephone Laboratories, 167
Bennett, Rawson, 136
Bering Strait, 175
Berkner, Lloyd, 147
Berlin, Germany, 186
Berlin Wall, 7
Bertram, Colin, 156
Best, Charles, 4
Bigelow, Wilfred G.: animal experiments, 4–5, 221; career achievements, 4; connection with DRB, 4–7, 216, 221; connection with DRML, 4; family, 6; hypothermia experiments, 4, 214; wartime research, 6
biological warfare: agents, 111; biodefence, 104–6; biowarfare, 105; and entomology, 98, 104–6; and insect vectors, 98, 104–6; and protective clothing, 111; and Western security, 98, 110–11, 117. *See also* Defence Research Board: Biological Warfare Research Panel; Reed, Guilford
biology: and Arctic research, 59–60, 86, 130, 165; of biting insects, 118; in cold-weather trials, 59–60, 124, 128; DDT, 120; human, 68, 124; Inuit, 68–71, 83, 88; *Journal of the History of Biology*, 181; marine, 96, 154; micro, 104; and nuclear

biology (*continued*)
fallout, 189; and Operation Hazen, 165; organic terrain, 118–19; and race, 68–71, 83, 88; student, 128. *See also* Law, Cecil; Pruitt, William "Bill"; Schenk, Carl

Bird, Peter: correspondence, 179–80, 192–3, 197, 201, 203; and human specimens, 192–3, 203, 221, 263n94; on ionizing radiation, 191, 193; on the nationwide fallout study, 180; on the Northern Radiation Study, 194–8, 201, 263n94

Black Brant (rocket), 52, 172
Blackadar, Robert, 159
Blouin, Arthur, 91–3
bombardier, 131
Booth, A.H., 191
Borisov, P.M., 175
botany, 51, 60, 97, 102, 155, 189. *See also* Porsild, Erling
Brandon, Manitoba, 6, 103
Brett, H.B., 197
Britain. *See* United Kingdom
British Army: cooperation with DRB, 113; Liaison Staff, 54; Parachute Regiment, 139; Royal Engineers, 146, 162; Royal Scots, 139; War Office, 147
British Columbia, 33, 40, 84, 99, 102, 104, 125, 148, 154, 203
British Imperial Chemical Industries, 92
Broad River, Manitoba, 187–8
Brown, A.W.A., 104
Brown, G. Malcolm: on acclimatization and race, 70–1, 164; acclimatization experiments, 67–9, 71, 164, 236n19, 238n35, 239n54; career at Queen's University, 73–4; connection with DRB, 67, 69, 74–5, 216, 221; and

human specimens, 68; and medical research ethics, 67, 70, 73; and military research, 68–9, 73, 221; wartime research, 73–4; Queen's University Arctic Expedition, 72, 75, 238n39

Bruce-Chwatt, Leonard J., 108
Bryan, J.W.P., 134
Buchan, W.R., 201
Bugher, John, 107
Bulletin of the Atomic Scientists, 175
Burton, Alan: on acclimatization, 80–1, 88; on biology and race, 88; cold-room experiments, 87; connection with DRB, 81, 86, 216; *Man in a Cold Environment* (Burton and Edholm), 81; and military research, 216
Burton Island (ship), 153
Burwash, Ontario, 115
Butler Lake, Ontario, 133

Cabinet: Advisory Committee on Northern Development, 76; Defence Committee, 43; Howard Green on radiation, 183; Lester Pearson on the Arctic frontier, 32
Calgary, Alberta, 200
Cameron, William, 154
Camp Borden, 120, 132
Camp Century, 152
Campney, Ralph, 137
Canadian Arctic Permafrost Expedition (1951), 151
Canadian Armed Forces. *See* Canadian Army; Royal Canadian Air Force (RCAF); Royal Canadian Navy (RCN)
Canadian Army: Artillery Guided Missiles Trials Troop, 170; *Canadian Army Journal*, 61, 142, 162, 219; Canadian Infantry

Brigade, 18, 131; Canadian
Regiment, 132–3; Corps of Signals,
132; Directorate of Staff Duties,
41; DRNL test team, 138, 140–1;
and entomology, 100–2; 1st Royal
Canadian Horse Artillery, 131; and
Fort Churchill, 29, 34, 36, 44, 53–
61, 134, 139, 142–4, 212–14, 219;
Medical Corps, 6 56, 73, 92; Mobile
Striking Force (MSF), 38–9, 72,
83, 113–14, 128–9, 132–3; non-
commissioned officer (NCO), 132,
212; and Operation Hazen, 160–2;
and operational research, 125–9;
Princess Patricia's Canadian Light
Infantry, 130, 132, 251n37; Provost
Corps, 56; Royal 22ᵉ Régiment,
132; School of Infantry, 132; in
the Second World War, 39–41;
Service Corps, 56; Staff College,
139; Survey Establishment, 162–3;
Western Command Joint Services
Operational Research Unit, 129;
and winter warfare, 39–41
Canadian Broadcasting Corporation
(CBC), 91
Canadian Institute of International
Affairs (CIIA), 216
Canadian Joint Staff, 110
*Canadian Journal of Biochemistry and
Physiology*, 75
Canadian Journal of Medical Sciences, 75
Canadian Medical Association Journal,
6, 198
Canadian Medical Hall of Fame, 4, 95
Canadian National Museum, 163
Canadian Pacific Air Lines, 53
Canadian Physiological Society, 86
Canadian Rangers, 39, 72, 129, 229n8
Canadian Wildlife Service, 180
cancer: and Inuit health, 75, 202;
from radiation experiments, 196;
from radiation-contaminated food,
197; from radioactive isotopes,
89, 108, 190–1. *See also* Institute of
Cancer Research (UK)
Canol Project, 37
Canso (aircraft), 162
Cape Aldrich, Nunavut, 159
Cape Columbia, Ellesmere Island, 159
Cape Dorset, Nunavut, 76
Cape Merry, Manitoba, 120
Cape Prince Alfred, Northwest
Territories, 153
Cape Race, Newfoundland, 148
Cape Thompson, Alaska, 185, 190
Carruthers, R.A.F., 92
Carson, Rachel, impact of *Silent
Spring* (1962), 96–7
Cavendish Laboratory. *See* University
of Cambridge
Central Intelligence Agency (CIA), 8
Chalk River, Ontario, 106, 108, 186.
See also National Research Council
(NRC)
Chandler Fiord, 161
Chapman, John, 170
Chapman, Sydney, 147
Charles Camsell Indian Hospital, 85,
178, 194. *See also* Schaefer, Otto
Château Laurier Hotel, 137
chemical warfare: agents, 119, 189; and
Cold War, 10, 108, 218; and DRB, 10,
101, 218; Porton Down, 101
Chesterfield Inlet, Nunavut, 205
Chicago, Illinois, 182
China, People's Republic of, 16
Chrétien, Jean, 119–20
Chukchi Sea, 159
Churchill, Manitoba: atomic testing
ground, 185–9; aurora borealis,
169; Churchill River, 29, 97;
Crimson Route, 13–14; Port of
Churchill, 36; Warkworth Lake, 97.

Churchill, Manitoba (*continued*)
  See also Churchill Research
  Range (CRR); Fort Churchill;
  International Geophysical
  Year (IGY)
Churchill Research Range (CRR),
  48–9, 56, 150, 169, 172, 175, 233n74
circumpolar region, 26, 178, 218
civil defence, 6, 260n49
Claxton, Brooke, 46
climate change, 93, 176, 184, 222
climatology, 21, 102, 135
Cockcroft, John, 186
Coffey, Mace, 63, 85, 128
Coke, William, 100–1
Cold Lake, Alberta, 48, 138
Cold Regions Research and
  Engineering Laboratory (US), 165
Cold War: arms race, 18, 177, 196;
  and Canadian security, 7–18,
  30–2, 49–52, 65–72, 94, 144–53,
  176–7, 208–10; and Canadian
  sovereignty, 155–9; colonialism,
  23, 205–8; definition of, 7–9, 17,
  219, 221; diplomacy, 183; and
  environmental impact, 121–2, 176,
  213–17, 220; and espionage, 7–8;
  and militarization of science, 5–12,
  21, 25–33, 62–6, 88, 98, 123–7,
  185–6, 212–22; and scientific
  intelligence, 50; space race, 170–3;
  and technological competition, 25,
  48, 92, 174–7, 180, 209–11, 216
Cole, H.I., 110
Collip, James, 74
Columbia River, 197
Commonwealth: Advisory Committee
  on Defence Science, 138, 167; and
  Canadian foreign policy, 18, 138;
  and Canadian scientists, 138, 167;
  Defence Conference on Clothing
  and General Stores, 135

communism, 9, 16, 185
Confederation, 21
consent. *See* human experiments
Coral Harbour, Nunavut, 76, 102, 205
Cornell University, 114
Cornwallis Island, Nunavut, 102,
  162–3
Cosmos 954 (satellite), 49, 172.
  See also Operation Morning Light
Cowie, Wilbert, 91–2
Crary, A.P., 165
Crimson Route, 14, 29
Croal, James Patrick, 57, 137, 150
Cronk, Bruce, 75–6
Crumlin, Bill, 57, 151
Cuba, 8, 183
Cuban Missile Crisis, 8, 48

Dakota (aircraft), 98–9, 103–4, 146,
  160–1, 182
Davies, E. Llewelyn, 13, 101
Dawson Creek, British Columbia, 102
Defence Research Board (DRB):
  Arctic Medical Research panel,
  74, 77, 85, 90, 239n65; Arctic
  Research Advisory Committee,
  51, 76–7; Arctic Section, 58, 64,
  137, 156; Biological Research
  Division, 101; Biological Warfare
  Research panel, 104; Canadian
  Armament Research and
  Development Establishment,
  61; Canadian Army Operational
  Research Establishment (CAORE),
  127; Civil Engineering Section,
  137; creation of, 13–15; Defence
  Medical Research Advisory
  Committee, 77; Directorate of
  Engineering Research, 137;
  Directorate of Physical Research,
  137; Environmental Protection
  Program, 58, 72, 102–3, 118,

122, 212, 214; Environmental
Protection Section, 72, 118;
Geophysics Section, 155;
Mechanical Section, 137; Medical
Research Advisory Committee,
77; Pacific Naval Laboratory,
154; Panel on Frostbite and
Immersion Foot, 4; Radio Physics
Laboratory, 168, 170; and secrecy,
7, 77, 224n15; symposium, 131;
Vehicle Experimental Proving
Establishment, 135
Defence Research Chemical
Laboratory (DRCL), 111
Defence Research Kingston
Laboratory (DRKL), 106
Defence Research Medical
Laboratories (DRML), 4, 149
Defence Research Northern
Laboratory (DRNL): closure of,
143–4; establishment of, 14, 29,
57–60; experiments at, 60–5, 86–7,
115, 124–37, 141, 169; illustration
of, 60, 87; official function of, 58–9,
149–50, 210–11; organizational
restructuring of, 62, 137–9,
149–50, 169
Defence Research
Telecommunications
Establishment (DRTE), 13, 168–9,
171–2
Dene, 122
Denmark, 153, 158, 165, 184
Department of Agriculture, 57, 62,
97, 101–2, 106, 114, 165, 217.
See also entomology: Entomology
Research Institute
Department of Citizenship and
Immigration, 196
Department of External Affairs
(DEA), 157
Department of Indian Affairs, 34, 193

Department of Indian Affairs and
Northern Development, 119
Department of Mines and Resources,
35, 57, 69, 76
Department of Mines and Technical
Surveys, 154
Department of National Defence
(DND): Cabinet Defence
Committee, 43; defence budget,
13, 16, 52, 143, 145, 210, 218,
250n15; defence policy, 11, 18,
47, 52; Directorate of Military
Training, 42; landownership and
expropriation, 28, 115–17; 122;
National Defence College, 48, 139;
National Defence Headquarters
(NDHQ), 13; Public Service
Surgical-Medical Insurance
Plan, 84. *See also* Canadian Army;
Gagetown; military research and
development; Royal Canadian Air
Force (RCAF); Royal Canadian
Navy (RCN)
Department of National Health and
Welfare, 67, 69, 75, 179–80, 192–3,
199, 202, 215, 217, 263n101
Department of Northern Affairs, 62,
152, 192
Department of Northern Affairs and
National Resources, 152, 196, 200
Department of Transport, 36, 53, 55,
150, 233n74
Devon Island, Nunavut, 163
dichloro-diphenyl-trichloroethane
(DDT), 96–9, 104, 111–12,
115–22
Diefenbaker, John, 143, 176, 182
Distant Early Warning (DEW) Line,
28, 48, 138, 151–2, 161, 166,
231n42
District of Keewatin, Northwest
Territories, 84, 162, 194

District of Mackenzie, Northwest Territories, 194–7. *See also* Mackenize Delta; Mackenzie River
Domashev, P.I., 46, 231n37
Dominion Observatory, 63
Doppler, 169
Downsview, 4
Duck Lake, 113
Dugal, Louis-Paul, 78–9, 81, 86
Dunbar, Moira, 137
DuPont, 92

Eagan, Charles, 85–7
Easterbrook, Jim, 128–9
Eastwind (ship), 161
ecologies, 117, 122
ecology, 33, 50, 97
Edholm, Otto, 81
Edmonton, Alberta, 91, 146, 200
Eisenbud, Merril, 182
Elk River, Minnesota, 182
Ellef Ringnes Island, Nunavut, 163
Ellesmere Island, Nunavut: Cape Columbia, 159; Gilman Glacier, 155, 160, 162; ice islands and, 156; Lake Hazen, 149–50, 155, 160–4; military value of, 174. *See also* Operation Hazen
Elliott, G.W., 191
Elliot Lake, Ontario, 23
Enewetak Atoll, Marshall Islands, 180
Engels, D., 160
engineering: Arctic, 55, 92–3, 151, 209; civil, 124, 137; human, 62, 92–3, 115, 137; material, 124; mechanical engineering, 60, 135, 137; radar, 151. *See also* Cold Regions Research and Engineering Laboratory
English Channel, 167
entomology: aerial spraying, 97–9, 104, 116, 120; and biological warfare, 110–11, 117; Entomology Research Institute, 165; and Fort Churchill, 60, 98, 103; insect control, 58, 62, 97–8, 100, 104, 119–22, 213; insect repellency, 111–15; insecticides, 96–9, 101, 104, 111–12, 118–21, 213, 221; Northern Insect Survey, 102–4; in northern Quebec, 106–7; Operation Bugbait, 97; Operation Insect, 103; radioactive mosquitoes, 106–11, 117; radioactive tagging, 107–8; in the Second World War, 96–7; vectors, 59, 98, 104–6, 110–11, 118. *See also* Department of Agriculture; dichloro-diphenyl-trichloroethane (DDT)
Environment Canada, 28
environment(al): activism, 28, 190; challenges for military training, 37, 50, 59, 92, 97, 102; contamination of, 23, 93–9, 115–17, 121–2, 152, 185–9, 197, 206, 213, 217, 222; control of nature, 57; knowledge of, 24, 51, 90, 152; and military research, 50, 135, 147, 153, 213, 217, 222; northern conditions, 54, 62, 65, 78, 100; protection of, 58, 99; protests, 190. *See also* Defence Research Board: Environmental Protection Program; dichloro-diphenyl-trichloroethane (DDT)
Environmental Health Laboratory (US), 65
epidemiology, 104
Eskimo Point (Arviat), 40, 205
Esquimalt, British Columbia, 154
Eureka, Nunavut, 151
Europe, 9, 11, 14, 19–20, 29, 38, 40, 74, 189
Explorer 1 (satellite), 171
expropriation. *See* Department of National Defence

Fairbanks, Alaska, 88, 124, 197
Falconer, W.L., 194
Falkland Islands Dependencies Survey, 156
Finland, 180, 184
1st Arctic Test Detachment (US), 54–6
First World War, 21
Fisher, K.C., 60
Fletcher's Island (T-3), 157–8
Flicek, Jerry, 171–2
Flying Boxcar (aircraft), 161
Fond du Lac, Saskatchewan, 194
Foothills, Alberta, 194, 201
Forsyth, Peter, 168
Fort Chimo (Kuujjuaq), Quebec, 102, 195
Fort Churchill: closure of, 24, 214; and Cold War secrecy, 45–6; 188; command structure, 53–6; establishment of, 14–15, 29, 36, 44–5, 53, 58–65; Joint Services Experimental Testing Station (JSES), 36, 44–5, 59; Naval Radio Station, 55; purpose of, 14–15, 29–37, 43–7, 58–65, 209–12, 219–20; and relocation of Inuit, 195. *See also* Arctic: warfare; Defence Research Northern Laboratory (DRNL); winter warfare
Fort Franklin (Délı̨nę), Northwest Territories, 194
Fort Nelson, British Columbia, 99
Fort Prince of Wales, Manitoba, 120
Fort Resolution, Northwest Territories, 49
Fort Smith, Northwest Territories, 193–4
Fort St. John, British Columbia, 99
Fort Yukon, Alaska, 192
Foulkes, Charles, 42, 230n24
Freeman, T.N., 102

Freshwater Biological Association (UK), 189
Frobisher Bay (Iqaluit), Nunavut, 102, 179, 205
frontier: and northern Canada, 18–22; 209–10, 214, 216, 220; theory and usage of term, 18–22. *See also Arctic Frontier, The*; Klondike–Alaska Gold Rush; Turner, Frederick Jackson
Frye, Northrop: garrison mentality, 20

Gagetown, New Brunswick, 115–17, 121
Galloway, Strome, 128, 134, 142–3, 171, 219
Gander, Newfoundland, 102
Geiger-Müller, 107, 181
Geneva, 183
geography: imaginary, 19, 22, 24–6, 32, 220; military, 42, 166
Geological Survey of Canada, 159
geology, 50, 137, 150, 155–6, 165
geophysics, 124, 147, 153, 155, 166
Germany, 10, 16, 18
Giant Mine, Yellowknife, 122
Gibbons, James, 76
Gilman Glacier, Ellesmere Island, 155, 160, 162
glaciology, 137, 148, 165, 177, 218
Goforth, Wallace, 17, 41–2
Goose Bay, Labrador, 99, 102, 104
Goose Creek, 97
Gorham, Eville, 189
Gouzenko, Igor, 8
Gray, G., 205
Gray, O.C., 198, 200
Great Bear Lake, Northwest Territories, 49, 102
Great Britain. *See* United Kingdom
Great Lakes, 29, 99

296 Index

Great Slave Lake, Northwest Territories, 49
Great Whale River, Quebec, 107
Greely, Adolphus, 160
Green, Howard, 182–3
Green, J.J., 137
Green, John, 76
Greenland, 11, 136, 139, 151–3, 157–9, 161, 189
Grinnell Peninsula, Nunavut, 163
Grosse-Île, St. Lawrence River, 16
Gulf Stream effect, 175

Hamilton, Joseph, 88
Hamilton, Ontario, 184
Hanford Laboratory, 190
Hansen, Godfred, 159
Harrow, Ontario, 148
Harvard University, 89, 104
Harwell, England, 186
Harwood, Trevor, 137, 147, 155, 157, 160
Hattersley-Smith, Geoffrey, 155–6, 159, 163–6
Health and Safety Laboratory (US), 182
Health Canada, 28
Hearne, Samuel, Massacre at Bloody Falls, 23
herbicides, 116–17
Hinton, Christopher, 186
Hiroshima, Japan, 8, 43, 186
Hiroshima Appeal, 191
Höhn, E.O., 146
Holman Island, Northwest Territories, 146
Honeyman, H.C., 162
Hope, E.R., 175
House of Commons (Canada), 108
Howe, William, 200
Hoy, T.L., 162

Hudson Bay: Arctic boundary, 33–4; atomic testing near, 187; drainage basin, 97; ecosystems, 188; Hudson Strait, 34; Indigenous peoples, 179; North Magnetic Pole, 21; northern Quebec, 106–7; radioactive fallout, 194; rail line, 36, 53, 125; skies over, 172; tide water, 125, 175–6. *See also* Churchill; Coral Harbour; Fort Churchill
Hudson's Bay Company (HBC), 22, 146, 178, 236n13
human experiments: autopsies, 179, 192–7, 201–4, 263n101; biopsies, 68, 80 221, 235n7; on blood, 67–8, 73–4, 78–87, 92, 100, 105; on bone, 179, 192–206, 215, 221, 263n94; 263n101; and consent, 27, 70, 83, 86, 193, 196, 202–3, 207; on frostbite, 4, 81–2, 85–6, 124; on hyperthyroidism, 80, 88; on malnourishment, 71; and medical research ethics, 70, 88, 93, 195, 222; on organs, 68, 205, 221; on placenta, 203, 205–6; race-based, 68, 71, 83, 85, 88, 115, 194; with radioisotopes, 88–9, 107–8, 258n8; on sensory deprivation, 216, 251n32; on teeth, 194, 203–4; on tissue, 68, 80, 87, 193, 204–6, 215, 221; on urine, 67, 73, 192, 197–8, 200–1, 205–6, 215. *See also* Defence Research Northern Laboratory (DRNL); Radiation Protection Division (RPD)
Hut Point Peninsula, Antarctica, 174
Hvinden, F.T., 189
Hyde Park Declaration, 37
hydrography, 137, 154

ice islands, 156–8
Iceland, 184

Idaho, 196
Igloolik, Nunavut, 67, 205, 236n19
India, 135
Indigenous peoples: agency of, 27; Alaskans, 88, 192; Anishinaabek, 23; and cancer, 202; and cultural appropriation, 90–3; and decolonization, 27; and depression, 202; as guides, 39, 79, 91, 179; and hysterectomies, 202; Inupiat, 190; Nenets, 183; and oral history, 23; and traditional knowledge, 23, 26, 92–3, 208–9; and tuberculosis, 202; and reconciliation, 26, 122; and relocation, 195; and residential schools, 22, 71; and resistance to colonialism, 23, 236n19; Serpent River First Nation, 23; as soldiers, 82–3; and sterilization, 202; and unmarked graves, 193. *See also* Canadian Rangers; human experiments: race-based; Inuit
industrial preparedness, 28, 38. *See also* civil defence; preparedness: military
insects. *See* entomology
Institute of Cancer Research (UK), 203–6
intelligence: on Arctic warfare, 111; on biological warfare, 110–11; Central Intelligence Agency (CIA), 8; and the Cold War, 11; Joint Intelligence Bureau, 138; Pentagon, 117; scientific, 49–50, 65, 83, 85, 105, 110–11; on Soviet miliary threat, 49, 111, 180; terrain, 150–1
International Atomic Energy Agency (IAEA), 189
International Committee on Radiological Protection (ICRP), 192

International Geophysical Year (IGY), 147–50, 153, 160, 162–7, 169–73, 177, 214. *See also* Operation Hazen
International Polar Year (IPY), 11, 147
Inuit: agency of, 28, 236n19; knowledge, 26, 28; patient transfers of, 179, 193, 196–7; and radioactive fallout, 179–80, 200–1, 204; relocation of, 195; sterilization of, 202–3. *See also* acclimatization; human experiments
Inuvik, Northwest Territories, 179, 194, 201
ionospheric research: and electromagnetic radiation signals, 170; magnetism, 148, 169; radio propagation, 168–71
Iowa, 197
Iqaluit. *See* Frobisher Bay
Iron Curtain, 10, 172, 174, 216
Irving, Laurence, 88
Isachsen, Nunavut, 151

Jacob, Eileen, 84
Jacob, Jack, 84
James, A.A., 110
James Bay, 102
JANET, 168
Jansky, Karl, 167
Japan. *See* Hiroshima; Nagasaki
Jenkins, Dale W., 106–9
Johns Hopkins University, 96
Johns Island, Nunavut, 161
Johnson, Jim, 128–9
Johnson, R.E., 78
Joint Arctic Weather Stations (JAWS), 151
Joint Intelligence Bureau, 138
Joint Services Experimental Testing Station (JSES). *See* Fort Churchill

298 Index

Jones, A.C., 58
Jones, H.M., 195
Jordan, Dennis, 76
Josiah Macy Foundation for Medical Research, 86
Justice Department, 8

Kazakhstan, 183
Keenleyside, Hugh, 51–2, 210, 243n114
Keewatin. *See* District of Keewatin
Kennedy, John F., 8, 48, 116
Khrushchev, Nikita, 8, 183–4
Kimble, George H.T., *Canadian Military Geography*, 42
King, William Lyon Mackenzie, 38, 44–6, 210
King George Island, 156
Kingscote, A.A., 111–14
Kingston, Ontario, National Defence College, 48. *See also* Brown, G. Malcolm; Defence Research Kingston Laboratory (DRKL); Queen's University
Kirkland Lake, Ontario, 133
Klondike–Alaska Gold Rush, 19
Kodak, 181–2
Koksoak River, Quebec, 195
Korea, 116
Korean conflict, 6, 42, 85, 105, 123–4, 131, 138, 144–5, 218, 249n3, 250n15
Kotzebue, Alaska, 192

Labrador, 33, 99, 102, 104
Labrador (ship), 154–5
Lagernoe, Novaya Zemlya, 183–4
Lake District, England, 189
Lake Hazen, Ellesmere Island, 149–50, 155, 160–4
LaMarsh, Judy, 200–1
Langford, Warren, 48–9
Langley, A.J.G., 110

Latin America, 11
Laval University, 78, 86
Law, Cecil, 128
LeBlanc, J.S., 83–4, 86, 240n68
Lent, Peter, 190
Lindsay, Ian, 118
Livingood, J.J., 88
Lloyd, Trevor, 51
Lompoc, California, 172
London, England, 10, 18, 38, 40, 47–9, 137, 147, 154, 169, 188–9, 204–5, 210
London, Ontario, 133
Lotz, Jim, 161
Lougheed Islands, Nunavut, 163
Lyon, Waldo, 154

Maass, Otto, 13
MacAulay, Ralph, 75
Macdonald, Sir John A., 22
Mackenzie. *See* District of Mackenzie
Mackenzie, Chalmers Jack, 108, 245n41
Mackenzie Delta, 179
Mackenzie River, 37, 102, 125
Mackenzie River Valley, 151
Macpherson, Andrew, 146
Mair, Winston, 128
malariology, 108
Manhattan Project, 181, 186, 190
Manitoba: atomic testing ground in, 186–9; barrens, 86; Chamber of Mines, 123; climate, 44, 57, 134, 143; ecology of, 33, 36, 187; geography of, 125; Hospital Commission, 84; Indigenous populations in, 213; meteorology in, 171; military training in, 15, 29, 100, 114, 124, 150, 219; science in, 15, 29, 81, 97, 124–8, 131, 138, 141–2; Territorial Hospital Insurance Services, 84

Mann, George, 89
Manning, Thomas, 146–7
Mar, P.G., 179
Maralinga Range, Australia, 138
Marier, Guy, 78
Marshall Islands, 180
Martin Marietta, 174
Matheson, D.M., 162–3
Maxwell, Moreau, 163
Mayne, John, 128
Mayneord, W.V., 203
McClure Strait, 154
McGill University, 51, 61. See also Arctic Institute of North America (AINA)
McMaster University, 118
McNamara, C.P., 186–8
McNamara, Robert, 48
McNaughton, Andrew, 44
McPherson, E.M., 92
medical examinations: on Inuit, 68, 79; X-rays, 76, 181, 258n8. See also *Nascopie*
medical research. See human experiments
meteorology, 21, 137, 147, 154, 157, 165, 177
Michigan, 196
microbiology, 104, 110
Mid-Canada Line (MCL), 48, 138, 150
Middle East, 9
militarization: in Alaska, 88; in the Arctic, 152, 184–5; of Arctic research, 10–11, 15, 25–7, 33, 214, 218; of Canadian science, 5; definition of, 10; of medical science, 94; in northern Canada, 217
Military Cooperation Committee (MCC), 38
military equipment: in Arctic conditions, 10, 28, 36, 40–9, 54–5, 72, 85, 112, 126–35, 139–41, 145, 214, 218; and the Canadian Rangers, 39; and DDT, 97–8, 111; and the DRB, 52–9, 102, 167–9, 214; at Fort Churchill, 14, 56–65; in high latitude, 38, 218; and Korea, 105; and Operation Hazen, 147, 160–1; for winter warfare, 40
military exercises: Bulldog II, 129; Deer Fly, 114; Deer Fly II, 114; Deer Fly III, 114; Eskimo, 40, 125; Lemming, 40, 125; Musk Ox, 46, 125; Polar Bear, 40, 125
military-industrial complex, 28, 216
military operations: in the Arctic, 40, 56, 62, 90, 165, 208; cold-weather, 39–40, 124, 126, 211; high-latitude, 62, 212; in northern Canada, 25, 29, 41, 62, 97, 115; in Vietnam, 175; warm-weather, 113. See also Mobile Striking Force (MSF)
military research and development: aircraft, 130; anti-submarine warfare, 154; anti-tank weapons, 132; civilian applications of, 6–7, 52–4, 65, 86, 91–7, 108, 112, 119, 125, 155, 172; guns, 130; kit, 28, 127, 133–5, 140; mines, 130; radar, 56, 126, 130, 167–8, 170; sensors, 130; sonar, 130, 138; tanks, 130; torpedoes, 130; Wapiti (oversnow vehicle), 135; warships, 130; weapons, 130, 170. See also Defence Research Board; missiles; nuclear
Millen, H.A., 64
Miller, Raymond, 120
Minneapolis, Minnesota, 182
Minnesota, 182
missiles: anti-aircraft, 170; ballistic, 5, 166; Bomarc, 48; guidance system, 49; guided, 62, 141, 158, 170; intercontinental, 117, 148, 166, 231n42; MGM-18 Lacrosse missile, 141; Soviet, 8, 147; tactical,

missiles (*continued*)
141; threat against North America, 137, 145. *See also* Canadian Army: Artillery Guided Missiles Trials Troop; Cuban Missile Crisis; Martin Marietta; Operation Frost Jet
Mitchell, D., 105
Mobile Striking Force (MSF). *See* Canadian Army
Mongolia, 183
Montebello Islands, Australia, 187
Montreal, Quebec, 201, 206. *See also* McGill University; University of Montreal
Moore, P.E., 193, 206
Moose Factory, Ontario, 102
Moose Jaw, Saskatchewan, 120
Moscow, Russia, 9, 38, 46–9, 118, 174, 180, 183–4
Mould Bay, Northwest Territories, 151
munitions: surplus disposal, 15; unexploded ordnance, 28
muskeg, 97, 112, 118–19, 188
Muskoka, Ontario, 99
mutual assured destruction (MAD), 9, 184

Nagasaki, Japan, 8, 43, 186, 190
Nanaimo, British Columbia, 154
Nanuk, 142
Nascopie (ship), 76, 236n13
National Aeronautics and Space Administration (NASA), 171–2, 175
National Defence College, 139
National Film Board (NFB): *Vigil in the North*, 87, 124, 136, 265n12
National Museum of Canada, 146, 163
National Parks, 119
National Reactor Testing Station, 196
National Research Council (NRC): Atomic Energy Project, 106; Division of Medical Research, 74;

and DRB, 16, 67, 237n30. *See also* Chalk River
Nebraska, 196
Nevada Test Site/Proving Grounds, 181–2, 187–8
New York City, New York, 37, 167, 182
New Zealand, 135, 185
Newfoundland, 102, 119, 148
Niagara Falls, Ontario, 184
Nike-Cajun (rocket), 170
Noble, H.R.R., 142
Nobel Prize, 185–6, 190
non-commissioned officer (NCO), 132–3, 139, 212
Norman Wells, Northwest Territories, 104, 146
North American Air/Aerospace Defense Command (NORAD), 28, 48
North Atlantic Treaty Organization (NATO): Advisory Group on Aeronautical Research and Development, 137; Canadian garrison in West Germany, 18; Canadian Infantry Brigade in Korea, 131; Canadian involvement in, 18, 137, 139, 144; North Atlantic Treaty, 123; and Western defence, 8, 136, 138, 182, 211
North Dakota, 182
North Magnetic Pole, 21, 167, 171, 208
North Polar Basin, 136
North Pole, 152, 156, 158, 185
North-West Mounted Police (NWMP), 22, 226n58
Northern Health Services, 180, 200
Northern Sea Route, 51
Northwest Passage, 154
Northwest Territories (NWT): Arctic boundary, 33–4; Bureau, 76–7; DDT spray trials in, 104; IGY research in, 147–8; Indigenous

soldiers from, 82; and Inuit relocation, 195; medical research in, 76, 194; military exercises in, 40; muskeg, 119; Northwest Territories Act, 34; Northwest Territories Council, 76; radioactive fallout in, 178. *See also* Keenleyside, Hugh
Northwind (ship), 154
Norway: effects of thermonuclear testing in, 184; Ministry of Industry, 51; Norwegian Polar Institute, 51; nuclear fallout in, 189
Novaya Zemlya, Russia, 183–5
nuclear bomb: Blue Danube, 187; Bravo, 184; Hiroshima and Nagasaki, 8, 43, 186, 191; Mk-28 thermonuclear weapons, 189; Tsar Bomba, 184–5. *See also* Nevada Test Site/Proving Grounds; Novaya Zemlya; Semipalatinsk Test Site; *Technical Feasibility of Establishing an Atomic Weapons Proving Ground in the Churchill Area*
nuclear fallout: in the Arctic, 184–5; atmospheric, 204, 218; effects on animals, 178–9, 188–9, 192, 195, 198, 200; effects on Beluga whales, 188; effects on people, 178–9, 206–7, 212, 221; effects on vegetation, 189; in food, 178, 198, 207; international research on, 180–3; in northern Canada, 174, 179–80, 188, 195–8, 202, 206–7, 212–14, 220; in the Soviet Union, 152, 184–5. *See also* radiation; Radiation Protection Division (RPD)
Nunavik, Quebec, 195
Nunavut, 34, 67, 125, 229n95, 236n19
Nuremberg Code, 70. S*ee also* human experiments
nursing, 6, 17, 205

oceanography, 90, 137, 147, 154, 157, 165
Official Secrets Act, 45–6, 77
Ogdensburg Agreement, 37
Old Crow, Yukon, 198
Ontario Veterinary College. *See* University of Guelph
Ontario Water Resources Commission, 120
Operation Frost Jet, 170
Operation Hazen: DRB involvement in, 155, 160–6; field research, 155, 163; Hazen Camp, 155, 161; location of, 149–50, 155, 160–4; military value of, 164–6, 174, 214; RCAF involvement in, 160–2. *See also* Ellesmere Island; Hattersley-Smith, Geoffrey; International Geophysical Year (IGY)
Operation Morning Light, 49
Operation Ski Jump, 156
operational research: and Arctic warfare, 41, 63, 125, 129–30, 133; Canadian Army Operational Research Establishment (CAORE), 127; at DRNL, 85, 126–9, 138–41, 145; Joint Services Operational Research Unit, Edmonton, 129; Korean War, 124, 131, 249n3; Operations Research Section (ORS), 62, 126; in the Second World War, 43
Oppenheimer, Robert, 186
Oshawa, Ontario, 184
Ottawa, Ontario, 196, 201
Ottawa River, 103, 106
Ottawa Valley, 186

Pacific Fisheries Centre, 154
Pacific Ocean, 37–8, 180
Page, Edward, 86
Pakistan, 135

Panama, 18
Pangnirtung, Nunavut, 178
Partial Test Ban Treaty, 191–2
Pauling, Linus, 190–1
Pavshoukov, Ivan, 46
Pearl Harbor, 37
Pearson, Lester: *Foreign Affairs* publication, 32; on northern Canada as a frontier, 32–3; on northern development, 51; Under-Secretary of State for External Affairs, 32
Peary, Robert E., 159, 255n56
Penfield, Wilder, 74
Penney, William, 186–8
Pennie, Archie, 13, 15, 57, 60–3, 130, 134
permafrost, 37, 57, 59, 62, 137, 150–1, 165, 188. *See also* Arctic Ice and Permafrost Project; Army Snow, Ice and Permafrost Research Establishment; Canadian Arctic Permafrost Expedition
Permanent Joint Board on Defence (PJBD), 37–8, 44–5
Petawawa, Ontario, 40
physiology, 7, 59, 87–8, 114, 124. See also *Canadian Journal of Biochemistry and Physiology*, 75
Pinetree Line, 48
Point Barrow, Alaska, 192
Point Hope, Alaska, 185, 190, 192
Polar Continental Shelf Project, 148, 165
Pope, Charles, 131, 140
Porsild, Erling, 51
Porton Down, 101
preparedness: Cold War, 149; emergency, 260n49; industrial, 28; military, 12, 42, 50, 100, 142, 208–14; and mobilization, 38; NATO, 211

Prince Albert, Saskatchewan, 48
Prince of Wales Strait, 146, 153
prisoner of war (POW). *See* Korean conflict
Privy Council, 75
Procter, H.A., 197, 201
procurement, of radioactive isotopes, 106
Project ABLE, 175
Project Chariot, 185, 190, 261n57
Project Iceworm, 152
Project Plowshare, 190
Pruitt, William "Bill," 178, 181, 189–90, 207
psychology, 59, 124, 162. *See also* Easterbrook, Jim; Hoy, T.L.; Zubek, John

Qausuittuq (Resolute Bay), Nunavut, 26
Quebec City, Quebec, 16, 182
Queen's Journal, 75
Queen's University, 67, 73–5, 80
Queen's University Arctic Expedition. *See* Brown, G. Malcolm

radar. *See* Distant Early Warning (DEW) Line; Mid-Canada Line; Pinetree Line
Radforth, Norman W., 118–19
radiation: atomic energy, 89, 106, 108, 181–2, 186, 189, 192, 245n41; body burden, 192; cesium-137 (Cs-137), 190–6, 200; cobalt-60, 108; half-life, 108–9; iodine-131 (I-131), 88, 190, 196, 198; polonium-210, 203–4; radiation decay, 108–10; radioactive contamination, 177–8, 182, 189, 212, 214; radioactive phosphorous (P-32), 107–9; strontium-90 (Sr-90), 189–93, 197, 201, 204;

tooth research, 194, 203–4; uranium mining, 23; whole-body counts, 178, 190; whole-body examinations, 196–7, 201. *See also* entomology; human experiments; nuclear fallout

Radiation Protection Division (RPD): and cooperation with Institute of Cancer Research (ICR), 203–6; creation of, 180, 214–15, 258n8; human bone survey, 198–203, 263n101; radiation examinations, 196–7; radiochemical analysis, 180, 193, 198, 204–5. *See also* human experiments; nuclear fallout

radio propagation, 110, 166–9

Rankin Inlet, Nunavut, 205

Rath, O.J., 194

Rea, R.F., 161

Reader's Digest (magazine), 92

Red Cross, 83

Reed, Guilford, 13, 104–6

Reindeer Station, Northwest Territories, 102

Research and Development Board (RDB), 49, 110

Research Department eXplosive (RDX), 61

residential schools, 22, 71

Resolute, Nunavut, 102, 151, 162, 179

Resolute Bay (Qausuittuq), 26

Richards Island, Northwest Territories, 102

Rideau Military Hospital, 58

Rivers, Manitoba, 128–9

Robertson, Owen, 155

Robinson, D.B.W., 60

Rockcliffe, 99

rocket: attacks, 29; and Canadian defence, 138; and the Cold War, 169–76; high-latitude, 49; sounding, 52, 150, 169–73, 233n74.

See also Churchill Research Range (CRR); International Geophysical Year (IGY)

Rockingham, J.M., 116

Rooks, Dr., 201

Ross, D.I., 139

Ross, Donald, 128

Ross, James, 61

Ross Island, Antarctica, 174

Rowley, Graham, 51–2, 156

Royal Air Force (RAF), 61, 147, 168

Royal Canadian Air Force (RCAF): aerial reconnaissance, 129, 147, 154, 157–8, 160, 177; Air Transport Command, 53, 64, 160–1, 164, 214; clothing research, 93, 97; at Fort Churchill, 44, 54–5, 58, 63, 125–8, 142, 173, 212; Joint Air Training Centre, 128–9; medical evacuations, 84; and Operation Hazen, 146–7, 160–2, 164–6; and Queen's University Arctic Expedition, 76, 79; Reconnaissance Squadron, 160, 166; *Roundel* (magazine), 141–2; and Second World War, 154; Station Cold Lake, 138; Station Downsview, 4; Survival Training School, 91; Tactical Air Command, 129. *See also* entomology: aerial spraying; entomology: Northern Insect Survey

Royal Canadian Engineers, 160

Royal Canadian Mounted Police (RCMP), 8, 22, 46, 56, 68, 93. *See also* North-West Mounted Police (NWMP)

Royal Canadian Navy (RCN), 4, 44, 53, 55, 57, 111, 142, 154

Royal Navy (RN). *See* Admiralty (UK)

Rupert's Land, 22

Rural Cooperative Power Association, 182

Rutherford, Ernest, 186

Saint John, New Brunswick, 115
Saint-Hubert, Quebec, 119
Salvation Army, 131
Samuels, E.R., 201
San Diego, California, 154
Sandspit, British Columbia, 148
Saskatchewan, 33, 40, 48, 119, 194, 229n95
Saskatchewan River, 117
Saskatoon, Saskatchewan, 117, 200
satellite reflectors, 175
Sawmill Bay, Northwest Territories, 102
Scandinavia, 178
Schaefer, Didi, 179
Schaefer, Otto, 178–80, 192–3, 198, 200
Schenk, Carl, 120, 248n82
Schmidlin, E.F., 64–5
Scholander, Per Fredrik, 88
Schulert, Arthur, 197
science: fallibility of, 25, 27; militarization of, 5, 10–11, 15, 25–7, 33, 94, 214, 218. *See also* human experiments; military research and development
Science (journal), 197
scientific intelligence, 49–50, 85, 105, 110
Scott Polar Research Institute (SPRI). *See* University of Cambridge
Scripps Institute of Oceanography, 154
Seaborg, T., 88
search-and-rescue, 55, 63, 159
Second World War: Allies, 8–10; Arctic research, 29, 39–41, 51, 167, 208; Arctic strategy, 37, 42; atomic bomb, 190; Battle of Britain, 126; biological warfare, 104, Canada's long Second World War (theory), 15–17; chemical production, 96–7; DDT, 96–7, 111, 117; demobilization in Canada,

15, 38; Fort Churchill, 65, 125; gas gangrene, 104; medicine, 6; militarization, 88, 217; military research, 124, 209, 211, 222; operational research, 42–3, 126, 131; radar, 126; relocation of people, 195; Royal Canadian Air Force (RCAF), 154; Russo-Finnish War, 39–40; toxoids, 104; transport, 14
security: Arctic, 11, 30, 35, 44, 69–72, 94, 124, 138, 158, 164, 176, 210, 220–2; Basic Security Plan (Canada–US), 38; Cold War, 5, 12–13, 18, 24, 32, 143, 210–12, 217; and continental defence, 38, 48, 51; Council (UN), 42; definition of, 35; international, 17, 37; and miliary research, 29, 77, 131, 188, 208; and national defence, 5–7, 28–30, 42, 64–5, 77, 100, 106, 121–3, 123, 131, 143, 159, 165, 218, 221–2; North Atlantic, 9–10, 18, 29–30, 38, 42–3, 47, 50–2, 66, 144, 148, 152, 167, 188, 209–11, 219; politics, 37, 44–5, 51, 148; regulations at Fort Churchill, 54; and sovereignty, 157; vs. sovereignty, 35; surveillance, 158; Western, 110, 117, 123, 182
Sellers, Edward A., 81–2, 86
Semipalatinsk Test Site, Kazakhstan, 183
sensory deprivation. *See* human experiments
Shilo, Manitoba, 40
Showler, J.G., 160
Siberia, 156–7
Silent Spring (Carson), impact of, 96–7
Sinclair, R.G., 75
Single Side Band, 168

Sivertz, B.G., 196
Snag, Yukon, 102
Solandt, Omond: on Arctic science, 42–3, 65, 126, 148, 167; British Mission to Japan, 43; on Canadian defence, 17, 65, 157–8; on the Cold War, 123; as Director General for Defence Research, 43; as DRB chair, 13, 131, 137–8, 148; and DRNL, 64; education, 42–3; on military research, 32, 63, 100–1, 123, 126, 131; on scientific intelligence, 50
South Africa, 9, 135
Southampton Island (Shugliaq), Nunavut, 55, 67, 75–6, 79–80, 89, 102
sovereignty: in the Arctic, 18, 32, 69, 126, 157–8, 164, 210, 220; definition of, 35; over Greenland, 153; territorial, 18, 32, 39, 48, 52, 148, 151; vs. security, 35
Soviet Union: Arctic, 136, 183–4, 219; Committee for State Security (KGB), 8; Eastern Bloc, 8, 12, 50, 183; Embassy in Ottawa, 8, 46; and espionage, 7–8, 49; *Izvestia* (newspaper), 46; *Krasnyi Flot* (newspaper), 108; Kremlin, 46, 183, 188; North Pole-1, 157; North Pole-2, 156; Soviet Navy, 108. *See also* Moscow; Northern Sea Route
space: age, 171; Alouette, 172; astronaut research, 172; atmosphere, 170; and climate amelioration, 175; Committee on the Peaceful Uses of Outer Space, 171; DRTE, 8, 13; as a frontier, 177; ionosphere, 169; junk, 172; Milky Way, 167; race, 170; weaponization in, 181. *See also* astronomy; astrophysics; aurora borealis; Churchill Research Range (CRR); National Aeronautics and Space Administration (NASA); *specific satellites*
Sparrow, I.M., 146–7
Spinks, John W.T., 31, 117, 216
Sputnik (satellite), 48, 117, 147, 170
St. Catharines, Ontario, 184
St. Lawrence River, 16
Stefansson, Vilhjalmur, 51
Sternberg, Joseph, 205–6
Stevens, Ward, 128
Stevenson, James, 86
Strategic Air Command (SAC), 56, 142
Suffield, Alberta, Suffield Experimental Station (SES), 13, 16, 61, 101
superpower: conflict, 48; diplomacy, 9, 11; ideological supremacy, 9; world dominance, 9
Sutherland, Robert J., 131
Sweden, 178, 184, 191

Tanquary Fiord, Ellesmere Island, 166
Taylor, Lauriston, 191
Taylor, Marjorie, 107
Technical Feasibility of Establishing an Atomic Weapons Proving Ground in the Churchill Area, 187
telecommunications, 25, 168, 172. *See also* Defence Research Telecommunications Establishment (DRTE)
Teller, Edward, 185–6
Tennessee, 196
Thor-Delta (rocket), 172
Thule, Greenland: airbase, 152, 159, 161; Arctic Night exercise in, 139; B-52 bomber crash near, 189; militarization in, 153

Tommy Hunter Show (TV program), 84
topography, 33, 41–2, 46, 65, 102, 116, 118, 125, 147, 150–1, 153, 162, 174, 177, 218
Toronto, Ontario: bomb threat to, 184; cold conference in, 81; Commonwealth Advisory Committee on Defence Science, 167; DRML, 3–4, 214; Dennis Jordan, 76; patients transferred to, 201; Toronto General Hospital, 6. *See also* Defence Research Medical Laboratories (DRML); University of Toronto
toxic chemicals: Agent Orange, 116, 122; DDT, 96–9, 104, 111–22. *See also* chemical warfare
tripartite (Canada–US–UK), 64–5
Truman, Harry, 44, 46
tuberculosis, 79, 104, 202
Tuktoyaktuk, Northwest Territories, 146
Turner, Frederick Jackson, frontier thesis, 19–20
Twinn, C.R., 97–8, 100, 104, 221

unexploded ordnance, 28
Ungava Bay, 102, 195
Union of Soviet Socialist Republics (USSR). *See* Soviet Union
United Kingdom: Arctic Expedition (1875–6), 160; and atomic energy research, 89, 180, 183, 186–9, 191; and biological warfare, 110–11; cooperation in defence research with, 8, 11, 18, 29–32, 37, 42, 52, 135, 144, 210–11, 216–19, 221; and Fort Churchill, 47, 63–6, 72, 130, 139–40; and the International Geophysical Year (IGY), 165–7; Lady Franklin Bay Expedition (1881–4), 160; Ministry of Supply, 168; origins of operational research in, 126. *See also* British Army
United Nations: Charter, 123; coalition forces in Korean War, 6, 42, 123–4; Committee on the Peaceful Uses of Outer Space, 171; Scientific Committee on the Effects of Atomic Radiation (UNSCEAR), 182, 192; Security Council, 42
United States: Air Force (USAF), 44, 53, 56, 142, 152, 159, 162, 169, 171, 177, 189, 195; Army Basic Arctic Manual, 130; Army Basic Combat Training, 105; Army Committee for Insect and Rodent Control, 97; Army Corps of Engineers, 47, 56–7, 140, 151, 169; *Army Information Digest* (magazine), 8; Army Ordnance Corps, 169; Army Signal Corps, 169; Army Snow, Ice and Permafrost Research Establishment, 165; Bureau of Entomology and Plant Quarantine, 97; Bureau of Fisheries, 96; Chemical Corps, 56; Congress, 48; Department of Defense (DOD), 55, 175, 196–7; Marine Corps, 142; Medical Department, 56; Military Air Transport Service, 53; Military Sea Transportation Service, 161–2; and military-industrial complex, 28, 216; National Bureau of Standards, 191; Naval Electronics Laboratory, 154; Navy (USN), 136, 153; Office of Aerospace Research, 169; Office of Naval Research, 136, 150; Ordnance Department, 56; Pentagon, 117, 185; Public Health Service, 197; Signal

Corps, 56; Transportation Corps, 56; War Department Medical Nutrition Laboratory, 78–9; War Department's Medical Division, 106; Weather Bureau, 151, 182
University of Alaska–Fairbanks, 178
University of Alberta, 146
University of British Columbia, 154
University of Cambridge: Cavendish Laboratory, 186; and DRB, 168; Scott Polar Research Institute (SPRI), 51, 155
University of Cincinnati, Medical Center, 196
University of Guelph: Department of Parasitology, 111; Ontario Veterinary College, 112
University of Iowa, 196
University of Manitoba, 101, 265n13
University of Michigan, 196
University of Montreal, 205
University of Ottawa, 146
University of Oxford, 73, 155–6
University of Toronto: Banting Institute, 89; Wilfred Bigelow, 6; Department of Medicine, 242n97; and DRB, 221; insulin trials, 4; and RPD, 189; Edward Sellers, 81, 86; Omond Solandt, 42; Keith Wightman, 89
University of Western Ontario, 80, 86, 104, 239n65

Valcartier, Quebec, 61–2, 149, 251n36
Van Allen, James, 147
Vancouver, British Columbia: and aircraft communications research, 168–9; Dental Health Services, 204; Lower Mainland, 99; Stanley Park, 99
Vandenberg Air Force Base, 172

Vanderbilt University, 196–7
Vanier, Georges, 173
Victoria, British Columbia, 84
Victoria Island, 153
Vietnam War: herbicides in the, 116–17; Project ABLE, 175
Vigil in the North, 87, 124, 136, 265n12

Warsaw Pact, 8
Washington, D.C., 8–12, 14, 18, 37–40, 44–9, 56, 98, 110, 117–22, 136–7, 150–4, 158, 174–6, 185
Washington State, 190
Watson Lake, Yukon, 99, 104
weapons: anti-tank, 132; atomic cannon, 64; atomic/nuclear bomb, 6, 8, 47, 49, 105, 173, 180, 184, 186–7; ballistics, 149; Heller, 132; MGM-18 Lacrosse missile, 141; thermonuclear bomb, 180, 184; Tsar Bomba, 184; unexploded ordnance, 28. *See also* biological warfare; chemical warfare; military research and development; missiles
Webb, Jean, 205–6
Webster, D.R., 6
Whillans, Morley, 75
White Sands, New Mexico, 173
Whitehorse, Yukon, 99, 102, 104, 178–9, 193–4, 205
Whitehorse General Hospital, 194, 201
Wightman, Keith, 89
Wilbush, Dr., 201
Williams, D.C., 101–2
Winegard, William "Bill," 221
Winmill, A.E., 120
Winnipeg, Manitoba: John Diefenbaker speech in, 176; District Aviation Forecast Office, 55; hospitals in, 193–4, 200; Hudson Bay rail line, 53; Andrew

Winnipeg, Manitoba (*continued*)
McNaughton visit to, 44; RCAF planes, 79; Omond Solandt speech in, 123

winter warfare: vs. Arctic warfare, 39–41; British observation of, 54, 135; Canadian military training, 40–1, 124, 134–5, 139–40, 143; and DRB, 42, 78, 124, 143; indoctrination courses, 139–40, 143; *Instructions for Winter and Ski Training*, 40; Inter-Service Committee on, 47; and operational research, 125–9, 134; school in Petawawa, 40; tactical, 40. *See also* Arctic: warfare; Petawawa; Shilo

Wisconsin, 182
World Health Organization, 108
Wrong, Hume, 17

X-ray, 57, 76, 78, 181, 258n8

Yellowknife, Northwest Territories, 122, 129, 179, 193, 205
York Factory, Manitoba, 187
Yukon: 21, 33–4, 37, 76–7, 99, 102, 104, 156, 178, 194, 229n95

Zimmerman, Adam Hartley, 137–8, 148–9, 171
zoology, 60, 154. *See also* Manning, Thomas
Zubek, John, 216